Advances in Comparative and Environmental Physiology 16

Volumes already published

Vol. 4: Animal Adaptation to Cold
Edited by *L. C. H. Wang* (1989)

Vol. 5: From the Contents: Osmoregulation – Chemoreception – Temperature – Spectroscopy – Metalloproteins – Prostaglandins (Eicosanoids) – Maternal-Fetal Relationships (1989)

Vol. 6: Vertebrate Gas Exchange: From Environment to Cell
Edited by *R. G. Boutilier* (1990)

Vol. 7: From the Contents: Protein Turnover – Xenobiotic Metabolism Lipoproteins (1991)

Vol. 8: From the Contents: Dormancy in Aquatic Invertebrates – Root Effect – Red Blood Cell Functions in Fish – Diving Mammals (1991)

Vol. 9: Volume and Osmolality Control in Animal Cells
Edited by *R. Gilles, E. K. Hoffmann,* and *L. Bolis* (1991)

Vol. 10: Comparative Aspects of Mechanoreceptor Systems
Edited by *F. Ito* (1992)

Vol. 11: Mechanics of Animal Locomotion
Edited by *R. McN. Alexander* (1992)

Vol. 12: Muscle Contraction and Cell Motility: Molecular and Cellular Aspects
Edited by *H. Sugi* (1992)

Vol. 13: Blood and Tissue Oxygen Carriers
Edited by *Ch. P. Mangum* (1992)

Vol. 14: Interaction of Cell Volume and Cell Function
Edited by *F. Lang* and *D. Häussinger* (1993)

Vol. 15: From the Contents: Salivary Glands – Nematocyst Discharge – Crustacean Physiology – Steroid-Sensitive Areas (1993)

Vol. 16: Ion Transport in Vertebrate Colon
Edited by *W. Clauss* (1993)

Volumes in preparation

Vol. 17: Effects of High Pressure on Biological System
Edited by *A. G. Macdonald* (1993)

Vol. 18: Biomechanics of Feeding in Vertebrates
Edited by V. L. Bels, M. Chardon and P. Vandewalle (1993)

Advances in

Comparative and Environmental Physiology 16

Ion Transport in Vertebrate Colon

Guest Editor: W. Clauss

With Contributions by

Martine Avella · Kim E. Barrett · H. J. Binder · W. Clauss
D. C. Dawson · W. von Engelhardt · R. Greger · J. P. Hayslett
R. Krattenmacher · K. Kunzelmann · B. Lahlou · M. A. Post
V. M. Rajendran · G. Rechkemmer · G. I. Sandle
E. Skadhauge · M. Tilman

With 73 Figures

Springer-Verlag
Berlin Heidelberg New York
London Paris Tokyo Hong Kong
Barcelona Budapest

Guest Editor:

Professor Dr. WOLFGANG CLAUSS

Institut für Tierphysiologie
Justus-Liebig-Universität
Wartweg 95
W-6300 Giessen, Germany

ISSN 0938-2673
ISBN-13:978-3-642-77120-0 e-ISBN-13:978-3-642-77118-7
DOI: 10.1007/978-3-642-77118-7

Softcover reprint of the hardcover 1st edition 1993

Typesetting: Best-set Typesetter Ltd., Hong Kong
31/3145-5 4 3 2 1 0 – Printed on acid-free paper

Foreword to the Series

The aim of the series is to provide comprehensive, integrated reviews giving sound, critical, and provocative summaries of our present knowledge in environmental and comparative physiology, from the molecular to the organismic level.

Living organisms have evolved a widespread range of basic solutions to cope with the different problems, both organismal and environmental, with which they are faced. A clear understanding of these solutions is of course of fundamental interest for all biologists, zoologically or medically oriented. They can be best comprehended in the framework of the environmental and/or comparative approaches. These approaches demand either wide surveys of animal forms or a knowledge of the specific adaptive features of the species considered. This diversity of requirements, both at the conceptual and technological level, together with the fact that physiology and biochemistry have long been mainly devoted to the service of medicine, can account for the fact that these approaches emerged only slowly amongst the other new, more rapidly growing disciplines of the biological sciences.

The field has now gained the international status it deserves and the organization of a series devoted to it appeared timely to me in view of its actual rapid development and of the interest it arouses for a growing number of biologists, physiologists, and biochemists, independently of their basic, major orientation.

Liège, Belgium, Spring 1988 *Raymond Gilles*

List of Editors

Those interested in contributing to this series should contact Professor R. Gilles or one of the Series Editors. Review proposals should include a short outline – and a Table of Contents – briefly defining the aims and scope of the review.

Preface

The absorptive and secretory capacities of the large intestine are of special importance in the refinement and termination of the digestive processes. For a long time colonic properties were poorly investigated, as the main interest focussed on the major exchange site of ions and nutrients, the small intestine. In textbooks colonic properties were usually dealt with in a small section, noting that the colon is the site of water and electrolyte conservation.

Whereas the overall absorptive and secretory features of the large intestine had been studied in vivo with perfusions or instillation techniques, details about the mechanisms of ion exchange have only been worked out since Ussing laid the theoretical background with his flux-ratio equation and double membrane model, and pioneered in vitro studies under controlled conditions with his famous chamber and short-circuit technique. Over the following years many of the new techniques in cellular physiology, such as intracellular microelectrodes, noise analysis, patch clamping, membrane vesicles, and reconstitution techniques, have also been applied to colonic epithelia and isolated enterocytes of various vertebrates and have yielded interesting results.

At first, researchers viewed the colon as a tube, and their interest focussed mainly on mammalian colonic epithelia (e.g. rat and rabbit). Later, they recognized segmental and diurnal differences, and functional properties similar to the organization of ion transport in kidney tubules. Comparative physiologists noted that the colon does not serve exclusively digestive functions, but is in some vertebrates involved in osmoregulation, and even in urine concentration. The regulation of colonic ion transport was investigated, and was found regulated not only be enteric gut hormones, but by the classical target hormones of the kidney such as adrenal steroids and antidiuretic hormones. Many of these fine studies have led to models and conclusions of general interest.

Therefore, the time seems ripe to collect some of this information, and to present an integrated view of ion transport in the vertebrate large intestine. Research on the colon not only serves medical interest or clinical purposes, but is a highly interesting subject of comparative physiology, from which we can learn much about the basic principles in physiology. The present book presents examples of research on colonic ion transport of each vertebrate group, not only the classical models of epithelia, rat and rabbit, but also on the special and unique features in fish, amphibian, reptiles, birds and selected mammals such as the guinea pig. Physiology and pathophysiology of ion transport in the human large intestine are as well covered as the use of colonic cell cultures as model systems for crypt cell properties.

I am especially grateful to the authors, each a specialist and leader in their field, who have contributed to this comparative approach to colonic ion transport, and hope that these interesting contributions will stimulate further research in this area.

Giessen, February 1993 WOLFGANG CLAUSS

Contents

Chapter 1

Electrolyte Transport Mechanisms in Fish Intestine

Brahim Lahlou and *Martine Avella*[1]

Contents

1 Introduction

The ability of fish to osmoregulate in freshwater, in seawater and, occasionally, in other salt media, is ensured by the transport activity of several epithelia. Among these tissues, the intestine is also involved in nutrition and the two major functions are largely linked together since the absorption of inorganic ions drives that of small organic solutes and water. In comparison to other organs more complex in morphology, such as gill and kidney, the intestine presents convenient features which make it appropriate for in vitro studies, namely:

- it is "flat", except for villi and ridges, and may be easily mounted in Ussing chambers,
- its cell population is relatively homogeneous,

[1]Laboratoire de Physiologie Cellulaire et Comparée, URA CNRS 651, Faculté des Sciences, Parc Valrose, 06034 Nice Cedex, France

Advances in Comparative and Environmental Physiology, Vol. 16

- it is possible to isolate viable cells from its mucosa and easy to prepare subcellular fractions, in particular membrane vesicles,
- it is directly accessible to electrophysiological measurements. Accordingly, the following techniques, some classical, others more recent, have been applied:
- everted sac of Wilson and Wiseman, which simultaneously permits measurements of ion and water transport,
- short-circuit current, together with flux measurements, in Ussing chambers,
- short-term apical uptake (measured over 1 min or less), by the method of Schultz et al. (1967),
- isolated cells and apical and basolateral membrane vesicles,
- intracellular recordings by way of microelectrodes (potential-sensitive and ion-selective) and patch-clamp.

As several reviews have appeared on the subject in the past (Robinson 1976; Gilles-Baillien and Gilles 1983; Kirsch and Lahlou 1987), this presentation will deal mostly with knowledge acquired since 1985. Not surprisingly, studies on fish intestine concerned almost exclusively teleosts. In elasmobranchs, the other fish group of common occurrence and major physiological interest, one should take into account an additional transporting organ associated with the gut: the "rectal gland", which secretes ions and water and releases a fluid highly concentrated in ions at the posterior end of the intestine. This tissue has become a favourite model of study in a small number of laboratories like the Mount Desert Island Marine Biological Laboratory, in Maine, USA. Although it is not an absorbing epithelium, it offers instructive transport mechanisms which are reported below and comparable to those involved in intestinal absorption.

2 Morphology

The intestine of teleosts is relatively uniform along its length but two or three regions may be distinguished in many species. No part can be adequately described as a colon. In trout it divides into an anterior, or proximal, and a posterior, or distal, segment, the latter with transverse thick ridges generating a thick mucous layer which hinder uptake measurements and other in vitro studies. Unlike the mammalian intestine, fish intestine does not present Lieberkuhn crypts, a difference stressed as physiologically essential. Its length relative to body size varies between species: it is greater in herbivorous than in carnivorous fish.

The main cells, enterocytes, are columnar and usually narrow: $60 \times 3.5\,\mu m$ in winter flounder *Pseudopleuronectes americanus* (Field et al. 1978) and $40 \times 5\,\mu m$ in plaice *Pleuronectes platessa* (Ellory et al. 1978), a size still

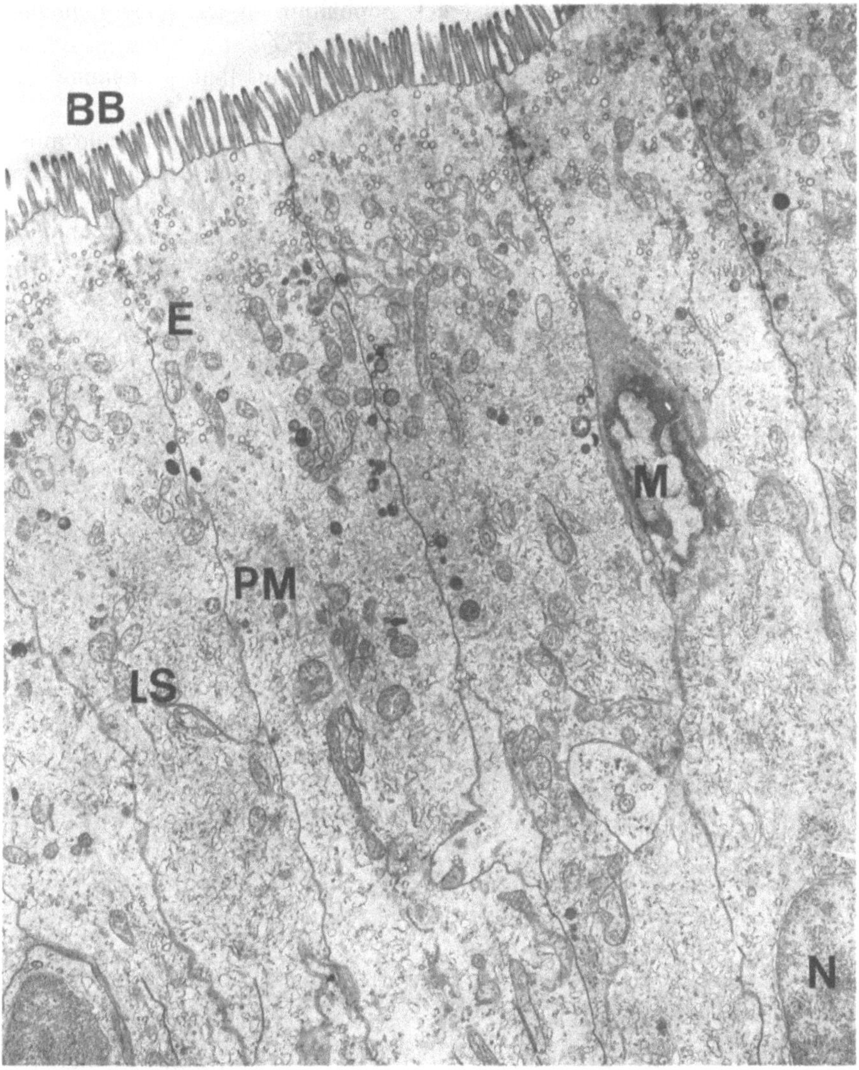

Fig. 1. Electron micrograph (×8750) of the apical part of enterocytes in the anterior intestine of juvenile *Liza ramada* (Mugilidae). *BB* Brush border; *E* enterocyte; *M* mucocyte; *PM* plasma membrane; *N* nucleus; *LS* lamellar structure. (Courtesy of J. Albertini-Berhaut)

sufficient for microelectrode impalements (Fig. 1). In flounder, a large proportion (95%) presents intercellular gaps which, according to Curtis et al. (1984), establish an electrical coupling between cells, a feature not encountered in the rabbit intestine. Presumably relevant to this fact, little variation is observed in membrane potential between regions. Mucous cells make up less than 10% of the cell population in the goby *Gillichthys mirabilis* (Loretz et al. 1985). In the mullet *Mugil* sp., Albertini-Berhaut (1987) reported that mucocytes are more numerous than in mammals and increase in number with age.

The intestinal mucosa is covered with a thick, very contractile muscle layer. In a number of species (flounder, plaice, tench, eel, goldfish) but not in others (rainbow trout), it is possible to strip this layer and therefore to remove a source of artifacts in electrical and flux measurements (see below).

Pyloric caeca have also been presented as a new absorptive organ involved in osmoregulation (Buddington and Diamond 1987).

In elasmobranchs like the small dogfish *Scyliorhinus canicula*, the intestine shows a well-developed spiral valve which has been successfully used to prepare membrane vesicles (Crane et al. 1979), but few transport studies have been made on this tissue.

With regard to ion and water absorption, it is worth mentioning that the oesophagus prepares proximally the intestinal transport in the European eel by allowing salts to diffuse across the gut wall towards the blood without accompanying water from swallowed seawater (Kirsch and Laurent 1975; Hirano and Mayer-Gostan 1976; Simmoneaux et al. 1987). In *Anguilla japonica*, Yamamoto and Hirano (1978) described three cell types in the oesophagus, namely numerous mucous cells, filament-rich (basal layer) and ribosome-rich cells. After 7 days in seawater, the stratified epithelium of freshwater animals is replaced by a simple epithelium made of columnar cells, without mucous cells, which are presumably more permeable.

In the intestine, morphological modifications may be induced by various factors. For example, acclimation to seawater leads to changes in the middle intestine of the rainbow trout. Metal pollutants (among others), such as cadmium, lead, mercury, arsenicals, copper, and aluminium, produce severe alterations and have a toxic effect on digestive tract function. These changes will not be discussed here in detail.

Temperature is an important factor. A change in temperature modifies the fluidity of the cell membranes by acting on the distribution and possibly the chemical nature of lipids. As described below, temperature acclimation of heterotherms like fish results in variations in transport efficiency. In carp, Lee and Cossins (1988) pointed out that changes occur in the gross anatomy as well as at the cellular level. The total intestinal volume in cold (10 °C)-acclimated fish is about double that of warm (30 °C)-acclimated fish. This holds for the entire length of the intestine.

The reason for this morphological involution is unclear but should certainly be considered as a regulatory process since acclimation to lower temperatures is the appropriate condition for fish life. Such changes should take place normally as a response to seasonal temperature variations. However, there may be a long-term adaptation, typically compensatory, so that it eventually provides independence from environmental temperature. The compensatory adjustment of membrane order by modifications in lipid composition has been called "homoeoviscous adaptation" presenting a "homoeostatic response" (Lee and Cossins 1990).

Even more important, dietary changes exert a striking control over intestinal morphology. In very active fish, like trout and goldfish, starvation produces thinning of mucosa, with electrical leakiness and reduction in active transport (see below). Pelletier et al. (1987) described in detail the dietary effects on the phospholipid composition of the brush border membrane in trout intestine. These authors distinguished the different phospholipid classes and the fatty acid composition of the brush border membranes when fish were fed an essential fatty acid-rich diet or a saturated diet.

For the above reasons, great care must be taken in comparative studies. In principle, transport capacities are corrected for surface area. This is not simple, however, for the intestine as the amplification factor between cylinder area and total villi surface is 5 to 15 times greater in an unstripped preparation (Lee and Cossins 1988). Gibson et al. (1985) showed that stripping the mucosa offsets the difference between 10 and 30 °C-acclimated carp with regard to transepithelial electrical resistance and serosa-to-mucosa Na^+ flux per unit nominal surface area. However, there was still substantial increase in sodium pump activity per unit surface area, indicating that the difference observed between media was physiologically consistent.

3 Solute-Linked Water Transport

Since the pioneering contributions of H. Smith (1930), it is known that the gills play a major role in fish ion regulation, in both freshwater and seawater animals. By contrast, absorption of water is the main intestinal function directly associated with osmoregulation. Classically, it is admitted that freshwater teleosts live in a hypo-osmotic medium and do not need to drink, while seawater teleosts have to obtain water by ingesting a hyperosmotic medium and to absorb it from intestinal lumen in the form of a salt solution isosmotic to the blood. The latter process is described as solute-linked water transport, which requires that sodium ions, which are transported actively by the Na^+/K^+ pump, be present at the luminal side.

In vivo experiments confirmed this relationship to environmental salinity. A simple approach made by several investigators was to label the external medium with radioactive particles (e.g., colloidal gold) or non-absorbed molecules (e.g., ^{131}I- phenol red), to sacrifice the fish after a definite time and to count the radioactivity accumulated in the various portions of the gut. The "drinking rate" was equated with the "absorption rate". A more elaborate technique was used in the European eel by Maetz and Skadhauge (1969). It consists of inserting a cannula at each end of the intestinal portion and circulating a known volume of fluid of definite composition through them, the changes of which were subsequently recorded. Skadhauge and Maetz found an absorption rate of 135 μl/100 g/h in freshwater, which increased to 325 in seawater and to 802 in double-strength seawater.

An increase in drinking rate with salinity was constantly observed in many teleosts from both media: *Anguilla anguilla*, *Anguilla japonica*, *Fundulus heteroclitus*, *Carassius auratus*, *Platichthys flesus*, *Xiphister atropurpureus*, *Serranus scriba*, *Salmo gairdneri*, *Opsanus tau*, etc. It appears as a typical feature of salt adaptation. In addition, a significant finding in the eel is that absorption takes place when the luminal solution is still consistently hyperosmotic with respect to blood: with a difference of 126 mOsmol/kg in seawater and 73 in freshwater. Thus, the "turning point", defined as the concentration at which the transmural osmotic movement of water is reversed from secretion to absorption, also varies with salinity.

The in vitro technique usually applied is the everted sac. It demonstrated the requirement of luminal sodium for water transport to take place, as in other vertebrates. The mechanism of Na^+-linked water is in accordance with the standing-gradient hypothesis proposed for epithelia (Diamond and Bossert 1967). In the Japanese eel, *Anguilla japonica*, Ando (1980) found that water absorption is linked to the coupled transfer of Na^+ and Cl^-. This finding and interpretation have not been confirmed since.

Since fish enterocytes are very narrow and adjacent, lateral space dilations may be rare or absent, a feature which should not favour transmural water movement. According to Field et al. (1978), fluid accumulates and hydrostatic pressure builds up only in preparations unstripped of their muscle layers.

The sac technique or perfused intestinal segments reveals that the absorption rate may be constant along the intestine (e.g., in goldfish: Smith 1964) or higher in one region: anterior intestine in rainbow trout (Bensahla-Talet et al. 1974) and European eel (Kirsch and Meister 1982), mid-intestine in Japanese eel (Ando 1980). In one case at least, *Tilapia*, which does not correspond to a typical euryhaline fish, water transport was found to be reduced after salt adaptation (Mainoya 1981). Similarly, in goldfish, a freshwater, stenohaline species, adaptation to a moderately hypertonic salt medium results in a decrease in water and ion transport (Ellory et al. 1972). By using the short-term exposure technique of Schultz et al. (1967), it was shown that the apical uptake of sodium was also lower following salt adaptation (Ellory et al. 1972). This reduction did not occur in the European flounder *Platichthys flesus*, which is a true euryhaline fish (Lahlou 1976).

Water transport is regulated by hormones. However, it is difficult to distinguish between an effect on water and a primary effect on ions. Long ago, Hirano and Utida (1968) demonstrated that cortisol injected into the Japanese eel promoted increased water transport by the isolated intestine. Porthé-Nibelle and Lahlou (1975) showed that this effect was displayed by cortisol and the synthetic compound dexamethasone but not by other corticosteroids, including aldosterone. This is of physiological relevance since cortisol is the major circulating interrenal hormone in teleosts, while the general presence of aldosterone is considered uncertain. Among recent findings, Balment and Carrick (1985) and Balment et al. (1987) showed in the European flounder and in the shark that mammalian angiotensin II and papaverine induce drinking behaviour in free-swimming animals. In addition, these findings support the presence of a renin-angiotensin system in elasmobranchs.

4 Ion Transport

As the most abundant electrolytes in extracellular fluids, sodium and chloride have received almost exclusive attention for many years in studies dealing with intestinal transport. Recent developments concerning coupled transport systems led investigators to examine other ions such as potassium, bicarbonate and calcium.

4.1 Transepithelial Properties

Ussing chambers have been widely used for unstripped and stripped preparations. Removal of the muscle layer is possible in a number of species, like tench, eel, plaice, flounder, but not in others, like trout, or may be difficult, as in goldfish. Stripping is of importance. In *Anguilla japonica*, it was shown to increase the transepithelial potential difference (PD) fourfold (Ando and Kobayashi 1978). Non-stripping implies generally a hindrance to diffusion and uncertainty regarding radioactive flux measurements. Fish intestine may be described as a typical leaky tissue by its electrical properties. The PD is low, in the range of a few millivolts, or may be nil, as in *Cottus* (House and Green 1965). On average, the total resistance is around $100\,\Omega \cdot cm^2$, with a range of 20 to 300.

Generally, the PD is serosa-positive in the intestine of freshwater fish, serosa-negative in that of seawater fish (Table 1). This striking feature led to the postulation that the intestine of seawater teleosts possesses a Na^+-independent Cl^- electrogenic pump (Huang and Chen 1971; Ando et al. 1975; Smith et al. 1975; Hirano and Mayer-Gostan 1976). In addition, chloride transport may take place in a Ringer solution containing relatively small concentrations of sodium, i.e., in the absence of a consistent electrochemical force driving sodium into the cells across the apical membrane. More compelling was the observation reported by Ando et al. (1975) according to which a large and sustained PD and short-circuit current (Isc) occur in *A. japonica* in Na-free Ringer (in choline chloride or TEA chloride medium). Finally, in the European flounder, *Platichthys flesus*, which is adapted to seawater (Smith et al. 1975) and in the plaice, *Pleuronectes*

Table 1. Selected electrical measurements in Ussing chambers on seawater fish intestine

Species	PD (mV)	Isc ($\mu A/cm^2$)	R ($\Omega \cdot cm^2$)	References
Anguilla anguilla	−5.4	−58	21−100	Schettino (pers. comm. 1991)
Anguilla japonica	−2.8	−82	36	Ando et al. (1975)
	−8.5	−120	65	Ando (1990)
Blennius parvicornis (at 50‰ salinity)	−1.0	−19	53	Lorenzo and Balanos (1989)
Cottus scorpius	None	–	–	House and Green (1965)
Gillichthys mirabilis	−4.6	−25	160−300	Loretz (1983)
Notothenia rossii (Antartic fish, at 4 °C)	−1.8	−20	94	Gibson and Ellory (1987)
Oncorhynchus kisutch	−2	(?)	41−191	Collie (1985)
Platichthys flesus	−1.9	−45	42	Smith et al. (1975)
	−2.0	−30	71	Gibson et al. (1987)
Pleuronectes platessa	−3.7	−100	56	Zeuthen et al. (1978)
	−5.5	−107	58	Ramos and Ellory (1981)
Pseudopleuronectes	−3.4	−22	155	Huang and Chen (1971)
americanus	−1.9	−36	54	Field et al. (1978)
	−8	−88	91	O'Grady et al. (1985)
	−4.5	−118	38	Smith et al. (1981)
Salmo gairdneri	−0.9	−25	43	Ando et al. (1975)

platessa, which lives in seawater (Ramos and Ellory 1981), Isc may be equated with chloride net flux.

Field et al. (1978) proposed an attractive alternative explanation which presented the active transport of sodium as the actual driving force for the transepithelial transfer of NaCl, that is for chloride as well, as in the intestine of higher vertebrates, despite serosal electronegativity. The model, as updated more recently (Rao and Field 1983), is depicted in Fig. 2. According to this model, Na^+ enters into the cell owing to its electrochemical gradient and is extruded towards the intercellular spaces by the classical basolateral Na^+/K^+ pump. Then Na^+ is refluxed to the mucosal side across the tight junctions, which are cation-selective. In contrast, Cl^- ions, which follow sodium passively, move in the opposite direction, towards the serosal side which acquires a negative potential. This explanation predicts that the net flux of chloride is higher than that of sodium, as observed in most cases. The merits of this interpretation are that it does not postulate a new type of ion pump and that it attempts to establish a unifying concept for all vertebrate intestine, the size and sign of the transepithelial PD depending on the ultimate diffusion of the two major ions. MacKay and Lahlou (1980) argued that 2,4,6-triaminipyrimidine (or TAP), which was believed to block the cation-selective junctions in leaky epithelia, did not reduce the serosal electronegativity but increased it instead in the isolated flounder intestine, a fact difficult to reconcile with Field's hypothesis. However, this drug was

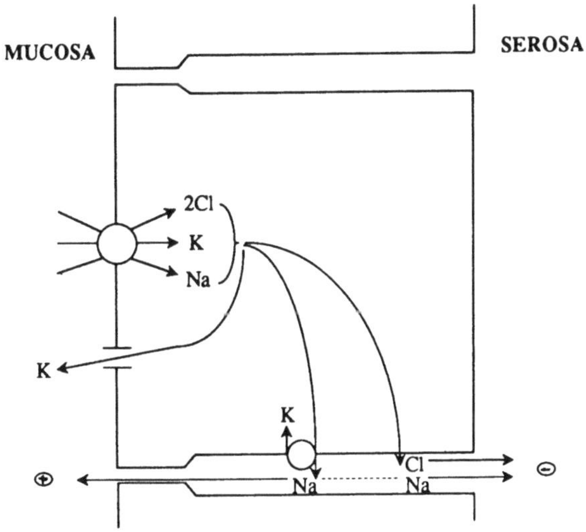

Fig. 2. Early version of ion transport by the intestine of seawater teleost. The figure is based on the model proposed by Field (itself presented as an amplification of the gallbladder mechanism of Machen and Diamond 1969), as completed by Rao and Field (1983)

shown to provoke alterations in the brush border membranes. In other fish epithelia, such as the opercular epithelium, the ciliary epithelium or the rectal gland, which show a negative PD on the trans-side, models have been similarly constructed to account for chloride secretion without postulating the intervention of a separate pump for this ion.

4.2 Basic Ion Couplings

Since the experiments of Diamond (1962) on gallbladder, it has been suggested that Na^+ and Cl^- move across epithelia as coupled ions by a compulsory, one-for-one, non-electrogenic mechanism. A straightforward demonstration of the coupling is shown by using ionic substitutions or pharmacological blockers.

Thus, in the intestine of the winter flounder, the uptake of sodium and chloride is reduced to the same extent by completely replacing the counterion by a less permeant ion, such as gluconate in place of chloride, choline or N-methyl D-glucamine (NMDG) in place of sodium, in the mucosal solution or on both sides.

Several "loop" diuretics given from the mucosal side provoke partial inhibition of sodium and chloride transports. Furosemide (Frizzell et al. 1979) and bumetanide (Halm et al. 1985a,b) were shown to be active on winter flounder. In plaice, piretanide inhibited Isc by 60%, and Cl^- net flux by 45% (Zeuthen et al. 1978); it also suppressed the coupled entry which corresponds to 20–30% of the uptake.

The double exchange, Na^+/H^+ and Cl^-/HCO_3^-, working in parallel represents a flexible alternative to NaCl absorption. H^+ and HCO_3^- are readily available as they result from intracellular metabolism of CO_2. The exchanges may be revealed separately by pharmacological inhibitors: amiloride for proton exchange, DIDS for bicarbonate transport.

Amiloride was generally described as ineffective on sodium transport in teleost intestine. This may imply lack of a Na^+/H^+ exchange mechanism or may simply result from competition at the exchange sites between amiloride and sodium ions (which are present in high concentration in the lumen). However, in winter flounder intestine (Charney and Pelkonen 1990) a consistent inhibition of sodium and chloride net fluxes by 1 mM amiloride was shown to occur, provided the pH in contact with the mucosa was high (7.6–7.8).

HCO_3^- exchange for Cl^- has been studied less frequently (Dixon and Loretz 1986; Ando 1990; Ando and Subramanyam 1990).

Regardless of its mechanism, NaCl absorption is maintained as long as there is a spontaneous apical inward movement of Na^+ into the cells along its electrochemical gradient. This gradient is maintained by the operation of the basolateral Na^+/K^+-ATPase. Therefore, the well-known blockade of this enzyme by addition of ouabain or removal of K^+ at the serosal side abolishes the coupled Na^+-Cl^- entry, among other effects.

4.3 Apical Cotransporters and Channels

Much progress has been made with the introduction of intracellular recordings by means of potential-sensing and ion-selective microelectrodes. The technique permitted one first to determine the apical (mucosa-to-cell) potential difference, Ea, and to compare it to the transepithelial (mucosa-to-serosa), Ems, the PD presented above. In winter flounder, Ea was found to be −60 mV (Musch et al. 1982) or −69 mV (Duffey et al. 1979), compared to an Ems of −3.3 mV. A value of −60 mV was also recorded in the tilapia *Oreochromis mossambicus* (Bakker and Groot 1988) and in the goby *Gillichthys mirabilis* (Loretz 1983). The apical potential was lower in plaice: −33.6 mV (Ellory et al. 1978) and −32.5 mV (Zeuthen et al. 1978) for a transepithelial PD of −5.7 mV.

When the chloride activity measured in cells was compared to the transepithelial gradients, there was obviously an intracellular accumulation of that ion above its expected equilibrium value. In plaice, the apparent concentration of chloride was 35 to 54 mM (Ellory et al. 1978; Zeuthen et al. 1978). In the winter flounder, the estimated Cl^- activity was 3.4 times greater than that predicted by passive distribution (Duffey et al. 1979).

More recently, it was shown that the NaCl-coupled apical mechanism should be extended to include K^+. This concept arose from electrophysiological studies and flux measurements in epithelia (thick ascending limb of

Henle's loop, diluting segment of amphibian kidney) and in isolated cells (erythrocytes, tumour cells, squid giant axon), indicating that K^+ had to be present in the luminal or extracellular medium for NaCl transport to occur. The new NaKCl cotransport so defined involved Na^+, K^+ and Cl^- at a stoichiometry of 1:1:2. It is also inhibited by loop diuretics, furosemide and bumetanide in particular. This modifies the earlier model of ion uptake proposed by Field and coworkers (Fig. 2).

A complication arises from the simultaneous presence of a conductive, outwardly directed K^+ channel. This channel, present in other cells and epithelia, is described as voltage-dependent and is blocked by Ba^{2+}. In winter flounder intestine, potassium movements measured by means of ^{86}Rb as radioactive tracer indicate the exit of K^+ at the apical membrane (Musch et al. 1982). Since K^+ enters across the basolateral membrane by the classical activity of Na^+/K^+-ATPase, this exit corresponds to the net secretion of potassium. In the presence of Ba^{2+}, the direction of K^+ movement reverses from secretion to absorption. If furosemide is then added to the mucosal medium, it suppresses K^+ absorption and provokes hyperpolarization of the apical membrane potential by 12–13 mV.

In the European eel *Anguilla anguilla*, T. Schettino and coworkers (pers. comm.) studied ion transport in the middle part of the intestine, using Ussing chambers, in seawater- and in freshwater-adapted animals. Ba^{2+} and bumetanide reduced Isc and PD, while amiloride and hydrochlorothiazide presented no effect, thus confirming the presence of an apical Na^+-K^+-$2Cl^-$ cotransport.

Not all authors have found this cotransport as the sole mechanism. In the European flounder, Lau (1985) suggested that the apical membrane is the site of the double exchange Na^+/H^+ and Cl^-/HCO_3^-, on the grounds that piretanide does not prevent accumulation of Cl^- above equilibrium in the enterocytes. This was also proposed for the goldfish *Carassius auratus* by Zuidema et al. (1985) from the effects of ion substitutions.

In addition to intracellular electrophysiological recordings, apical and basolateral properties have been conveniently explored by means of membrane vesicles (MV) prepared from brush border (BBMV) or from basolateral (BLMV) membranes. Although vesicles do not reproduce the actual fluxes which take place across cell membranes in situ, they proved very useful in analyzing absorption or exit mechanisms. One of the first works on fish intestine vesicles was performed by Crane et al. (1979) on the spiral valve of the elasmobranch *Scyliorhinus canicula*. To mention but a few teleosts, the species most frequently used for BBMV have been the winter flounder (Eveloff et al. 1980), the sea bass *Dicentrarchus labrax* (Bogé et al. 1982; Drai et al. 1990), the rainbow trout (Di Costanzo et al. 1983), the European eel (Storelli et al. 1986), the tilapia *Oreochromis mossambicus* (Reshkin and Ahearn 1987).

In winter flounder vesicles, O'Grady et al. (1985) confirmed the apical localization of NaKCl cotransport by using the binding of radiolabelled bumetanide.

In trout intestine, Colin and coworkers (Aguenaou et al. 1987, 1989a,b) made a survey of uptake for several ions and described a large number of pathways. NaKCl-coupled entry was found in seawater fish only. In contrast, both freshwater and seawater animals are equipped with Na^+/H^+ and Na^+/Na^+ exchangers. There is also a Na^+ conductance which accounts for Na^+ entry along its gradients. Cl^- entry was shown to take place independently of Na^+ as it was of the same amplitude regardless of the associated cation (Na^+, K^+ or Li^+) even if the cation was replaced by TMA^+ or mannitol. The authors ruled out the presence of Cl^-/OH^- exchange as a Cl^- gradient did not affect the intravesicular pH (pHi). They also observed a Cl^-/Cl^- exchange and separate Cl^-, H^+ and K^+ conductances. The authors warn, however, due to probable interactions, that this complex set of conductances and exchanges should be interpreted with caution.

In salt-adapted tilapia, Howard and Ahearn (1988) found no evidence for the occurrence of apical NaCl or NaKCl cotransport, since an inwardly directed Cl^- gradient did not stimulate Na^+ uptake and Na^+ uptake itself was not sensitive to furosemide. On the contrary, Na^+/H^+ and Cl^-/HCO_3^- exchanges were present.

In other words, while the apical K^+ conductance is a constant feature in brush border membranes, not all teleosts possess the NaKCl cotransport mechanism.

With regard to HCO_3^-, which is transferred across both apical and basolateral membranes, Ando (1990) and Ando and Subramanyam (1990) showed that its omission from bathing solutions (without changing the pH) inhibits NaKCl cotransport in seawater-adapted Japanese eel. DIDS is similarly inhibitory, with a more pronounced effect when given from the serosal side. The authors propose that bicarbonate ions raise the intracellular pH, which itself controls the apical K^+ conductance. Alkalinization favours K^+ backflux to mucosa, thus increasing Na^+ and Cl^- transfer to serosa and transepithelial PD and Isc. This is in agreement with previous studies of Field and coworkers showing that winter flounder intestine transports ions more efficiently when the pH of bathing Ringer is elevated to 7.8 or more.

4.4 Basolateral Cotransports and Channels

At the serosal side, the ouabain-sensitive Na^+/K^+ active exchange maintains low intracellular sodium concentration. This induces Na^+ entry at the apical side and, ultimately, the transepithelial transfer of NaCl. K^+ taken up in exchange for Na^+ is recycled towards the serosal medium owing to the permeability of the basolateral membrane for potassium.

Other mechanisms were shown to be present however, namely non-conductive KCl symport in winter flounder and goby (Stewart et al. 1980; Loretz 1983; Halm et al. 1985a,b), Cl^-/HCO_3^- antiport in goby (Dixon and

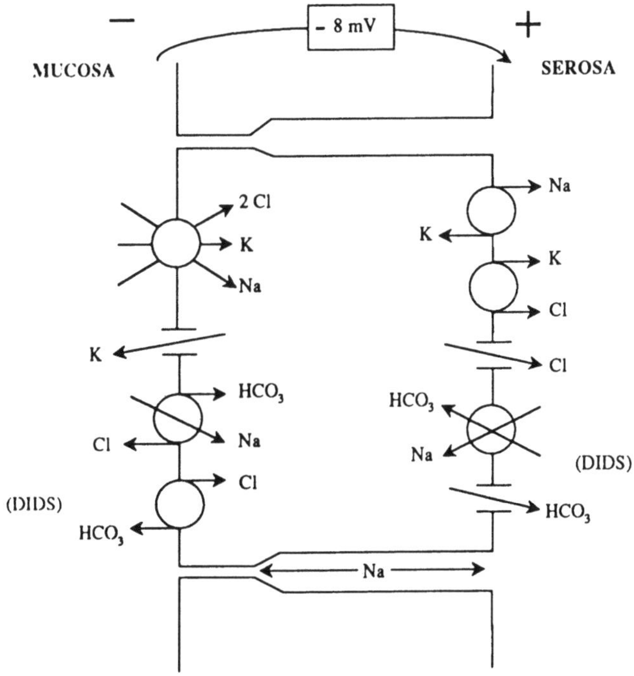

Fig. 3. Ion pathways in seawater eel intestine. (After data from Ando and Utida 1986; Ando and Subramanyam 1990)

Loretz 1986) and conductive Cl^- exit in winter flounder (Halm et al. 1985a,b), goldfish (Zuidema et al. 1985) and European eel (T. Schettino, pers. comm.). In addition, Ando and Subramanyam (1990) considered HCO_3^- movements in seawater-adapted Japanese eel and suggested the presence of a Na^+-HCO_3^- symport at the basolateral membrane. These results have been obtained on intact epithelium.

Recently (1988), Loretz and Fourtner applied for the first time the patch-clamp technique to the intestinal basolateral membrane of the goby. They isolated a calcium-insensitive, voltage-gated anion channel which may function as a conductive pathway for chloride. This confirmed directly the Cl^- conductance found in winter flounder intestine by classical electrophysiology (Halm et al. 1985a,b). The channel described by Loretz presents a conductance of 20–90 pS, well below the "maxi-chloride channel" conductance of 300–500 pS observed in several epithelia (like rabbit urinary bladder, toad nephron or MDCK cells). It is similar in size to the apical conductance found in some secretory epithelia (20–50 pS, in T84 and HT29 colonic cells and human tracheal epithelium). In addition, Loretz proposed that a functional link adjusts the basolateral conductance when the apical NaKCl entry is modified (by loop diuretics or other treatments). Figure 3 illustrates most of the transport and exchange pathways mentioned above.

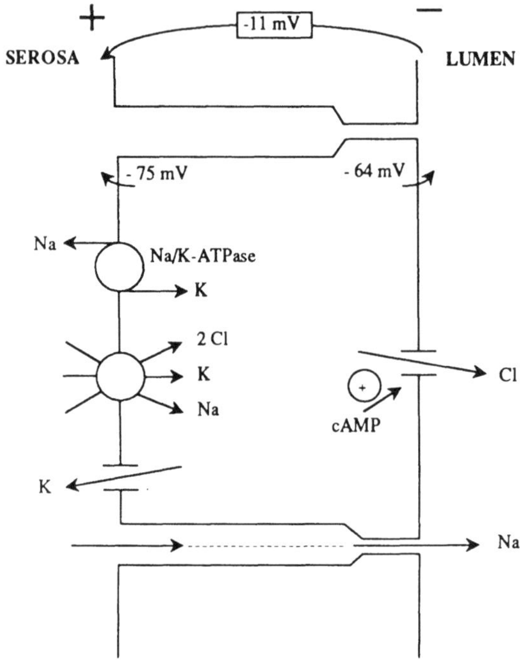

Fig. 4. Main ion pathways and secondary active transport of chloride in stimulated elasmobranch rectal gland. (After Epstein et al. 1983; Greger et al. 1984, 1987)

It is interesting to compare these features of intestinal absorption to those established recently for the elasmobranch rectal gland (Fig. 4). In this tissue, Cl^- is secreted from blood to duct lumen against its electrochemical gradient, resulting in a lumen-negative PD which is low (2 mV) at rest but increases to about 10 mV in the gland perfused with cyclic AMP (cAMP)-containing solution (see below). This secretion depends on the presence of sodium in the perfusate and is inhibited by ouabain and loop diuretics. The intracellular potential ranges between −60 and −90 mV with respect to blood. Cl^- activity measured by selective microelectrodes is around 70 mM, which is two to four times that predicted by the Nernst equation. Using the patch-clamp technique in the excised-patch configuration, Gögelein et al. (1987) and Greger et al. (1985, 1987, 1988) found a small (11 pS) and large (40–50 pS) chloride-selective channels in the apical membrane. The latter, which is functionally the more important, resembles the teleostean channel by its conductance but differs obviously from it by its hormonal sensitivity (see below).

4.5 Divalent Ion Absorption

For a long period after Smith's observations (1930), it was accepted that fish intestine is only slightly permeable to divalent ions. In vivo studies on seawater-adapted animals showed that magnesium and sulphate ions are retained, while 30–70% of ingested calcium is absorbed. It was proposed that magnesium remaining in the posterior intestine could be used to estimate the volume of water absorbed across the gut. More recently, in vitro techniques have been devised to measure transepithelial fluxes of these ions.

In seawater fish, intestinal lumen contains a five times higher Ca^{2+} concentration than in blood (ca. 10 mM compared to 2). It is therefore admitted that calcium moves to the lumen (efflux) by an active process to compensate passive entry (influx) down its concentration gradient. In the Atlantic cod, *Gadus morhua*, Sundell and Björnsson (1988) used a perfused intestine preparation which permits the determination with ^{45}Ca of the amount of calcium transferred between lumen and circulation under conditions imitating the in vivo situation. They found that Ca^{2+} influx, which is partially inhibited by the Ca^{2+}-ATPase inhibitor chlorpromazine, presents a saturable component with a Km of 8.4 mM and a Vm of 0.6 $\mu mol\,kg^{-1}\,h^{-1}$. Similarly, Ca^{2+} efflux had a saturable component with a Km of 6.15 mM and a Vm of 3.8 $\mu mol\,kg^{-1}\,h^{-1}$. The authors estimated that intestinal absorption (net flux) represents about 30% of the whole body influx.

In tilapia *Oreochromis mossambicus*, Flik et al. (1990) determined calcium fluxes in Ussing chambers on stripped epithelium. Calcium absorption was sodium-dependent, indicating the presence of a Na^+/Ca^{2+} exchanger, as in the basolateral membrane of mammalian intestine. Active Ca^{2+} extrusion was then analyzed using basolateral membrane vesicles to compare ATP-and Na^+ gradient-driven calcium fluxes. In fish, unlike mammals, it appears that the Na^+/Ca^{2+} exchange accounts for the major part of calcium transport across the basolateral border of enterocytes.

Studies dealing with magnesium transport mechanisms are scarce. Nakamura and Hirano (1986) suggested that magnesium crosses the intestinal wall of *Anguilla japonica* by solvent drag. Recently, Van der Velden et al. (1990) measured radioactive fluxes of magnesium in freshwater-adapted *O. mossambicus*, using ^{28}Mg as tracer. They proposed that the solvent-drag effect cannot account for the transport of this ion. In addition, considering the estimated electrochemical gradients, magnesium uptake across the apical membrane could be passive, while basolateral transfer requires an active process. The latter step necessitates Na^+ and Na^+/K^+-ATPase activity and suggests the presence of a Na^+/Mg^{2+} exchange mechanism which therefore operates in parallel with the Na^+/Ca^{2+} exchanger.

5 Control of Ion Transport by Second Messenger Systems

Induction of chloride secretion was discovered as a striking response of mammalian intestine to cholera toxin and cyclic AMP. Since the intestine of marine teleosts presents a serosa-negative potential initially attributed to electrogenic chloride absorption, it was of interest to determine the effects of this and other secondary messengers on this tissue.

Table 2. Some teleost species responding to cAMP (isolated intestine in Ussing chambers)

Species	Authors
Pseudopleuronectes americanus	Field et al. (1980), Rao et al. (1984), O'Grady (1989)
Platichthys flesus	Mackay et al. (1978), Mackay and Lahlou (1980)
Gillichthys mirabilis	Loretz (1983)
Carassius auratus	Groot (1985), Bakker and Groot (1989)
Tilapia mossambica	Groot and Bakker (1987)

In vitro studies in relation to cAMP levels were performed initially in Ussing chambers. They concerned mainly the teleost species listed in Table 2.

In the presence of cAMP or its derivatives dB-cAMP and 8Br-cAMP and of theophylline or forskolin, there is a large decrease in PD and Isc. This indicates an inhibition of ion transport, an increase in Cl^- unidirectional fluxes with a reduction in net flux, and an increase in total conductance. However, unlike that found in mammals and in other tetrapods, chloride secretion is never observed in teleost intestine and Na^+ fluxes are not affected (with one exception: O'Grady and Wolters 1990, for the intestine of the winter flounder). It is interesting that freshwater and seawater fish respond in the same way (the species in Table 2 are from both media). These observations led Field and coworkers (1978, 1980) to point out that the absence of chloride secretion is directly related to the lack of Lieberkuhn crypts in fish intestine.

In elasmobranchs, no similar work has been done on the intestine, but the NaCl secretion of the perfused rectal gland is largely stimulated (up to 30 times) by cAMP plus theophylline (Epstein et al. 1983). Even more meaningful, the rectal gland hardly secretes any notable quantities of water and salt unless cAMP is added in the perfusion fluid. This response is associated with enhanced steady-state binding of ^{3}H-ouabain and increased ^{86}Rb uptake (Shuttleworth and Thompson 1980; Marver et al. 1986), indicating that cAMP acts on Na^+/K^+-ATPase activity located at the serosal side. However, from detailed intracellular electrophysiological recordings, Greger et al. (1984) showed that cAMP increases first the apical Cl^- conductance and second the Na^+/K^+-ATPase activity.

The question of whether the ion permeability increase is located in the cell membrane or in the paracellular pathway has been studied on stripped goldfish intestine by Groot and Bakker (1987). By associating hypotonic treatment of the tissue and dilution and streaming potentials, which are supposed to reflect the properties of the paracellular pathway, they proposed that cAMP increases the Cl^- conductance of the tight junctions. However, they did not exclude an effect on a definite cell population, owing to the presence of conductive Cl^- channels which may be regulated.

Fewer studies have been done with cyclic GMP (cGMP). The results are as with cAMP but the effects are presented rather as cellular. In winter flounder (Halm et al. 1985a,b), mucosal bumetanide or serosal cGMP induces cell hyperpolarization and decreases the apical fractional resistance, with an increase in basolateral conductance. Similarly, ANP (the atrial natriuretic peptide), which acts specifically to increase cGMP in tissues, and 8Br-cGMP reduce PD and Isc and abolish Cl^- net flux (O'Grady et al. 1985). According to O'Grady (1989), atriopeptin III, like bumetanide and ionomycin, inhibits the initial rate of Na^+ and Cl^- uptake across the brush border membrane and thus intervenes presumably by blocking the NaKCl cotransport.

Intracellular calcium is now widely described as an essential component of the second messenger system. In goby intestine (Loretz 1987), its elevation (by the ionophore A23187) produced hyperpolarization of the apical membrane and reduced Cl^- absorption. Calmodulin is a likely mediator since the antagonist of the Ca^{2+}-calmodulin system trifluoperazine induced opposite effects. Mucosa-to-serosa flux and net flux of calcium are reduced, without a significant effect on Isc and tissue resistance, suggesting a calcium effect on a non-conductive pathway. The NaKCl cotransport may be that pathway, considering the above effect of ionomycin, which acts also as a calcium ionophore.

Further studies are necessary to determine connections between the different second messengers and to localize their respective target cells. At the present stage, it is striking that they all present inhibitory effects. In other words, the short-term acting hormones only produce a decrease in ion and water transport in fish intestine. In contrast, long-term acting hormones like cortisol induce an increase which appears as an adaptive response of the tissue.

In addition, the intestinal wall of fish (teleosts or elasmobranchs) contains a large number of endocrine cell types, the secretory products of which may act as circulating hormones, local hormones or neuromediators. Glucagon, glucagon-like, substance P-like immunoreactivity, serotoninergic and chromaffin cells, gastrin-releasing peptide and VIP have been found but their actual involvement has not been demonstrated. In fact, in teleostean intestine, few hormones have been shown to affect ion transport. For example, urotensin II stimulates active Na and Cl absorption across the posterior intestine of the goby adapted to 5% seawater.

By contrast, the hormonal control of the shark rectal gland has been explored in depth. As with cAMP, secretion of the perfused gland is largely stimulated by adenosine or VIP (Stoff et al. 1979). A similar effect obtained with the atrial natriuretic peptide (ANP) was shown to be indirect and mediated by the release of VIP from neural stores in the gland (Silva et al. 1987). Conversely, the secretion produced by cAMP, forskolin, adenosine or ANP is inhibited by somatostatin. The mechanism involved in these actions is obviously not limited to cAMP production since these agents produce a release of intracellular calcium.

6 Sodium-Linked Transport of Organic Solutes

Absorption of sugars and amino acids is a fundamental function of fish intestine. As in mammals, it has been demonstrated that this absorption takes place in many cases by a sodium-dependent process. A survey of relevant literature can be found in the review of Ferraris and Ahearn (1984). We report briefly some recent results concerning these organic solutes, in view of their relationships with Na^+ transport.

6.1 Sugar Transport

Relatively few studies have dealt with sugar transport. For example, by using everted sacs of goldfish intestine, Smith (1964) showed that the glucose concentration is increased three-fold or more with respect to the mucosal solution and that glucose added to the mucosal medium provokes an immediate rise in transepithelial PD. This transport has also been studied in catfish (Chen and Huang 1972), toadfish (Farmanfarmaian et al. 1972), tench *Tinca tinca* (Cartier et al. 1979) and winter flounder (Naftalin and Kleinzeller 1981).

More recently, Reshkin and Ahearn (1987) and Reshkin et al. (1989) examined glucose transport by brush border vesicles in the tilapia *Oreochromis mossambicus* with regard to salt adaptation. Growth stimulation of the intestine by thyroid hormones (T3) or by 17a-methyltestosterone induces an increase in this transport. They reported for the first time a change in stoichiometry of the Na-glucose cotransport which shifts from 1:1 to 1:2 under hormone treatment and suggested that higher ratios should help to produce higher intracellular sugar concentrations.

In goldfish, Groot and coworkers (Groot et al. 1983a,b) described relations between sugar transport and intracellular electrical phenomena. Glucose evoked a rapid drop in E_a in a few seconds, followed by slow repolarization. Assuming that resistances were not altered significantly, they considered that the membrane electromotive forces were not modified. They reported that glucose transport is rheogenic and involves two transport systems, with low and high affinities respectively.

6.2 Amino Acid Transport

Several techniques have been used: everted sacs, apical short-term uptake according to the technique of Schultz et al. (1967) as adapted for goldfish intestine (Ellory et al., 1972) and, more recently, membrane vesicles. However, until now, the amino acid transport systems involved have not been defined for fish intestine in general.

In goldfish, Smith (1970) showed that everted sacs transport 18 amino acids against their concentration gradients. The work focused on the effect of the temperature to which the animals were previously adapted (between 8 and 30 °C). The ability to be concentrated for some amino acids was independent of the acclimation temperature. These "non-adapting" amino acids (Met, Leu, Tyr, Phe, His) were more lipophilic than the others. With reference to the A- and L-systems characterized in Ehrlich cells, the L-system would act in the goldfish for "non-adaptors" and the A-system (Ala, Ser, Thr, Val) for "adaptors". However, there is cross-affinity between the two systems and there are differences between species since, for example, this distinction does not apply to the same amino acids in the sea bass *D. labrax* (Avella et al. 1992).

Under short-circuit conditions, mutual interference in transport of amino acid and sodium is reflected by effects on transepithelial PD. In the Japanese eel, Ando (1988) showed that L-cysteine and D-glutamine provoked large increases in serosa positivity in freshwater animals, while L- and D-alanine and glutamine and L-glutamate enhanced serosa negativity in seawater fish. According to Ando (1988), intestinal cells metabolize these amino acids to enhance ATP production which, in turn, stimulates Na^+/K^+-ATPase activity, ion transport and transepithelial PD.

The brush border membrane vesicles have been used routinely in recent years to explore relationships with Na^+ entry into the cells. In elasmobranchs, vesicles were prepared from the spiral valve of *Mustelis canis* and *Scyliorhinus canicula* (see Crane et al. 1979). In teleosts, observations of a large number of amino acids were made on the European eel by Storelli and coworkers and on the tilapia by Ahearn and coworkers. In the European eel, Storelli et al. (1989) recently characterized some specific features. All amino acids examined exhibited carrier-mediated, Na-dependent and Na-independent processes, plus diffusion, except that glutamic acid and proline did not appear to be driven by a Na-independent mechanism. The authors proposed four Na-dependent transport systems: anionic (Glu, Asp), cationic (Lys, Arg), neutral-specific (Pro, a Methyl AIB), neutral-non-specific (others). There is also a Na-independent transport which concerns Ala, Gly, Lys. The scheme so defined for a carnivorous fish resembles somehow that of mammalian intestine. Strikingly, however, considerably lower concentrations of luminal substrates are sufficient to saturate the system in fish than in mammals, as the affinity constant is below 2 mM in most cases. In addition, for L-glutamate and L-lysine, there was a sigmoidal dependence in relation to external sodium, so that two Na^+ or more are taken up with one amino acid.

Isolation of enterocytes in higher vertebrates (birds and mammals) has permitted the separation of cells from villi and crypts in order to analyze, among other functions, their respective contributions to amino acid transport. It was convincingly demonstrated that neutral amino acids are taken up preferentially by the villus tip (see Smith et al. 1983). Concern about the precise location in the case of fish intestine is illustrated by the work of Lee et al. (1991) on carp showing that the initial entry of labelled Ala or Lys, as measured by autoradiography, is restricted to the distal part of the villus. This entry is sensitive to acclimation temperature of the animals but, in this respect, the Na-dependent uptake of the two amino acids varies in opposite directions, Ala influx being enhanced at 10 °C and Lys influx at 30 °C. This aspect of different cell localizations of transport is important and deserves further exploration in relation to ion transport.

In addition to temperature, amino acid and sugar transports are necessarily under the influence of various external factors, such as the quantity (including the effects starvation; Avella et al. 1992) and composition of the diet, pollutants (hydrocarbons, heavy metals) and seasons. Although striking effects have been observed, their detailed analysis is beyond the scope of the present review. In contrast, environmental salinity is relevant but as yet little investigated.

7 Conclusion

Fish intestine has proved over decades to be an interesting and reliable model in physiological investigations. Among recent technical developments, intracellular electrophysiology and membrane vesicles have become routine procedures for this tissue. However, with the exception of the particular case of the elasmobranch rectal gland, new approaches such as cell culture, patch-clamp technique, molecular aspects in endocrine control, and cell volume regulation have not been applied often to fish intestine. In this presentation, we have not considered the pharmacological or toxicological effects of many environmental substances, like environmental pollutants, which have begun to produce precise and interesting insights at the cellular level. All these areas of research offer fruitful perspectives.

References

Aguenaou H, Hubsch A, Duportail G, Colin DA (1987) Cl^- transport through the intestinal brush border of the trout (*Salmo gairdneri* R.) Ichtyophysiol Acta 11: 111–126

Aguenaou H, Boeuf G, Colin DA (1989a) Na^+ uptake through the brush border membranes of intestine from fresh water and sea water adapted trout (*Salmo gairdneri* R.) J Comp Physiol B 159: 275–280

Aguenaou H, Husch A, Colin DA (1989b) Is there a Cl^--OH^- exchange (Cl^- H^+ cotransport) mechanism in the brush border membrane of the freshwater trout (*Salmo gairdneri*)? J Membr Biol 108: 13–20
Albertini-Berhaut J (1987) L'intestin chez les Mugiladae (Poissons; Téléostéens) à différentes étapes de leur croissance. I. Aspects morphologiques et histologiques. J Appl Ichthyol 3: 1–12
Ando M (1980) Chloride dependent sodium and water transport in the sea water eel *Anguilla japonica* intestine. Comp Biochem Physiol 138: 87–91
Ando M (1988) Amino acid metabolism and water transport across the seawater eel intestine. J Exp Biol 138: 93–106
Ando M (1990) Effects of bicarbonate on salt and water transport across the intestine of the seawater eel. J Exp Biol 150: 367–379
Ando M, Kobayashi M (1978) Effects of stripping off the outer layers of the eel intestine on salt and water transport. Comp Biochem Physiol 61: 497–503
Ando M, Subramanyam MVV (1990) Bicarbonate transport systems in the intestine of the seawater eel. J Exp Biol 150: 381–394
Ando M, Utida S (1986) Effects of diuretics on sodium, potassium, chloride and water transport across the seawater eel intestine. Zool Sci 3: 605–612
Ando M, Utida S, Nagahama H (1975) Active transport of chloride in eel intestine with special reference to sea water adaptation. Comp Biochem Physiol 51A: 27–32
Avella M, Blaise O, Berhaut J (1992) Effects of starvation on valine and alanine transport across the intestinal mucosal border in sea bass, *Dicentrarchus labrax*. J Comp Physiol B 162: 430–435
Bakker R, Groot JA (1988) Induction of bumetamide-sensitive Na(K)Cl transport and K^+ permeability in the apical membrane of intestinal epithelium of *Oreochromis mossambicus* due to seawater adaptation. Comp Biochem Physiol 90A: 824
Bakker R, Groot JA (1989) Further evidence for the regulation of the tight junction ion selectivity by cAMP in goldfish intestinal mucosa. J Membr Biol 111: 25–35
Balment RJ, Carrick S (1985) Endogenous renin-angiotensin system and drinking behavior in flounder. Am J Physiol 248: R157
Balment RJ, Hazon N, Perrot MN (1987) Control of corticosteroid secretion and its relation to osmoregulation in lower vertebrates. In: Kirsch R, Lahlou B (eds) Adaptations to salinity and dehydration. Karger, Basel, pp 92–102
Bensahla-Talet A, Porthé-Nibelle J, Lahlou B (1974) Le transport de l'eau et du sodium par l'intestin isolé de la truite *Salmo irideus* au cours de l'adaptation à l'eau de mer. C R Acad Sci 278: 2541–2544
Bogé G, Rigal A, Pérès G (1982) The use of intestinal brush border membrane vesicles for comparative studies of glucose and 2-amino-isobutyric acid transport by four species of marine teleosts. Comp Biochem Physiol 72A: 85–89
Buddington RK, Diamond JM (1987) Pyloric ceca of fish: a "new" absorptive organ. Am J Physiol 252: G65–76
Cartier M, Buclon M, Robinson JWL (1979) Preliminary studies on the characteristics of the phenylalanine and β-methylglucose transport in the tench intestine in vitro. Comp Biochem Physiol 62A: 363–370
Charney AN, Pelkonen AM (1990) Effects of pH, amiloride and bumetanide on intestinal sodium and chloride transport in the winter flounder (*Pseudopleuronectes americanus*). Bull Mt Desert Isl Biol Lab 29: 56
Chen TST, Huang KC (1972) Structural specificity in the intestinal transport of hexoses, tyrosine derivatives and electrolytes in freshwater catfish. J Pharm Exp Ther 180: 777–783
Collie N (1985) Intestinal nutrient transport in coho salmon, *Oncorhynchus kisutch*, and the effects of development starvation and seawater adaptation. J Comp Physiol 156: 163–179
Crane RK, Bogé G, Rigal A (1979) Isolation of the brush border membranes in vesicular form from the intestinal spiral value of the small dogfish (*Scyliorhinus canicula*). Biophys Biochim Acta 554: 264–267

Curtis RL, Trier JS, Frizell RA, Lindem NM, Madara JL (1984) Flounder intestinal absorptive cells have abundant gap junctions and may be coupled. Am J Physiol 246: C77
Diamond JM (1962) The mechanism of solute transport by the gallbladder. J Physiol 161: 474–502
Diamond JM, Bossert WH (1967) Standing gradient osmotic flow. J Gen Physiol 50: 2061–2083
Di Costanzo G, Duportail G, Florentz A, Leray C (1983) The brush-border membrane of trout intestine: influence of its lipid composition on ion permeability, enzyme activity and membrane fluidity. Mol Physiol 4: 279–290
Dixon JM, Loretz CA (1986) Luminal alkalinization in the intestine of the goby. J Comp Physiol B 156: 803–811
Drai P, Albertini-Berhaut J, Lafaurie M, Sudaka P, Giudicelli J (1990) Simultaneous preparation of basolateral and brush border membrane vesicles from sea bass intestinal epithelium. Biophys Biochim Acta 1022: 251–259
Duffey ME, Thompson SM, Frizzell RA, Schultz SG (1979) Intracellular chloride activities and chloride absorption in the intestinal epithelium of the winter flounder *Pseudopleuronectes americanus*. J Membr Biol 50: 331–342
Ellory JC, Lahlou B, Smith MW (1972) Changes in the intestinal transport of sodium induced by exposure of goldfish to a saline environment. J Physiol 222: 497–509
Ellory JC, Ramos M, Zeuthen T (1978) Cl^- accumulation in the plaice intestinal epithelium. J Physiol 287: 12–13
Epstein FH, Stoff JS, Silva P (1983) Mechanism and control of hyperosmotic NaCl-rich secretion by the rectal gland of *Squalus acanthias*. J Exp Biol 106: 25–41
Eveloff J, Field M, Kinne R, Murer H (1980) Sodium-cotransport systems in intestine and kidney of the winter flounder. J Comp Physiol B135: 175–182
Farmanfarmaian A, Ross A, Mazal D (1972) In vivo intestinal absorption of sugar in the toad-fish (marine teleost *Opsanus tau*). Biol Bull 142: 427–445
Ferraris RP, Ahearn GA (1984) Sugar and amino acid transport in fish intestine. Comp Biochem Physiol 77A: 397–413
Field M, Karnaky KJ, Smith PL, Bolton J, Kinter WB (1978) Ion transport across the isolated intestinal mucosa of the winter flounder, *Pseudopleuronectes americanus*. I. Functional and structural properties of cellular and paracellular pathways for Na and Cl. J Membr Biol 41: 265–293
Field M, Smith PL, Bolton JE (1980) Ion transport across the isolated intestinal mucosa of the winter flounder *Pseudopleuronectes americanus*. 2. Effects of cyclic AMP. J Membr Biol 55: 153–163
Flik G, Schoenmakers TJM, Groot JA, Van Os CH, Bonga SEW (1990) Calcium absorption by fish intestine: the involvement of ATP- and sodium-dependent calcium extrusion mechanisms. J Membr Biol 113: 13–22
Frizzell RA, Smith PL, Vosburgh E, Field M (1979) Coupled sodium chloride ion flux across brush border of flounder intestine. J Membr Biol 46: 27–39
Gibson JS, Ellory JC (1987) Intestinal salt transport in the stenothermic Antarctic fish *Notothemia rossii*. J Exp Biol 130: 437–442
Gibson JS, Ellory JC, Cossins AR (1985) Temperature acclimation of intestinal Na^+ transport in the carp (*Cyprinus carpio*). J Exp Biol 114: 355–364
Gibson JS, Ellory JC, Lahlou B (1987) Salinity acclimation and intestinal salt transport in the flounder: the role of the basolateral cell membrane. J Exp Biol 128: 371–382
Gilles-Baillien M, Gilles R (1983) Intestinal transport: fundamental and comparative aspects. Springer, Berlin Heidelberg New York
Gögelein H, Schlatter E, Greger R (1987) The "small" conductance chloride channel in the luminal membrane of the rectal gland of the dogfish (*Squalus acanthias*). Pflügers Arch Eur J Physiol 409: 122–125
Greger R, Schlatter E, Wang F, Forrest JN (1984) Mechanism of NaCl secretion in rectal gland tubules of spiny dogfish (*Squalus acanthias*). 3. Effects of stimulation of secretion by cyclic AMP. Pflügers Arch Eur J Physiol 402: 376–384

Greger R, Schlatter E, Gögelein H (1985) Cl^- channels in the apical cell membrane of the rectal gland "induced" by cAMP. Pflügers Arch Eur J Physiol 403: 446–448
Greger R, Schlatter E, Gögelein H (1987) Chloride channels in the luminal membrane of the rectal gland of the dogfish (*Squalus acanthias*). Properties of the "larger" conductance channel. Pflügers Arch Eur J Physiol 409: 114–121
Greger R, Gogelein H, Schlatter E (1988) Stimulation of NaCl secretion in the rectal gland of the dogfish *Squalus acanthias*. Comp Biochem Physiol 90A: 733–738
Groot JA (1985) Modulation of transepithelial Cl^- permeability. In: Gilles R, Gilles-Baillien M (eds) Transport processes, ions and osmoregulation, Springer, Berlin Heidelberg New York, pp 206–217
Groot JA, Bakker R (1987) Cellular and paracellular transport pathways. In: Kirsch R, Lahlou B (eds) Adaptations to salinity and dehydration. Karger, Basel, pp 56–66
Groot JA, Albus H, Bakker R, Dekker K (1983a) Changes in sugar transport and in electrophysiological characteristics of the intestinal preparation of temperature-acclimated goldfish (*Carassius auratus* L.). J Comp Physiol 151: 163–170
Groot JA, Albus H, Bakker T, Segenbeek Van Heukelom J, Zuidema Th (1983b) Electrical phenomena in fish intestine. In: Gilles-Baillien M, Gilles R (eds) Intestinal transport. Springer, Berlin Heidelberg New York, pp 321–340
Halm D, Krasny EJ, Frizzell RA (1985a) Electrophysiology of flounder intestinal mucosa. I. Conductance properties of the cellular and paracellular pathways. J Gen Physiol 85: 843–864
Halm D, Krasny EJ, Frizzell RA (1985b) Electrophysiology of flounder intestinal mucosa. II. Relation of the electrical potential profile to coupled NaCl absorption. J Gen Physiol 85: 865–883
Hirano T, Mayer-Gostan N (1976) Eel oesophagus as an osmoregulatory organ. Proc Natl Acad Sci USA 73: 1348–1350
Hirano T, Utida S (1968) Effects of ACTH and cortisol on water movement in isolated intestine of the eel *Anguilla japonica*. Gen Comp Endocrinol 11: 373–380
Howard JW, Ahaern GA (1988) Parallel antiport mechanisms for Na^+ and Cl^- transport in herbivorous teleost intestine. J Exp Biol 135: 65–76
House CR, Green K (1965) Ion and water transport in isolated intestine of the marine teleost, *Cottus scorpius*. J Exp Biol 42: 177–189
Huang KC, Chen TST (1971) Ion transport across intestinal mucosa of winter flounder, *Pseudopleuronectes americanus*. Am J Physiol 220: 1734–1738
Kirsch R, Lahlou B (1987) Adaptations to salinity and dehydration. Karger, Basel
Kirsch R, Laurent P (1975) L'oesophage, organe effecteur de l'osmorégulation chez un téléostéen euryhalin, l'Anguille (*Anguilla anguilla* L.). C R Acad Sci 280: 2013–2015
Kirsch R, Meister MF (1982) Progressive processing of ingested water in the gut of sea water teleosts. J Exp Biol 98: 67–81
Lahlou B (1976) Ionic permeability of fish intestinal mucosa in relation to hypophysectomy and salt adaptation. In: Robinson JWL (ed) Intestinal ion transport. MTP, London, pp 318–387
Lau KR (1985) The effect of salinity adaptation on intracellular chloride accumulation in the European flounder. Biophys Biochim Acta 818: 105–108
Lee JAC, Cossins AR (1988) Adaptation of intestinal morphology in the temperature-acclimated carp, *Cyprinus carpio* L. Cell Tissue Res 251: 451–546
Lee JAC, Cossins AR (1990) Temperature adaptation of biological membranes: differential homeoviscous responses in brushborder and basolateral membranes of carp intestinal mucosa. Biophys Biochim Acta 1026: 195–203
Lee JAC, James PS, Smith MW, Cossins AR (1991) Amino acid transport in the intestine and mucosa of temperature acclimated carp. J Therm Biol 16: 7–11
Lorenzo A, Balanos A (1989) Efectos de la salinidad sobre el transporte de Na-Cl a través del intestino del pez *Blennius parvicornis*. Bol Inst Esp Oceanogr 5: 37–42
Loretz CA (1983) Ion transport by the intestine of the goby, *Gillichthys mirabilis*. Comp Biochem Physiol 75A: 205–210

Loretz CA (1987) Regulation of goby intestinal ion absorption by the calcium messenger system. J Exp Zool 244: 67–78
Loretz CA, Fourtner CR (1988) Functional characterization of a voltage-gated anion channel from teleost intestinal epithelium. J Exp Biol 136: 383–403
Loretz CA, Howard ME, Siegel AJ (1985) Ion transport in goby intestine: cellular mechanism of urotensin II stimulation. Am J Physiol 249: G284–293
Machen TE, Diamond JM (1969) An estimate of the salt concentration in the lateral intercellular spaces of rabbit gallbladder during maximal fluid transport. J Membr Biol 1: 194
Mackay WC, Lahlou B (1980) Relationship between Na^+ and Cl^- fluxes in the intestine of the European flounder, *Platichthys flesus*. In: Lahlou B (ed) Epithelial transport in the lower vertebrates. Cambridge University Press, Cambridge, pp 151–162
Mackay WC, Lahlou B, Porthé-Nibelle J (1978) AMP cyclique et contrôle des échanges ioniques au niveau de l'intestin de poisson. C R Acad Sci 287: 1239–1242
Maetz J, Skadhauge E (1968) Drinking rate and ionic gill turnover in relation to external salinities in the eel. Nature 217: 371–373
Mainoya JR (1982) Water and NaCl absorption by the intestine of the tilapia, *Sarotherodon mossambicus*, adapted to fresh water or sea water and the possible role of prolactin and cortisol. J Comp Physiol 146B: 1–8
Marver D, Lear S, Marver LT, Silva P, Epstein FH (1986) Cyclic AMP-dependent stimulation of Na,K-ATPase in shark rectal gland. J Membr Biol 94: 205–216
Musch MW, Orellana S, Kimberg L, Field M, Halm DR, Krasny E Jr, Frizzell R (1982) Na^+-K^+-Cl^--co-transport in the intestine of a marine teleost. Nature 300: 351–353
Naftalin RJ, Kleinzeller A (1981) Sugar absorption and secretion by the winter flounder intestine. Am J Physiol 240: G392–400
Nakamura Y, Hirano T (1986) Effect of hypophysectomy on absorption of organic phosphate in eel intestine. Comp Biochem Physiol 84A: 595–599
Nonotte L, Nonotte G, Leray C (1986) Morphological changes in the middle intestine of the rainbow trout, *Salmo gairdneri*, induced by hyperosmotic environment. Cell Tissue Res 243: 609–617
O'Grady SM (1989) Cyclic nucleaotide-mediated effects of ANF and VIP on flounder intestinal ion transport. Am J Physiol 256: C142–146
O'Grady SM, Wolters PJ (1990) Evidence for chloride secretion in the intestine of the winter flounder. Am J Physiol 258(2): C243–247
O'Grady SM, Field M, Nash NT, Rao MC (1985) Atrial natriuretic inhibits Na-K-Cl cotransport inteleost intestine. Am J Physiol 249: C531–534
Pelletier X, Duportal G, Leray C (1987) Dietary modifications of phospholipid composition and biophysical properties of the brush border membrane along the trout intestine. Membr Biochem 7(1): 55–56
Porthé-Nibelle J, Lahlou B (1975) Effects of corticosteroid hormones and inhibitors of steroids on sodium and water transport by goldfish intestine. Comp Biochem Physiol 50A: 801–805
Ramos MMP, Ellory JC (1981) Sodium and chloride transport across the isolated anterior intestine of the plaice *Pleuronectes-platessa*. J Exp Biol 90: 123–142
Rao MC, Field M (1983) Role of calcium and cyclic nucleotides in the regulation of intestinal ion transport. In: Gilles-Baillien M, Gilles R (eds) Intestinal transport. Springer, Berlin Heidelberg New York, pp 227–239
Rao MC, Nash NT, Field M (1984) Differing effects of cGMP and cAMP on ion transport across flounder intestine. Am J Physiol 246: C167–171
Reshkin SJ, Ahearn GA (1987) Intestinal glucose transport and salinity adaptation in a euryhaline teleost. Am J Physiol 252: R567–578
Reshkin SJ, Grover ML, Howerton RD, Grau EG, Ahearn GA (1989) Dietary hormonal modification of growth, intestinal ATPase and glucose transport in tilapia. Am J Physiol 256: G10–G18
Robinson JWL (1976) Intestinal ion transport. MTP, London
Schultz SG, Curran PF, Chez RA, Fuisz RE (1967) Alanine and sodium fluxes across mucosal border of rabbit ileum. J Gen Physiol 50: 1241–1260

Shuttleworth TJ, Thompson JL (1980) The mechanism of cyclic AMP stimulation of secretion in the dogfish rectal gland. J Comp Physiol 140: 209–216
Silva P, Stoff JS, Solomon RJ, Lear S, Kniaz D, Greger R, Epstein FH (1987) Atrial natriuretic peptide stimulates salt secretion by shark rectal by releasing VIP. Am J Physiol 252: F99–103
Simonneaux V, Barra JA, Hmbert W, Kirsch R (1987) The role of mucus in ion absorption by the oesophagus of the sea-water eel (*Anguilla anguilla* L). Electrophysiological, structural and cytochemical investigations. J Comp Physiol B 157: 187–200
Smith H (1930) The absorption and excretion of water and salts by marine fishes. Am J Physiol 93: 480–505
Smith MW (1964) The in vitro absorption of water and solutes from the intestine of goldfish *Carassius auratus*. J Physiol 175: 38–49
Smith MW (1970) Selective regulation of amino acid transport in the intestine of goldfish (*Carassius auratus* L). Comp Biochem Physiol 35: 387–401
Smith MW (1983) Membrane transport in fish intestine. Comp Biochem Physiol 75A: 325–335
Smith MW, Ellory JC, Lahlou B (1975) Sodium and chloride transport by the intestine of the European flounder *Platichthys flesus* adapted to fresh or sea water. Pflügers Arch Eur J Physiol 357: 303–312
Smith MW, Sepulveda FV, Petersen JVF (1983) Cellular aspects of aminoacid transport. In: Gilles-Baillien M, Gilles R (eds) Intestinal transport. Springer, Berlin Heidelberg New York, pp 46–63
Smith PL, Welsh MG, Stewart CP, Frizzell RA, Orellana SA, Field M (1981) Chloride absorption by the intestine of the winter flounder: mechanism of inhibition by reduced pH. Bull Mt Desert Isl Biol Lab 20: 96–101
Stewart CP, Smith PL, Welsh MJ, Frizzell RA, Musch MW, Field M (1980) Potassium transport by the intestine of the winter flounder, *Pseudopleuronectes americanus*: evidence for KCl cotransport. Bull Mt Desert Isl Biol Lab 20: 92–96
Stoff JS, Rosa R, Hallac R, Silva P, Epstein FH (1979) Hormonal regulation fo active transport in the dogfish rectal gland. Am J Physiol 237: F138–144
Storelli C, Vilella S, Cassano G (1986) Na-dependent D-glucose and L-alanine transport in eel intestinal brushborder membrane vesicles. Am J Physiol 251: 463–469
Storelli C, Vilella S, Romano MP, Maffia M, Cassano G (1989) Brushborder amino acid transport mechanisms in carnivorous eel intestine. Am J Physiol 257: R506–510
Sundell K, Björnsson BT (1988) Kinetics of calcium fluxes across the intestinal mucosa of the marine teleost *Gadus morrhua*, measured using an in vitro perfusion method. J Exp Biol 140: 171–186
Van der Velden JA, Groot JA, Flick G, Polak P, Kolar ZI (1990) Magnesium transport in fish intestine. J Exp Biol 152: 587–592
Yamamoto M, Hirano T (1978) Morphological changes in the oesophageal epithelium of the eel, *Anguilla japonica*, during adaptation to sea water. Cell Tissue Res 192: 25–38
Zeuthen T, Ramos M, Ellory JC (1978) Inhibition of active chloride transport by piretanide. Nature 273: 678–680
Zuidema T, Van Riel JW, Siegenbeek Van Heukelom J (1985) Cellular and transepithelial responses of goldfish intestinal epithelium to chloride substitutions. J Membr Biol 88: 293–304

Chapter 2

Mechanisms of Electrogenic Sodium Transport in the Amphibian Colon

Rolf Krattenmacher[1,2] and *Wolfgang Clauss*[1]

Contents

1 Introduction

In the early 1950s Ussing and Zerahn (1951) introduced the technology of the short-circuit current as a method to study active ion transport through isolated epithelia. They could show that the electrical potential difference across the isolated frog skin is caused by a primary active absorption mechanism for sodium ions, located in the tissue itself. Since that time this technology as well as amphibian tissues became more and more established in many "epithelial ion transport laboratories" all over the world. In the meantime hundreds if not thousands of investigations dealing with this active sodium transport system have been published, wereby the toad urinary bladder and the frog skin developed into classical model systems. In contrast, ion transport processes in the amphibian intestine, especially those in the large intestine (colon), have been only relatively poorly investigated.

[1] Institut für Tierphysiologie, Justus-Liebig-Universität Giessen, Wartweg 95, W-6300 Giessen, Germany
[2] Present address: Dept. of Exp. Gynecology, Schering AG, P.O. Box 650311, W-1000 Berlin 65, Germany

Advances in Comparative and Environmental Physiology, Vol. 16

Table 1. The zoological system of the amphibians

Order	Suborder[a]	Representative
Anura	* Amphicoela	
	* Aglossa	*Xenopus* sp.
	* Opistocoela	
	* Anomocoela	
	Displasiocoela	*Rana* sp.
	Procoela	*Bufo* sp.
Urodela	* Cryptobranchnoidea	
	Sirenoidea	
	Salamandroidea	*Necturus* sp., *Amphiuma* sp.
Gymnophiona	–	

[a] * Ancient suborders.

The first results concerning colonic sodium transport in amphibians were reported by Ussing and Andersen (1955). They showed that sodium ions in the toad colon, similarly as in the frog skin, are also actively absorbed from the mucosal to the serosal side of the tissue. Since then different laboratories have focussed again and again on a more detailed characterization of this intestinal sodium transport and the mechanisms of its regulation. The aim of this chapter is to summarize the essential results of these studies in order to compare the mechanisms of electrogenic sodium transport in the amphibian colon with those in other tissues.

Among transport physiologists confusion sometimes arises from the position of the different amphibian species such as frogs, toads, or salamanders within the zoological system. Therefore the following section will give a brief summary of the amphibians system.

2 The Zoological System of Amphibians

An animal often used in physiological investigations is the African clawed toad *Xenopus laevis*. The English name indicates that this species is a toad. The German name (Glatter Krallenfrosch) and the French name (Grenouille onglée) indicate that *Xenopus* is a frog. Therefore, the question can arise whether *Xenopus laevis* is a frog or a toad. This problem can be solved by considering the zoological system of the amphibians (Grzimek 1970) which is composed of three orders with several suborders each, (Table 1). These representative amphibian species, which are most commonly used in epithelial ion transport studies, are assigned to the corresponding suborder.

Anurans lose their tail during metamorphosis,while urodeles keep it as adult animals. Gymnophiones are worm-like animals which have neither tails nor legs. One of the main criteria to divide the anurans into six suborders is the shape and number of the vertebrae, whereby those suborders which are indicated by an asterisk in Table 1 are regarded as ancient groups. The most dominant and widespread anurans are the Displasiocoela to which the "typical toad" (*Bufo* sp.) belongs, and the Procoela with the "typical frog" (*Rana* sp.) as one representative. The criteria to divide the urodeles into three suborders are different from those in the anurans. In the Cryptobranchnoidea the lower jaw is more ancient (*) than that of the two other groups. The Sirenoidea are permanent larvae with only fore- but no hindlegs, whereas the "typical salamanders" with more than 400 species are the Salamandroidea.

Obviously, *Xenopus laevis* is neither a real toad nor a real frog but a representative per se of an ancient group of amphibians, showing that systematic considerations might be helpful in comparative studies.

3 The Function of the Amphibian Colon

Osmoregulation in amphibians is of special interest since amphibians inhabit both aquatic and terrestrial environments. Although amphibians were the first vertebrate group on land (Bogart 1981), most species must inhabit an aquatic environment during at least one stage of their life. Thus, extreme demands are made on the osmoregulatory capability of this vertebrate group. Exposed to air amphibians lose a considerable amount of water by evaporation through the skin, which is comparable to that from a free water surface (Bentley 1971). Therefore the behaviour of terrestrial amphibians is one of the most important osmoregulatory factors. Exposed to fresh water, which is hypoosmotic to the body fluid, amphibians accumulate water through their skin. This water uptake may be equivalent to 30–40% of their body weight per day and must be excreted by the kidneys. Although this urine is very hypoosmotic, a continuous electrolyte excretion occurs. The loss of some of these electrolytes, especially sodium, may be physiologically significant (Bentley 1971). The organs which regulate water and ion balance are mainly the skin, the kidney, the urinary bladder, and the large intestine. The movements of water and ions are controlled by the action of mainly hypophyseal, thyroid, and adrenocortical hormones (Warburg and Rosenberg 1990). In the ion balance, the regulation of sodium uptake plays a predominant role. Especially aquatic anurans can take up sodium actively from solutions that contain socium only at a concentration of 10^{-5} mol/l. In contrast to the skin, which is in contact with freshwater containing such low sodium concentrations, the renal tubule, the urinary bladder, and the colon are exposed to higher sodium concentrations at the luminal surface of the epithelium. During the aquatic stage, the glomerular filtrate is for different species about 100–160 mmol/l (for review, see Bentley 1971). During the passage through the nephron, sodium is progressively reabsorbed from the glomerular filtrate, so that the sodium concentration in the ureteral urine is lowered to about 5 mmol/l (Walker et al. 1937). Because of the close relationship, both anatomically and physiologically, the urinary bladder, the cloaca and the large intestine can be considered as a "functional syncytium", which is not merely a provisorial short-term storage place for urine or feces. In many amphibians the urinary bladder and the colon can store large volumes of urine and feces for extended periods of time. During such storage periods, the epithelia lining these organs are known to exchange water and salts with the body fluids. Thus, the colonic epithelium is a major site of osmoregulatory processes which play a significant role in water and ion balance (Deyrup 1964; Rosoff et al. 1983).

4 Electrogenic Sodium Absorption

4.1 Basic Electrophysiological Parameters

This section will give an overall view of the basic electrophysiological parameters of the amphibian colon derived from transepithelial in vitro measurements in Ussing chambers. Basic parameters are the spontaneous transepithelial potential difference (V_t), the transepithelial resistance (R_t) and the short-circuit current (I_{sc}), which is required to compensate V_t to zero. Table 2 summarizes the results obtained from different species. Furthermore, the electrogenic sodium absorption (I_{Na}) is also given as a percentage of short-circuit current. I_{Na} was either measured by isotope fluxes (^{22}Na or ^{24}Na as tracers), sodium substitution by organic cations (mostly choline), or by using sodium channel blockers (amiloride).

As can be seen in Table 2, the transepithelial resistance of the colonic tissue is relatively high and almost comparable to that of the frog skin. According to Nagel (1976), the amphibian colon must therefore be regarded as a tight epithelium because R_t is higher than $0.35\,k\Omega \cdot cm^2$. Thus, the amphibian colon seems to be considerably tighter than the colon of other vertebrates such as turtles (Dawson 1977), rabbits (Frizzell et al. 1976), rats (Schultzke et al. 1986) or hens (Clauss et al. 1984), in all of which R_t does not exceed $0.5\,k\Omega \cdot cm^2$. This makes the amphibian colon well suited for electrophysiological ion transport studies. However, Table 2 also shows that large intestinal ion transport was so far only studied in anuran amphibians, but not in urodeles or gymnophiones. In urodeles, especially in *Necturus maculosus* (Gunter-Smith et al. 1982) and *Amphiuma* sp. (White 1977), intestinal ion transport was mainly studied in the small intestine. In contrast to R_t, the transepithelial potential difference and the short-circuit current show a wide scatter which does not reflect species differences but rather individual variations within the species. I_{sc} and V_t also depend on the composition of the Ringer's solutions used by the investigators. These

Table 2. Basic electrophysiological parameters of ampibian colonic epithelium

Species	V_t (mV)	I_{sc} ($\mu A/cm^2$)	I_{Na} (% of I_{sc})	R_t ($k\Omega \cdot cm^2$)	Reference
Rana catesbeiana	45	107.2	92	0.40	Cooperstein and Hogben (1959)
Rana esculenta	29.5*	19.8	85–90	1.49	Krattenmacher and Clauss (1988)
Rana escul./temp.	46.3*	33.1	85	1.4	Krattenmacher et al. (1988)
Xenopus laevis	25.7*	24.2	60	1.06	Krattenmacher et al. (1990)
	9.4*	14.6	–	0.64	Krattenmacher et al. (1991)
Bufo marinus	10.7	32.8	–	0.33*	Cofré and Crabbé (1965)
	5.8	21.5	93	0.27*	Cofré and Crabbé (1967)
	10–40	5.3–37.4	–	0.9–1.9	Cuthbert (1973)
	38.6	61.6	88	0.53	Dawson and Curran (1976)
	70.8	61.5	87	0.74–1.2	Macchia and Helman (1979)
Bufo marinus/bufus	–	14.1	87	–	Ferreira and Smith (1968)
Bufo arenarum	35	65.9	71	0.53*	Lew and Carlisky (1967)
	15–60	75	62–100	–	Lew (1970)

Values are given as mean values or range, * calculated by the authors using Ohms law.
V_t = transepithelial potential difference, I_{sc} = short circuit current.
I_{Na} = sodium-dependent current, R_t = transepithelial resistance.

solutions are in most cases very similar but show nevertheless slight variations. However, in all investigations listed in Table 2 the major part of the short-circuit current (60–100%) is explained by the active absorption of Na^+ ions. The residual fraction of I_{sc} can therefore only be explained on the basis of an anion transport from serosa to mocosa and/or a cation transport in the opposite direction. There is evidence that the non-Na^+ component is bicarbonate, which is formed in the epithelial cells and secreted into the gut lumen, at least in *Bufo arenarum* (Carlisky and Lew 1970). Transepithelial electrogenic bicarbonate or chloride fluxes were not measured (Cooperstein and Hogben 1959; Lew and Carlisky 1967; Lew 1970). Electrogenic Ca^{2+} absorption in the colon of *Bufo marinus* was reported by Rosoff et al. (1983). However, the amount of this transport, which was found to be lower than $0.1\,\mu A/cm^2$, is negligible when it is compared to the much higher sodium and bicarbonate transport rates.

4.2 Pathways of Electrogenic Sodium Absorption

In many tight epithelia, such as the amphibian skin or the vertebrate urinary bladder and large intestine, sodium ions are actively absorbed by generating a transepithelial potential difference of several millivolt. Due to the net movement of positive charges, this type of sodium transport is called electrogenic sodium absorption. Here, Na^+ ions diffuse along an electrochemical gradient from the outer medium into the epithelial cells. This may occur by different transport mechanisms which are located in the apical membrane of the epithelial cell. The Na^+-entry step may be coupled to the transport of substrates such as glucose or amino acids, or Na^+ ions enter the cell via ion channels which may be more or less specific for sodium ions. In contrast to these differences in the apical Na^+-transport system, the basolateral mechanism by which Na^+ leaves the cell towards the blood side seems to be in all cases the same: 3 Na^+ ions are actively exchanged against $2K^+$ ions by a Na^+/K^+-ATPase located in the basolateral membrane. Potassium ions, which have been transported into the cell, diffuse back to the blood side along their chemical gradient through potassium channels. In the amphibian large intestine the apical Na^+-transport systems are, to the best of our knowledge, exclusively ion channels. So far, no evidence was found that an electrogenic Na^+-substrate cotransport mechanism exists in the apical membrane of amphibian colonic enterocytes. However, since ion transport studies in the amphibian large intestine have been performed so far only in a small number of species, the results summarized here are not necessarily representative for the amphibians per se.

4.2.1 The Amiloride Sensitive Sodium Channel

Undoubtedly, one of the most important and most investigated epithelial transport systems is a highly specific sodium channel which is located in the apical membrane of various epithelial cells and which can be blocked by micromolar concentrations of the diuretic amiloride (for review, see Benos 1986; Palmer 1986; Garty and Benos 1988). Amiloride, which can be used as a tool in the study of ion transport (Bentley 1968; Cuthbert and Shum 1974; Kleyman and Cragoe 1988) was also used for characterization of large intestinal electrogenic Na-absorption in amphibians. However, early reports have already suggested that the toad colon is almost *one decade* less sensitive to amiloride than other amphibian epithelia such as the skin or the urinary bladder (Bentley 1968; Ehrlich and Crabbé 1968). In a recent study this lower sensitivity was also described for the colon of the frog *Rana esculenta* (Krattenmacher and Clauss 1988). In this tissue a half-maximal inhibition of the electrogenic sodium transport was achieved by 0.83 μmol/l mucosal amiloride (Fig. 1), indicating, however, a higher sensitivity to amiloride as e.g., in the large intestine (coprodeum) of the hen, where 1.8 μmol/l, amiloride is needed to block 50% of the sodium current.

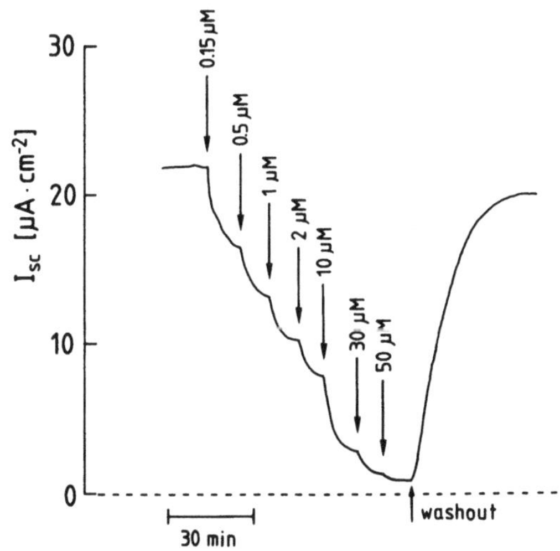

Fig. 1. Recording of the response of the short circuit current (I_{sc}) to graded doses of the sodium channel blocker amiloride in an isolated colon of *Rana esculenta*. (Krattenmacher and Clauss 1988)

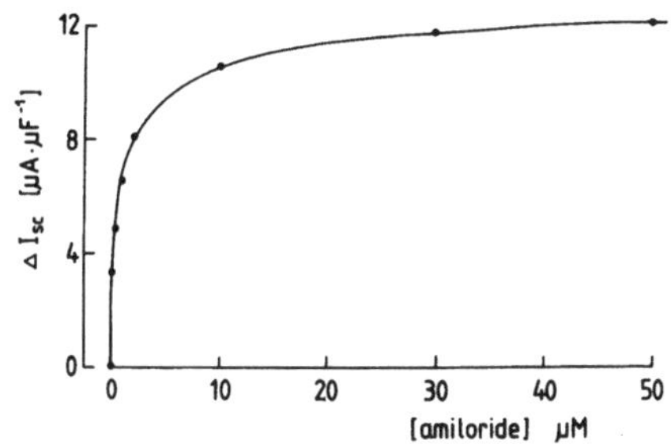

Fig. 2. The amiloride-sensitive current in an isolated colon of *Rana esculenta* is plotted against the concentration of mucosal amiloride, and shows Michaelis-Menten kinetics. (Krattenmacher and Clauss 1988)

Although Fig. 2 indicates that the amiloride-induced inhibition of the sodium current shows Michaelis-Menten kinetics, this type of kinetics possibly represents a simplified model of the interaction between amiloride and its receptor site. While amiloride is generally regarded to block the sodium channel one to one, Cuthbert (1973) could show that at least in the colon of *Bufo marinus* two different receptor sites with different affinities to amiloride are likely: a high affinity with an apparent affinity constant (K_a) of about 0.02 µmol/l and a low affinity with a K_a of 1.31 µmol/l. Following this model, Cuthbert (1973) showed that the binding of amiloride to one of these two receptor types would induce a 50% inhibition of the sodium current. The existence of two different receptor sites may possibly be the explanation for the incomplete blockage of the sodium current in the colon of *Rana* sp. with the amiloride analogue CDPC (6-chloro-3,5-diamino-pyrazine-2-carboxamide), which blocked only 41–46% of the amiloride sodium current (Krattenmacher et al. 1988). In the same study the number of sodium channels per unit area was calculated from current fluctuations which were induced by either amiloride or CDPC (for details concerning the analysis interested readers are referred to Lindemann and Van Driessche 1977; Lindemann 1984). Using amiloride, 23×10^6 channels were calculated for $1\,cm^2$ tissue area, whereas with CDPC only about half as many ($12 \times 10^6/cm^2$) were found. The corresponding single channel current was in both cases found to be about 1 pA. These results are in good agreement with those of Cuthbert (1973) and possibly also indicate two different binding sites for amiloride in the colon of frogs. On the other hand, the anaylsis of the amiloride-induced inhibition revealed a Hill coefficient (n_H) of 1.03 indicating that amiloride has only one binding site. A deviation of n_H from one would indicate that Michaelis-Menten kinetics could not be assumed or that multiple binding sites are involved.

However, amiloride remains a suitable blocker of electrogenic sodium transport in the amphibian colon, which allows a rapid estimation of the amount of transported sodium ions. It should, however, be kept in mind that, as pointed out in the following section, electrogenic sodium transport is not necessarily in all cases sensitive to mucosal amiloride.

4.2.2 Non-Selective Cation Channels

In the basic electrophysiology of the anuran colon, the transepithelial potential difference, transepithelial resistance and short-circuit current are very similar in all species investigated so far (Table 1). In all these species also the amount of electrogenically absorbed sodium ions (given as the percentage of short-circuit current in Table 1) is almost identical. At first glance, these similarities suggest that in all these species the electrogenic sodium transport follows the same mechanism. However, whereas in *Bufo* sp. and *Rana* sp. electrogenic sodium transport can be totally blocked by mucosal amiloride, sodium transport in the African clawed toad *(Xenopus laevis)* is completely insensitive to this sodium channel blocker (Fig. 3). It was found that in the colon of *Xenopus laevis* apical Na^+ entry into the epithelial cell does not occur through amiloride-sensitive Na^+ channels, but through non-selective cation channels, which are sensitive to mucosal Ca^{2+} ions (Krattenmacher et al. 1990, 1991). The authors demonstrated that electrogenic sodium transport (I_{Na}) can be increased by more than 100%, when Ca^{2+} ions are removed from the mucosal Ringer's solution. This Ca^{2+}-sensitive increase in I_{Na} could be stepwise reversed by readdition of increasing micromolar amounts of mucosal Ca^{2+} ions. Figure 4 shows that already Ca^{2+} concentrations of about 1 µmol/l were sufficient to induce a half-maximal blocking effect.

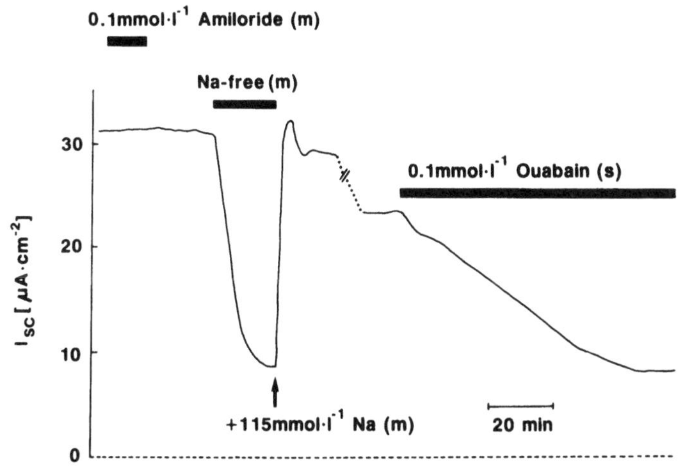

Fig. 3. Time course of the short circuit current (I_{sc}) in the colon of the African clawed toad (*Xenopus laevis*) when all mucosal (*m*) sodium is substituted by impermeant organic cations (choline). The specific sodium channel blocker amiloride (0.1 mmol/l, mucosal) shows no effect on I_{sc}. The sensitivity of I_{sc} to serosal ouabain (0.1 mmol/l) indicates the presence of a Na/K-ATPase in the basolateral membrane of the epithelial cells (Krattenmacher et al. 1990)

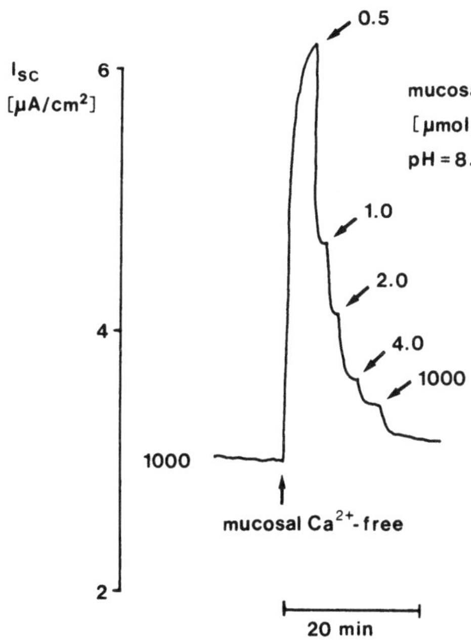

Fig. 4. Time course of the short circuit current (I_{sc}) in the colon of the African clawed toad (*Xenopus laevis*) after removal and subsequent stepwise readdition of mucosal Ca^{2+} ions. The increase in I_{sc} fully accounts for the increased electrogenic Na^+ absorption (Krattenmacher et al. 1991)

The analysis of the current noise revealed that this Ca^{2+}-sensitive sodium-transport system must be an ion channel in the apical membrane, because the mucosal Ca^{2+} removal induced a Lorentzian component in the power density spectrum that is typical for ion channels (Krattenmacher et al. 1990). From transepithelial selectivity experiments it was concluded that the ion selectivity of this non-selective cation channel is, at least in the absence of mucosal Ca^{2+} ions, $K^+ > Na^+ = Rb^+ > Cs^+ > Li^+$ (Krattenmacher et al. 1991). However, the idea that *Xenopus laevis*, in general, may not provide amiloride-sensitive sodium channels must be rejected, because their existence was demonstrated in the lung epithelium (Fischer et al. 1989). This Ca^{2+} sensitivity of electrogenic sodium transport is not a peculiarity of the colon of *Xenopus laevis*, but a mechanism which was so far found in various other amphibian epithelia such as the frog skin (Hillyard et al. 1982; Van Driessche and Zeiske 1985) or the toad urinary bladder (Van Driessche et al. 1987; Aelvoet et al. 1988). Also, for the intestinal epithelia of other vertebrate groups such as mammalians (Martens et al. 1990) or birds (Heinz et al. 1991), a similar Ca^{2+}-sensitive absorption of Na^+ was observed. The physiological function of this Ca^{2+} sensitivity is, however, in all cases still unclear.

At least for the intestinal epithelia, a regulation of electrogenic sodium transport by the Ca^{2+} concentration of the gut lumen must be excluded. Luminal Ca^{2+} concentrations are much too high as to allow an increased Na^+ transport. Thus, in the colon of *Bufo marinus* the luminal Ca^{2+} concentration was found to be about 5 mmol/l (Rosoff et al. 1983). To regulate the amount of Na^+ absorption, however, luminal Ca^{2+} concentrations should be more in the micro- than in the millimolar range. Thus far only speculations about the physiological role of calcium sensitivity are possible. For instance, it was considered that the binding of Ca^{2+} at or near the outside of the channel has no regulatory function, but rather determines the properties (e.g., conductance, selectivity) of the channel (Das and Palmer 1989; Krattenmacher et al. 1991). Van Driessche (1987) even speculated that the Ca^{2+}-sensitive channel in the toad urinary bladder may represent an epithelial calcium channel, because similar properties have been described for calcium channels in muscle fibres (Almers and McCleskey 1984; Almers et al. 1984; Hess et al. 1986). Interestingly, in the colon of *Bufo marinus*, Ca^{2+} is actively absorbed from the gut lumen (Rosoff et al. 1983). For final assertions, however, further investigations on the physiological role of calcium-sensitive sodium absorption are required.

4.3 Hormonal Regulation

4.3.1 Neurohypophyseal Hormones

In contrast to most fish, the neurohypophysis of amphibians is enlarged posteriorly to form a neural bulb. This persists throughout the tetrapods.

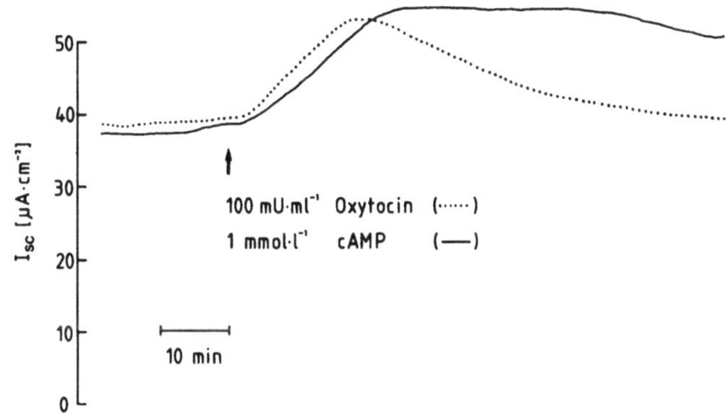

Fig. 5. Time course of short circuit current (I_{sc}) after serosal stimulation with 100 mU/ml oxytocin (----) or 1 mmol/l cAMP (——) in the colon epithelium of *Rana esculenta*. (Krattenmacher et al. 1988)

In mammals, the prevailing neurohypophyseal hormones are arginine-vasopressin and oxytocin. In amphibians, the predominant hormone that is released by this gland is arginine-vasotocin, a peptide that is also present in reptiles, birds and most of fish, but is absent in mammals. A second substance, with properties similar to those of oxytocin, was also found to be present in amphibians and was called mesotocin. Mesotocin is also present in reptiles, but is absent in most fish (with the interesting exception of some lung fish). Oxytocin has been identified in the neurohypophysis of a single amphibian, *Rana pipiens*. Vasotocin, mesotocin and oxytocin can all promote water retention in various amphibians, but vasotocin is by far the most active of these peptides (Bentley 1971).

It is well known that in various epithelia antidiuretic hormones not only increase the water permeability of epithelia, but are also potent stimulators of electrogenic sodium absorption (Hays 1983; Lewis 1983). However, there is evidence that both cellular actions are caused by different and independent processes. While the increase in water permeability is the result of a recruitment of water channels by fusion of cytoplasmic vesicles with the apical cell membrane (Wade 1985, 1986), the increased sodium uptake seems to be caused by an activation of preexisting apical sodium channels (Palmer et al. 1980; Palmer 1984, 1986). However, our understanding of these hormonal actions is derived from experiments on the toad urinary bladder. Moreover, in intestinal epithelia, such as the colon, different responses of ion transport in reaction to neurohypophyseal hormones were found. In the amphibian colon antidiuretic hormones (ADH) and their second messenger cyclic adenosine monophosphate (cAMP) induce a 20–80% increase in electrogenic sodium absorption as demonstrated in Fig. 5 (Cofré and Crabbé 1967; Krattenmacher and Clauss 1988; Krattenmacher et al. 1988). In contrast, in the mammalian colon (rat) ADH induces a secretion of chloride into the intestinal lumen (Bridges et al. 1984). In

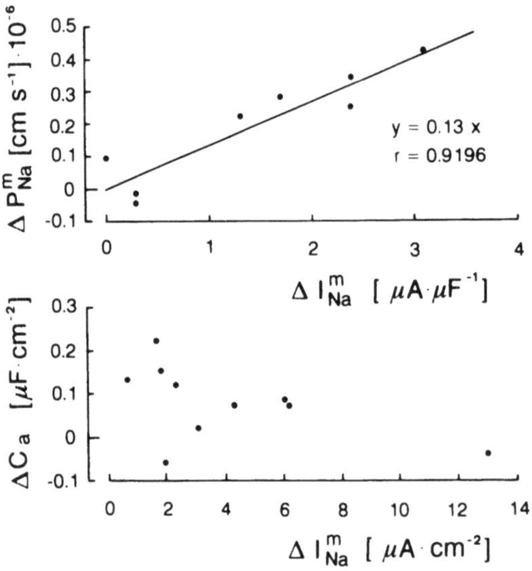

Fig. 6. Arginine-vasotocin (10^{-6} mol/l) induced changes in apical Na^+ permeability (P^m_{Na}; *upper part*) and apical membrane capacitance (C_a) as a function of the corresponding increase in Na^+ current (I^m_{Na}). Apical membrane capacitance can be used as a parameter of apical membrane area. A regression analysis between the increase in sodium permeability and sodium current shows a linear relationship (*solid line*). No correlation exists between changes in apical membrane capacitance (membrane area) and sodium current, as it would be expected if sodium channels were recruited into the apical membrane by fusion of cytoplasmic vesicles. (Krattenmacher and Clauss 1988)

reptiles (lizzard, turtle) no response of electrogenic ion transport to ADH was found (Bentley 1962; Bentley and Bradshaw 1972). Thus, compared to other vertebrate groups, only the amphibian colon exhibits a similar ADH action as described for the toad urinary bladder.

Concerning the mechanism of action of ADH in the amphibian colon, Krattenmacher and Clauss (1988) showed that the sodium permeability of the apical membrane of the colonic epithelium of *Rana esculenta* increased significantly by about 20% after application of 10^{-6} mol/l vasotocin on the serosal side in vitro. The increased apical sodium permeability correlated very well with the concomitant increase in electrogenic sodium absorption (I_{Na}), whereas no correlation was found between I_{Na} and the hormone-induced increase in apical membrane capacitance (Fig. 6). Membrane capacitance is an indicator of apical membrane area (Lewis and Diamond 1976). Furthermore, noise analysis experiments clearly showed that the number of conducting amiloride-sensitive apical sodium channels in the colon epithelium of *Rana esculenta* and *R. temporaria* increased from 23 to 49 million channels/cm^2 epithelium after stimulation with 1 mmol/l cAMP. In contrast, the current through a single sodium channel remained unaffected by cAMP and was about 1 pA (Krattenmacher et al. 1988).

However, the sensitivity of sodium absorption in the colon to antidiuretic hormones is not common to all amphibians. For example, in *Xenopus laevis*, electrogenic sodium transport remains unaffected by oxytocin or cAMP in vitro (Krattenmacher et al. 1990). The reason for this difference may be the different kind of "sodium channel" in *Xenopus*. The described ADH effects on sodium absorption in the amphibian colon were only described for epithelia in which sodium is transported via the amiloride-sensitive Na^+ channel. In the colon epithelium of *Xenopus*, however, this channel is not present, but Na^+ passes the apical cell membrane through a Ca^{2+}-sensitive non-selective cation channel (see also Sect. 4.2.2). Obviously, this channel type is not regulated by the cAMP system and/or the colonic epithelial cells have no receptors for neurohypophyseal hormones. Interestingly, electrogenic ion transport through a similar non-selective cation channel in the toad urinary bladder can be stimulated by both oxytocin and cAMP (Van Driessche et al. 1987). Whether intracellular Ca^{2+} ions have a regulatory function in this channel is so far unclear.

4.3.2 Adrenocortical Hormones

Amphibian adrenocortical tissue appears to be capable of secreting principally two steroids, aldosterone and corticosterone, both of which are also found in reptiles, birds and mammals. The plasma concentration of corticosterone is about ten times that of aldosterone (Hanke et al. 1990), but the functional diversity of corticosteroids does not seem very pronounced in Amphibia (Hanke 1990). Cortisol is of minor importance in amphibians, but was found in *Xenopus laevis* and the urodele *Amphiuma tridactyla*. Phylogenetically, the mineralocorticoid aldosterone, which is secreted by the zona glomerulosa of the adrenal gland, appears for the first time in amphibians, whereas in fish aldosterone is not yet present (Bentley 1971). There seems to be a close interaction between the neurohypophyseal hormones (see Sect. 4.3.1) and the adrenocortical hormones, because arginin-vasotocin is able to increase the plasma concentration of both corticosterone and aldosterone within 1 h.

Although epithelial ion transport in vertebrates is affected by both gluco- and mineralocorticoids (Clauss et al. 1985; Halevy et al. 1988), aldosterone is the predominant regulating hormone. From a variety of epithelia it is known that aldosterone increases electrogenic Na^+ absorption by activation of apical amiloride-sensitive Na^+ channels (Lindemann 1984; Clauss et al. 1988; Palmer 1988), and by stimulation of the basolateral Na/K-ATPase (Petty et al. 1981; Geering et al. 1982; Doucet 1988). While the regulation of electrogenic sodium transport by aldosterone in the large intestine has been well investigated in mammals and birds, studies on the amphibian colon are scarce. To the best of our knowledge only two studies were performed on the regulation of sodium transport by adrenocortical steroids in the amphibian colon so far. Cofré and Crabbé (1965, 1967) demonstrated in the colon of the toad *Bufo marinus* that Na^+ absorption can be stimulated by aldosterone by about 100%. This increase was measured after both subcutaneous injection of 10 µg aldosterone or serosal application of 10^{-5} mol/l aldosterone in vitro. This stimulation showed a typical latency period of about 1 h between addition of the steroid to the incubating fluid and the detection of the hormonal effect. From other tissues it is known that the reason for the delayed hormonal action is the mechanism of action of aldosterone: the hormone diffuses through the basolateral membrane into the epithelial cell and binds to a cytoplasmic receptor. The aldosterone receptor complex diffuses to the nucleus where it binds to a nuclear receptor. This latter binding step leads to a specific gene expression and mRNA formation which induces the synthesis of aldosterone-induced proteins (Benjamin and Singer 1974; Scott and Sapirstein 1975; Cox and Geheb 1984; Rossier et al. 1985). Sodium transport does not increase before the synthesis of these proteins, which then effect specific changes in the transport proteins of the apical and basolateral membrane. Trials to achieve a similar stimulation of Na^+ absorption in vitro in *Rana esculenta* and *Xenopus laevis* have failed so far (Krattenmacher, unpubl. observ.).

4.4 Sodium Self-Inhibition

In the colon of the frog *Rana esculenta* the sodium absorption rate saturates with increasing luminal sodium concentrations, with a half-maximal transport capacity at a luminal sodium concentration of about 27 mmol/l. Concomitantly, the sodium permeability of the apical membrane of the colonic epithelium decreases (Fig. 7). This type of kinetics has the advantage that the Na^+-absorption rate is buffered especially at high luminal Na^+ concentrations, allowing a sensible regulation of sodium absorption by endogenous factors like aldosterone or antidiuretic hormone (Krattenmacher and Clauss 1989).

Similar saturation kinetics of electrogenic sodium absorption was described for various other epithelia. In all cases the apical sodium permeability is downregulated by the luminal sodium concentration (Fuchs et al. 1977; Li et al. 1982, 1986; Palmer et al. 1982; Thomas et al. 1983; Turnheim et al. 1983). Therefore, this mechanism was called sodium self-inhibition. This downregulation is caused by a reduction in the number of conducting apical sodium channels, presumably due to a binding of sodium ions to a receptor at or near the mouth of the sodium channel. As a consequence of this binding process, the sodium channel is transformed to a non-conducting state (Van Driessche and Lindemann 1979; Christensen and Bindslev 1982; Lewis et al. 1984). However, it is also conceivable that changes in the intracellular ionic composition are responsible for the regulation of the apical sodium permeability. Thus, calculations of the intracellular sodium activity $[(Na)_c]$ from current-voltage relationships showed that in colonic cells of *Rana esculenta* $(Na)_c$ increases with increasing luminal sodium con-

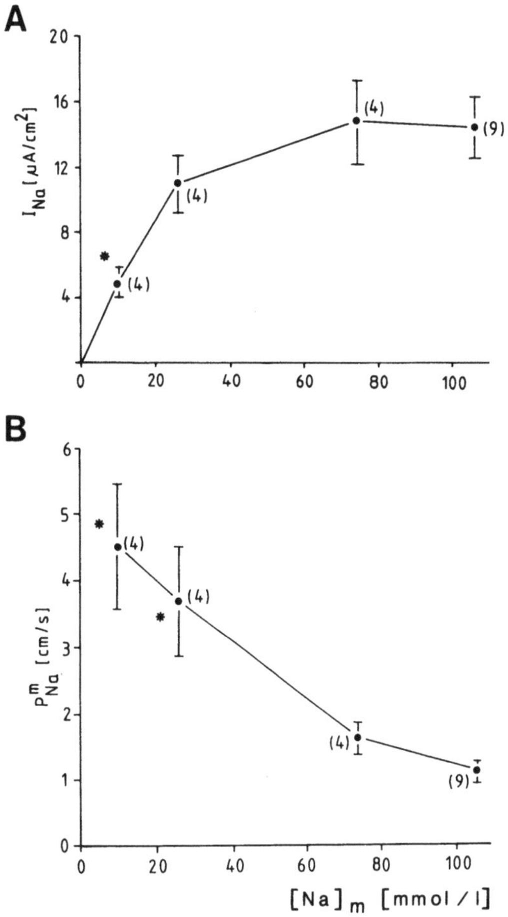

Fig. 7A,B. Sodium absorption (I_{Na}) and apical sodium permeability (P'''_{Na}) in the colon epithelium of *Rana esculenta* at different luminal Na concentrations in vitro (Krattenmacher and Clauss 1989)

centrations (Fig. 8; Krattenmacher and Clauss 1989). Whether this inhibition of the apical sodium entry is due to a direct effect of the increased intracellular sodium concentration (Lewis et al. 1976; Turnheim et al. 1978) or due to a secondary effect via a changed intracellular calcium concentration (Taylor and Windhager 1979; Windhager and Taylor 1983) is unclear. Increased intracellular sodium would reduce the driving force for a Na^+/Ca^{2+} exchanger and would therefore result in an increased intracellular calcium concentration. That the amiloride-sensitive sodium transport is inhibitable by intracellular Ca^{2+} ions was demonstrated for the toad urinary bladder (Chase and Al-Awqati 1983; Garty 1984; Garty et al. 1987).

Interestingly, in the colon of *Xenopus laevis*, no saturation kinetics of electrogenic sodium transport occurs. In this tissue the sodium absorption

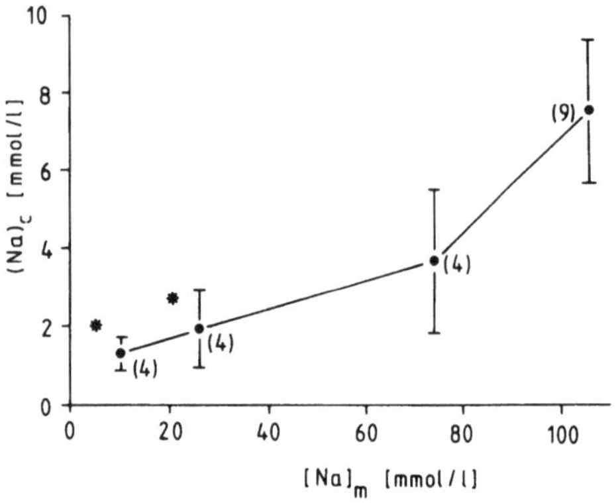

Fig. 8. Calculations of the intracellular sodium activity [$(Na)_c$] from current-voltage relationships in the colon epithelium of *Rana esculenta* at different luminal Na concentrations in vitro (Krattenmacher and Clauss 1989)

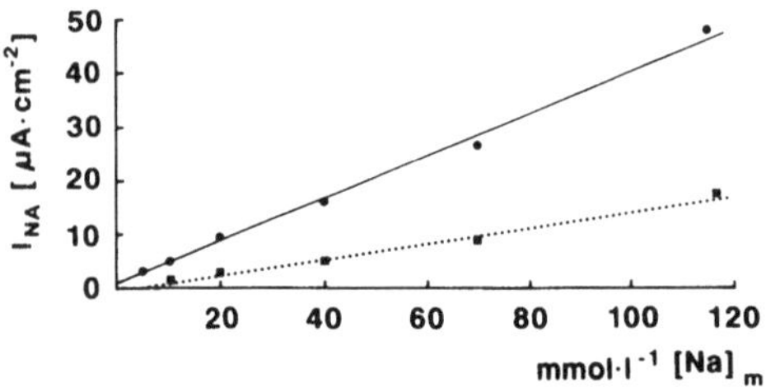

Fig. 9. Relationship between sodium current (I_{Na}) and mucosal sodium concentration ($[Na]_m$) in two experiments (-----, ——) in the colon epithelium of *Xenopus laevis*. The regression lines through the data points reveal a linear relationship between the parameters. Experiments were performed in tissues in which sodium transport was stimulated by mucosal Ca^{2+} removal (Krattenmacher et al. 1990)

rate increases linearly with the luminal sodium concentration (Fig. 9), at least in the absence of mucosal Ca^{2+} ions, a condition, in which electrogenic sodium absorption is stimulated. This is in good agreement with findings in a similar Ca^{2+}-sensitive cation channel in the urinary bladder of the toad *Bufo marinus*, where a similar linear relationship was found (Aelvoet et al. 1988). Thus, the phenomenon of sodium self-inhibition (increasing luminal sodium concentrations inhibit the sodium entry step at the luminal cell membrane) is obviously not common to the amphibian large intestine. It seems to be restricted to colonic epithelia that absorb sodium ions through the amiloride-sensitive pathway.

5 Conclusions

In contrast to other epithelia, ion transport in the amiphibian large intestine, especially of urodeles and gymnophiones, has been poorly investigated. So far it is clear that electrogenic ion transport in the amphibian colon does not follow a general and uniform mechanism, but differs considerably among different species. Regarding the electrogenic sodium transport, two different transport pathways were characterized, the classical amiloride-sensitive pathway and a calcium-sensitive pathway via non-selective cation channels. Neither pathway is typical for the amphibian colon, but they were also described for intestinal epithelia of other vertebrate groups. However, the type of regulation of electrogenic sodium absorption in the colon, at least in amiloride-sensitive tissues, seems to be amphibian-specific: antidiuretic hormones stimulate sodium transport via cyclic AMP, whereas in higher vertebrates similar hormonal action in the large intestine has not been described. On the contrary, in these tissues cyclic AMP stimulates electrogenic chloride secretion. This phylogenetic change in hormonal action in the large intestine from regulation of sodium absorption to regulation of chloride secretion is of great interest.

Compared to the amphibian skin, the colon may be of minor importance in the regulation of salt and water exchange with the body fluids, but nevertheless it is organ that plays a significant role in water and ion balance.

Acknowledgment. Research in the authors' laboratory was supported by grants from the Deutsche Forschungsgemeinschaft, the European Community, and the H. Wilhelm Schaumann Stiftung zur Förderung der Agrarwissenschaften.

References

Aelvoet I, Erlij D, Van Driessche W (1988) Activation and blockage of a calcium-sensitive cation-selective pathway in the apical membrance of toad urinary bladder. J Physiol (Lond) 398: 555–574

Almers W, McCleskey EW (1984) Non-selective conductance in calcium channels of frog muscle: calcium selectivity in a single file pore. J Physiol (Lond) 353: 585–608

Almers W, McCleskey EW, Palade PT (1984) A non-selective cation-conductance in frog muscle membrane blocked by micromolar external calcium ions. J Physiol (Lond) 353: 565–583

Benjamin WB, Singer I (1974) Aldosterone-induced protein in toad urinary bladder. Science 186: 269–272

Benos DJ (1986) Amiloride-sensitive epithelial sodium-channels. In: Latorre R (ed) Ionic channels in cells and model systems. Plenum Press, New York, pp 401–420

Bentley PJ (1962) Studies on the permeability of large intestine and urinary bladder of the tortoise (*Testudo graeca*) with special reference to the effects of neurohypophysial and adrenocortical hormones. Gen Comp Endocrinol 2: 323–328

Bentley PJ (1968) Amiloride: a potent inhibitor of sodium transport across toad bladder. J Physiol (Lond) 195: 317–330

Bentley PJ (1971) Endocrines and osmoregulation. A comparative account of the regulation of water and salt in vertebrates. Springer, Berlin Heidelberg New York, pp 161–197

Bentley PJ, Bradshaw SD (1972) Electrical potential difference across the cloaca and colon of the Australian lizzard *Amphibolurus ornatus* and *A. inermis*. Comp Biochem Physiol 42A: 465–471

Bindslev N, Cuthbert AW, Edwardson JM, Skadhauge E (1982) Kinetics of amiloride action in the hen coprodaeum in vitro. Pflügers Arch Eur J Physiol 392: 340–347

Bogart JP (1981) How many times has terrestrial breeding evolved in anuran amphibians? Monit Zool Ital Suppl XV 3: 29–40

Bridges RJ, Rummel W, Wollenberg P (1984) Effects of vasopressin on electrolyte transport across isolated colon from normal and dexamethasone-treated rats. J Physiol (Lond) 355: 11–23

Carlisky NJ, Lew VL (1970) Bicarbonate secretion and non-Na component of the short-circuit current in the isolated colonic mucosa of *Bufo arenarum*. J Physiol (Lond) 206: 529–541

Chase HS Jr, Al-Awqati Q (1983) Calcium reduces the sodium permeability of luminal membrane vesicles from toad bladder. Studies using a fast reaction apparatus. J Gen Physiol 81: 643–666

Christensen O, Bindslev N (1982) Fluctuation analysis of short-circuited current in warm-blooded sodium-retaining epithelium: site current, density and interaction with triamterene. J Membr Biol 65: 19–30

Clauss W, Arnason SS, Munck BG, Skadhauge E (1984) Aldosterone-induced sodium transport in lower intestine. Effects of varying NaCl intake. Pflügers Arch Eur J Physiol 401: 354–360

Clauss W, Dürr JE, Skadhauge E, Hörnicke E (1985) Effects of aldosterone and dexamethasone on apical membrane properties and Na-transport of rabbit distal colon in vitro. Pflügers Arch Eur J Physiol 403: 186–192

Clauss W, Dürr JE, Krattenmacher R, Hörnicke H, Van Driessche W (1988) Circadian rhythm of apical Na-channels and Na-transport in rabbit distal colon. Experientia 44: 608–610

Cofré G, Crabbé J (1965) Stimulation by aldosterone of active sodium transport by the isolated colon of the toad, *Bufo marinus*. Nature 207: 1299–1300

Cofré G, Crabbé J (1967) Active sodium transport by the colon of *Bufo marinus*: stimulation by aldosterone and antidiuretic hormone. J Physiol (Lond) 188: 177–190

Cooperstein IL, Hogben CAM (1959) Ionic transfer across the isolated frog large intestine. J Gen Physiol 42: 461–473

Cox M, Geheb M (1984) Aldosterone-induced proteins in renal epithelia. In: Bronner F, Kleinzeller A (eds) Current topics in membranes and transport, vol 20. Academic Press, New York, pp 271–293

Cuthbert AW (1973) Evidence for multiple forms of receptors for amiloride in transporting epithelia. Eur J Pharmacol 23: 187–190

Cuthbert AW, Shum WK (1974) Binding of amiloride to sodium channels in frog skin. Mol Pharmacol 10: 880–891

Das S, Palmer LG (1989) Extracellular Ca^{2+} controls outward rectification by apical cation channels in toad urinary bladder: patch-clamp and whole-bladder studies. J Membr Biol 107: 157–168

Dawson DC (1977) Na and Cl transport across the isolated turle colon: parallel pathways for transmural ion movement. J Membr Biol 37: 213–233

Dawson DC, Curran PE (1976) Sodium transport by the colon of *Bufo marinus*: Na uptake across the mucosal border. J Membr Biol 28: 295–307

Deyrup IJ (1964) Water balance and kidney. In: Moore JA (ed) Physiology of the Amphibia. Academic Press, New York, pp 251–328

Doucet A (1988) Function and control of Na-K-ATPase in single nephron segments of the mammalian kidney. Kidney Int 34: 749–760

Ehrlich EN, Crabbé J (1968) The mechanism of action of amipramizide. Pflügers Arch Ges Physiol Menschen Tiere 302: 79–96

Ferreira HG, Smith MW (1968) Effect of a saline environment on sodium transport by the toad colon. J Physiol (Lond) 198: 329–343
Fischer H, Van Driessche W, Clauss W (1989) Evidence for apical sodium channels in frog lung epithelial cells. Am J Physiol 256: C764–C771
Frizzell RA, Koch MJ, Schultz SG (1976) Ion transport by rabbit colon. I. Active and passive components. J Membr Biol 27: 297–316
Fuchs W, Hviid Larsen E, Lindemann B (1977) Current voltage curve of sodium channels and concentration dependence of sodium permeability in frog skin. J Physiol (Lond) 267: 137–166
Garty H (1984) Amiloride blockable Na-fluxes in toad bladder membrane vesicles. J Membr Biol 82: 269–279
Garty H, Benos DJ (1988) Characteristics and regulatory mechanisms of the amiloride-blockable Na^+-channel. Physiol Rev 68: 309–373
Garty H, Asher C, Yeger O (1987) Direct inhibition of epithelial Na^+ channels by a pH-dependent interaction with calcium, and by other divalent ions. J Membr Biol 95: 151–162
Geering K, Giradet M, Bron C, Kraehenbühl JP, Rossier BC (1982) Hormonal regulation of (Na^+K^+)-ATPase biosynthesis in the toad bladder. Effects of aldosterone and 3,5,3-triiodo-L-thyronine. J Biol Chem 257: 10338–10343
Grzimek B (ed) (1970) Grzimeks Tierleben-Enzyklopädie des Tierreichs (Vol 5-Fische 2/Lurche). Kindler Zürich
Gunter-Smith PJ, Grasset E, Schultz SG (1982) Sodium coupled amino-acid and sugar transport by *Necturus* small intestine. An equivalent electrical circuit analysis of a rheogenic cotransport system. J Membr Biol 66: 25–40
Halevy J, Boulpaep E, Budinger ME, Binder HJ, Hayslett JP (1988) Glucocorticoids have a different action than aldosterone on target tissues. Am J Physiol 254: F153–F158
Hanke W (1990) Corticosteroid function – evolutionary aspects. Prog Comp Endocrinol 342: 445–452
Hanke W, Kaltenhuser U, Maser-Gluth C, Pathak V (1990) Regulation of interrenal activity by electrolytes and peptide hormones. Fortschr Zool 38: 241–255
Hays RM (1983) Alteration of luminal membrane structure by antidiuretic hormone. Am J Physiol 245: C289–C296
Heinz M, Krattenmacher R, Hoffmann B, Clauss W (1991) Different modes of electrogenic Na-absorption in the coprodeum of the chicken embyo: role of extracellular Ca^{2+}. J Comp Physiol B 161: 363–370
Hess P, Lansman JB, Tsien RW (1986) Calcium channel selectivity for divalent and monovalent cations. Voltage and concentration dependence of single channel current in ventricular heart cells. J Gen Physiol 88: 293–319
Hillyard SD, Zeiske W, Van Driessche W (1982) Poorly selective cation channels in the skin of the larval frog (stage XIX). Pflügers Arch Eur J Physiol 394: 287–293
Kleyman TR, Cragoe EJ Jr (1988) Amiloride and its analogs as tools in the study of ion transport. J Membr Biol 105: 1–21
Kloas W, Hanke W (1990) Neurohypophysial hormones and steroidogenesis in the interrenals of *Xenopus laevis*. Gen Comp Endocrinol 80: 321–333
Krattenmacher R, Clauss W (1988) Electrophysiological analysis of sodium-transport in the colon of the frog (*Rana esculenta*). Modulation of apical membrane properties by antidiuretic hormone. Pflügers Arch Eur J Physiol 411: 606–612
Krattenmacher R, Clauss W (1989) Autoregulation of apical sodium entry in the colon of the frog (*Rana esculenta*). Comp Biochem Physiol 93A: 593–596
Krattenmacher R, Fischer H, Van Driessche W, Clauss W (1988) Noise analysis of cAMP-stimulated Na current in frog colon. Pflügers Arch Eur J Physiol 412: 568–573
Krattenmacher R, Voigt R, Clauss W (1990) Ca-sensitive sodium absorption in the colon of *Xenopus laevis*. J Comp Physiol B 160: 161–165
Krattenmacher R, Voigt R, Heinz M, Clauss W (1991) Electrolyte transport through a cation-selective ion channel in large intestinal enterocytes of *Xenopus laevis*. J Exp Biol 155: 275–290

Lew VL (1970) Short-circuit current and ionic fluxes in the isolated colonic mucosa of *Bufo arenarum*. J Physiol (Lond) 206: 509–528
Lew VL, Carlisky NJ (1967) Evidence for a special type of bicarbonate transport in the isolated colonic mucosa of *Bufo arenarum*. Biochim Biophys Acta 135: 793–796
Lewis SA (1983) Control of Na and water absorption across vertebrate tight epithelia by ADH and aldosterone. J Exp Biol 106: 9–24
Lewis SA, Diamond JM (1976) Na transport by rabbit urinary bladder, a tight epithelium. J Membr Biol 28: 1–40
Lewis SA, Eaton DC, Diamond JM (1976) The mechanism of Na^+ transport by rabbit urinary bladder. J Membr Biol 28: 41–70
Lewis SA, Ifshin MS, Loo DDF, Diamond JM (1984) Studies of sodium channels in rabbit urinary bladder by noise analysis. J Membr Biol 80: 135–151
Li JHY, Palmer LG, Edelmann IS, Lindemann B (1982) The role of sodium channel density in the natriferic response of the toad urinary bladder to an antidiuretic hormone. J Membr Biol 76: 77–89
Li JHY, Zuzack JS, Kan ST (1986) Effects of detergents on sodium transport in toad urinary bladder. J Pharmacol Exp Ther 238: 415–421
Lindemann B (1984) Fluctuation analysis of sodium channels in epithelia. Annu Rev Physiol 46: 497–515
Lindemann B, Van Driessche W (1977) Sodium-specific membrane channels of frog skin are pores: current fluctuations reveal high turnover. Science 195: 292–294
Macchia DD, Helman SI (1979) Transepithelial current-voltage relationships of toad urinary bladder and colon. Biophys J 27: 371–392
Martens H, Rübbelke M, Gäbel G (1990) Evidence for a divalent cation sensitive short circuit current across the isolated rumen of sheep. Pflügers Arch Eur J Physiol 415(1): R28 (Suppl)
Nagel W (1976) The intracellular electrical potential profile of the frog skin epithelium. Pflügers Arch Eur J Physiol 365: 135–143
Palmer LG (1984) Use of potassium depolarization to study apical transport properties in epithelia. In: Bronner F, Kleinzeller A (eds) Current topics in membranes and transport, vol 20. Academic Press, New York, pp 105–121
Palmer LG (1986) The epithelial sodium channel. In: Poste G, Crooke ST (eds) New insights into cell and membrane transport processes. Plenum Press, New York, pp 327–344
Palmer LG (1988) Regulation of NaCl transport in tight epithelia. In: Greger R (ed) NaCl transport in epithelia. Advances in Comparative and Environmental Physiology, vol 1. Springer, Berlin Heidelberg New York, pp 292–309
Palmer LG, Edelman IS, Lindemann B (1980) Current voltage analysis of apical sodium transport in toad urinary bladder: effects of inhibitors of transport and metabolism. J Membr Biol 57: 59–71
Palmer LG, Li JHY, Lindemann B, Edelmann IS (1982) Aldosterone control of the density of sodium channels in the toad urinary bladder. J Membr Biol 64: 91–102
Petty KJ, Kokko JP, Marver D (1981) Secondary effect of aldosterone on Na-K-ATPase activity in the rabbit cortical collecting tubule. J Clin Invest 68: 1514–1521
Rosoff CJ, Baldwin GF, Bentley PJ (1983) Active transport of calcium across the urinary bladder and colon of the toad *Bufo marinus*. Am J Physiol 245: R91–R94
Rossier BC, Geering K, Atkinson J, Roch-Ramel F (1985) Renal receptors. In: Seldin DW, Giebisch G (eds) The kidney: physiology and pathophysiology. Raven Press, New York, pp 775–806
Schultzke JD, Fromm M, Hegel U (1986) Epithelial and subepithelial resistance of rat large intestine: segmental differences, effect of stripping, time course, and action of aldosterone. Pflügers Arch Eur J Physiol 407: 632–637
Scott WN, Sapirstein VS (1975) Identification of aldosterone-induced proteins in the toad urinary bladder. Proc Natl Acad Sci USA 72: 4056–4060
Taylor A, Windhager EE (1979) Possible role of cytosolic calcium and Na-Ca-exchange in regulation of transepithelial sodium transport. Am J Physiol 236: F505–F512

Thomas SR, Suzuki Y, Thompson SM, Schultz SG (1983) Electrophysiology of *Necturus* urinary bladder. I. Instantaneous current voltage relations in the presence of varying mucosal sodium concentrations. J Membr Biol 73: 157–175

Turnheim K, Frizzell RA, Schultz SG (1978) Interaction between cell sodium and the amiloride-sensitive sodium entry step in rabbit colon. J Membr Biol 39: 233–256

Turnheim K, Thompson SM, Schultz SG (1983) Relation between intracellular sodium and active sodium transport in rabbit colon: current-voltage relations of the apical sodium entry mechanism in the presence of varying luminal sodium concentrations. J Membr Biol 76: 299–303

Ussing HH, Andersen B (1955) The relation between solvent drag and active sodium transport of ions. Proc 3rd Int Congr Biochem Brussels, pp 434–441

Ussing HH, Zerahn K (1951) Active transport of sodium as the source of electric current in the short-circuited isolated frog skin. Acta Physiol Scand 23: 110–127

Van Driessche W (1987) Ca^{2+} channels in the apical membrane of the toad urinary bladder. Pflügers Arch Eur J Physiol 410: 243–249

Van Driessche W, Lindemann B (1979) Concentration dependence of currents through single sodium-selective pores in frog skin. Nature 282: 519–520

Van Driessche W, Zeiske W (1985) Ca^{2+}-sensitive, spontaneously fluctuating cation channels in the apical membrane of the adult frog skin epithelium. Pflügers Arch Eur J Physiol 405: 250–259

Van Driessche W, Aelvoet I, Erlij D (1987) Oxytocin and cAMP stimulate cation movements through a Ca^{2+}-sensitive, amiloride-insensitive channel in the apical membrane of toad urinary bladder. Proc Nate Acad Sci USA 84: 313–317

Wade JB (1985) Membrane structural studies of the action of vasopressin. Fed Proc 44:2687–2692

Wade JB (1986) Role of membrane fusion in hormonal regulation of epithelial transport. Annu Rev Physiol 48: 213–224

Walker AM, Hudson CL, Findlay T, Richards AN (1973) The total molecular concentration and the chloride concentration of fluid from different segments of the renal tubule of Amphibia. Am J Physiol 118: 121–129

Warburg MR, Rosenberg M (1990) Ion and water balance and their endocrine control in the aquatic amphibians. In: Hanke W (ed) Progress in zoology, vol 38. Biology and physiology of amphibians. Fischer Stuttgart, pp 385–403

White JF (1977) Alterations in electrophysiology of isolated amphibian small intestine produced by removing the muscle layers. Biochim Biophys Acta 467: 91–102

Windhager EE, Taylor A (1983) Regulatory role of intracellular calcium ions in epithelial Na transport. Annu Rev Physiol 45: 519–532

Chapter 3

Ion Transport by Turtle Colon: A Role for Volume-Sensing Transporters in the Basolateral Membrane

David C. Dawson and *Marc A. Post*[1]

Contents

1 "Throughput": The Other Problem for Epithelial Cells

It is now widely appreciated that cells have evolved a variety of mechanisms for regulating intracellular composition. Prominent among these are a wide range of so-called volume regulatory mechanisms which permit the cell to adjust its *solute* composition, and hence its water content, in the face of potentially injurious changes in extracellular osmolarity. Lymphocytes, for example, possess well-characterized transport mechanisms that, when activated, promote solute loss or gain and resist osmotically induced cell swelling or shrinking (Grinstein et al. 1984). Epithelial cells face an additional hazard, the "other problem" which is encountered as a direct result of high rates of *transcellular* solute movement. The cell must be protected from the potentially enormous changes in cell composition which could result if solute influx and efflux become "unbalanced". Schultz (1981) has used the graphic term "flush through" to describe the problem confronting salt-transporting epithelial cells. It is likely, given the extremes of throughput encountered by epithelial cells, that they have evolved a variety of mechanisms for avoiding the disaster of "flush through". In this review we will consider two components of a potential volume regulatory mechanism which we have found to be prominent in the cells of the reptilian colon, a basolateral K^+ conductance (Dawson 1987; Dawson and Chang 1990) and a

[1] Department of Physiology, The University of Michigan Medical School, Ann Arbor, MI 48109, USA

Advances in Comparative and Environmental Physiology, Vol. 16

basolateral Na/H antiporter (Post and Dawson 1992). Although it is a bit presumptuous to refer to these as volume "sensors", it is clear that both are activated (or inactivated) by changes in cell volume. Cell swelling activates a specific subpopulation of basolateral K^+ channels and inactivates basolateral Na/H exchange. Cell shrinkage, in contrast, activates Na/H antiport and inactivates the volume-sensitive potassium conductance (g_K). The basolateral membrane of reptilian colonic cells thus contains two volume regulatory elements similar to those identified in a variety of symmetric cells. This seems appropriate in that the basolateral membrane is generally regarded as the "housekeeping" membrane of the polar epithelial cell. The possibility has been raised, however, that these housekeeping functions might be intimately linked to the regulation of transcellular transport (Harvey and Ehrenfeld 1988; Harvey et al. 1988). In particular, volume-sensitive elements may play a role in the coordinate modulation of apical and basolateral membrane transport that acts to minimize changes in cell composition (Schultz and Hudson 1986; Palmer and Sackin 1988; Dawson and Richards 1990).

2 Functional Signatures and the Importance of Polarity

It is interesting to note in retrospect that at about the same time Hodgkin and Huxley (1952) produced what might be thought of as the first definitive characterization of Na^+ and K^+ channel *function*, Watson and Crick (1953) were sowing the seeds of a revolution in biology which would ultimately provide the tools required to associate specific transport functions with specific *gene products*. In the last decade these two lines of inquiry have collided in a spectacular way, and the combination of the tools of molecular biology, cell biology, and biophysics now offers powerful insights into the molecular basis for the function of specific transport proteins. Among physiologists who were trained predominantly in the use of *function* as an analytical tool this revolution has produced no small amount of anxiety, having mainly to do with the need to become fluent in the newly emerging language of recombinant DNA technology and overcoming the well-known, uneasy feeling that if one had a clone one wouldn't know what to do with it! As the dust settles, however, it is amusing to find that cloning, and the consequent emphasis on molecular structure and processing, have led us straight back to the measurement of function as a primary analytical tool. As more and more transport proteins yield the secrets of their primary structure it becomes increasingly apparent that the limiting factor in our ability to understand them will not be in the molecular biology, but rather in the analysis of function. Progress in understanding the relation between structure and function will be limited by our ability to characterize transport processes in a detailed, quantitative fashion.

The notion of a "functional signature" emerges as a primary objective of any experiment designed to identify a particular transporter in a particular physiological setting. Insights gained from molecular biology have reinforced the emerging notion that multiple isoforms of the same transporter may be present, even within the same cell, and one goal of a biophysical analysis is to develop a *functional taxonomy* of transporters based on these "signatures". Indeed, one hopes that a functional profile will emerge that will permit the transport process to be identified unambiguously in an alternative expression system so that it can be associated with the product of a specific gene. Epithelial channels generally appear to lack the distinctive neurotoxin sensitivity and voltage dependence that have been so important in the functional identification of cloned neuronal and muscle channels expressed in, for example, *Xenopus* oocytes. However, each of the basolateral transporters considered here has a specific transport signature that could provide a useful tag. Certain aspects of this signature, such as the details of ion-ion interaction (see below), may not be of obvious significance in terms of salt transport mechanisms, but may be of crucial importance in the process of correlating transport function with molecular structure.

Implicit in the notion of a "functional signature" for a transporter is the question of where and how to look for it. The arsenal of weapons in the struggle to decipher transport function has increased enormously in the last two decades. The development and refinement of single channel recording techniques has brought us into direct contact with the functional properties of single proteins – ion channels – and has permitted profound insights into the relation between channel structure and function (Catterall 1988; Hoshi et al. 1990; Zagotta et al. 1990). Similarly, the spectacular success in the development of fluorescent reporter molecules provides the basis for a detailed portrait of the dynamics of cellular composition (Tsien 1988; Foskett 1990; Lechleiter et al. 1991). In the application of these and other techniques epithelial cells present a special challenge because of their polarity; specifically, because the development of polarity appears to be linked to the expression of differentiated function. Techniques involving patch clamping and fluorescent probes are most easily accomplished using isolated, single cells, but it has become apparent that many epithelial cells do not express transport properties associated with the highly differentiated state unless they are allowed to form a confluent sheet and develop functional polarity. The turtle colon, for example, possesses a higher density of apical epithelial Na^+ channels than any other tissue examined by noise analysis (Wilkinson and Dawson 1990), yet these channels are not detectable in freshly isolated cells either as single channels or whole cell currents. Interestingly, however, the epithelial Na^+ channel from these cells is readily expressed in *Xenopus* oocytes after injection with poly(A+)-RNA prepared from colonic mucosa (Richards et al. 1991). In the case of A6 cells it is now well known that the expression of apical Na^+ channels characteristic of the highly differentiated state requires that the cells be grown to confluence on permeable supports (Sariban-Sohraby et al. 1983). Similarly, the small conductance apical Cl^- channel which appears to be responsible for cAMP-activated Cl^- secretion in HT29 cells is only expressed in cells which are a part of a confluent cell layer (Morrison and Frizzell, pers. comm., in press).

This trick of nature, which decrees that epithelial cells may only reveal their differentiated properties when they are part of a cell layer, creates an impetus to design methods for assaying the transport properties of epithelial cell membranes while the cells are a part of a polarized, confluent sheet. One approach to investigating the properties of the basolateral membrane which has been particularly successful with the turtle colon is the use of various agents to permeabilize the *apical* membrane and remove it as a barrier to solute transport. Digitonin, for example, was used to render the apical membrane freely permeable to molecules as large as lactate dehydrogenase and permitted the study of calcium-proton interactions in the activation of basolateral conductances for K^+ and Cl^- (Chang and Dawson 1988; Chang et al. 1991). The studies reviewed here made use of epithelial sheets apically permeabilized with amphotericin-B, a pore-forming antibiotic. The amphotericin-permeabilized apical membrane is highly permeable to Na^+ and K^+, moderately permeable to Cl^-, and impermeable to Ca^{2+} and gluconate. The more selective permeabilization effected by exposure to amphotericin-B had several advantages for these studies. First, the high apical conductance to Na^+ and K^+ facilitated the study of *basolaterally* located transport mechanisms for these two cations. Second, the limited solute permeability of the amphotericin-treated membrane made it relatively easy to design solutions which would cause the cells to swell or shrink by simply adjusting the amount of the *permeant* anion. In the presence of mucosal Cl^- the finite Cl^- permeability of the amphotericin-B pore permits solute entry and cell swelling, whereas in the presence of the impermeant anion, gluconate, the cells shrink (Germann et al. 1986a).

For both the swelling-activated K^+ channel and the shrinking-activated Na/H antiporter the permeabilized cell layer permitted us to measure isotope fluxes and voltage clamp currents simultaneously. It will be apparent that in both cases these measurements provided information which would have been difficult or impossible to obtain any other way, and which was crucial in defining a unique functional signature for each transporter. In both cases the identification of the transport event was contingent upon the emergence of a particular type of anomalous, nondiffusional tracer behavior. In the case of the K^+ channels this was *single-filing*, the positive coupling between the flow of tracer and unlabeled K^+. For the Na/H antiporter the fingerprint was based on *counterflow* or *trans* acceleration of tracer flow by an opposing gradient of unlabeled ion, in this case Na^+ or H^+. Both of these phenomena result in anomalous tracer flux ratios (Dawson 1991) which are readily discerned, even against a background of parallel, diffusional transport events, because they result from the coupling of the free energy inherent in the gradient of the unlabeled ion to the flow of tracer.

3 Basolateral K^+ Channels Activated by Cell Swelling

Kirk and Dawson (1983) carried out a detailed study of K^+ currents and $^{42}K^+$ fluxes across the basolateral membrane of turtle colon and concluded that the basolateral membrane was the site of a barium-sensitive K^+ conductance (g_K) which was characterized by a highly anomalous tracer flux ratio of the type identified by Hodgkin and Keynes (1955) as characteristic of "single-file diffusion" (Fig. 1). The fluxes behaved as if tracer and unlabeled K^+ could occupy the channel at the same time. Subsequent experiments by Germann et al. (1986a,b) revealed that the apparent simplicity of the basolateral membrane in the earlier experiments was the fortuitous consequence of the choice of benzene sulfonate$^-$ as the anion to replace Cl^-. If sheets of colon were bathed by a Na-benzene sulfonate Ringer's solution the cells swelled and the basolateral g_K was dominated by a population of swelling-activated K^+ channels. These particular K^+ channels could be distinguished from other K^+ channels present by their sensitivity to quinidine and lidocaine. Richards and Dawson (1986) identified these channels in isolated colonic cells using the patch-clamp method and Dawson et al. (1988) showed that this population of channels could also be identified in permeabilized cell layers by analyzing the lidocaine-induced fluctuations in the swelling-activated basolateral K^+ current. Because the anomalous fluxes reported by Kirk and Dawson (1983) suggested that single-filing might be a property of the swelling activated g_K, Germann et al. (1986a) conducted an experiment in which $^{42}K^+$ fluxes were determined across apically permeabilized cell layers in the presence of either mucosal (M)-to-serosal (S) or S-to-M, K^+ gradients. As illustrated in Fig. 2 (above), the results were very striking. In the presence of a steep, M-to-S K^+ gradient, permeabilization was associated with a dramatic increase in the M-to-S rate coefficient for $^{42}K^+$ flow (λ_{MS}), whereas the rate coefficient for S-to-M $^{42}K^+$ flow (λ_{SM}) was virtually unchanged. Reversing the orientation of the K^+ gradient (Fig. 2, below) reversed the result in the presence of an S-to-M K^+ gradient $\lambda_{SM} > \lambda_{MS}$. It is worth emphasizing that this result is strictly anomalous; it is thermodynamically incompatible with simple diffusional ion flow through a channel in which tracer and unlabeled K^+ do not interact. The result is, however, compatible with the notion that swelling activates a population of basolateral K^+ channels which share an important conduction property: the pore can be occupied by more than one K^+ ion at a time. The possibility of double occupancy gives rise to positive coupling between the flows of tracer and unlabeled K^+ that can be envisioned in terms of a kinetic and conformational state diagram, illustrated in Fig. 1.

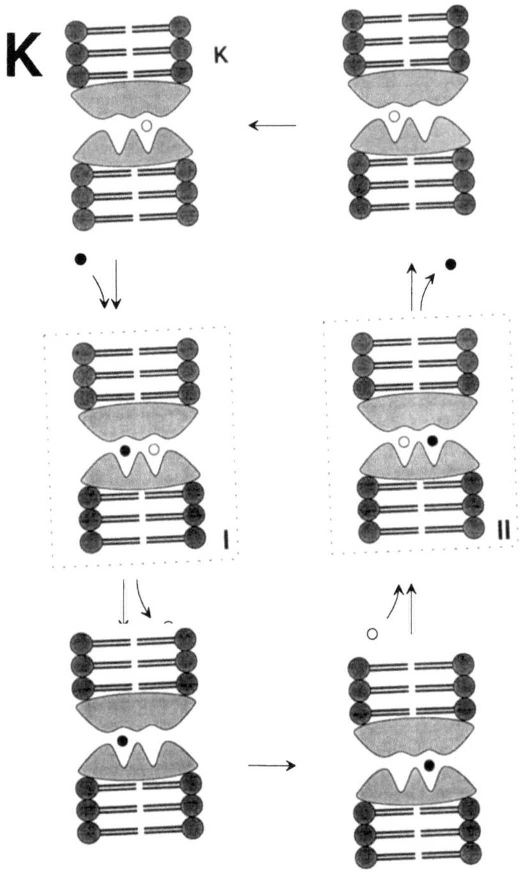

Fig. 1. Left-to-right K^+ fluxes in a channel which can be occupied by two ions at the same time. The diagram shows only those states of the channel involving $^{42}K^+$ (*open circle*) and its interaction with unlabeled K^+ (*solid circle*). The exit of tracer K^+ from the channel is strongly influenced by the presence of an unlabeled K^+ ion in the second site (i.e., the conformational states within the *dashed boxes*). The presence of a K^+ ion in the left-hand site enhances the probability that tracer K^+ will exit toward the right. Imposition of a left-to-right gradient of unlabeled K^+ enhances the time-average probability of the doubly occupied state with abundant K^+ in the left-hand site and tracer K^+ in the right-hand site (i.e., *box I*) occurring. This will be detected as a larger rate coefficient for left-to-right tracer flow than for right-to-left tracer flow. For completeness an entire cycle for left-to-right $^{42}K^+$ and abundant K^+ channel translocation are shown. Note that all of the reactions are completely reversible, although for clarity the "reverse" arrows have been omitted

The diagram depicts the cycle of kinetic and conformational states which are created as labeled ($^{42}K^+$) and unlabeled K^+ transit a two-site K^+ selective channel. Note that only one direction of the cycle is depicted, although all reactions are fully reversible. The positive coupling of the two K^+ flows is directly related to the probability of finding the channel occupied

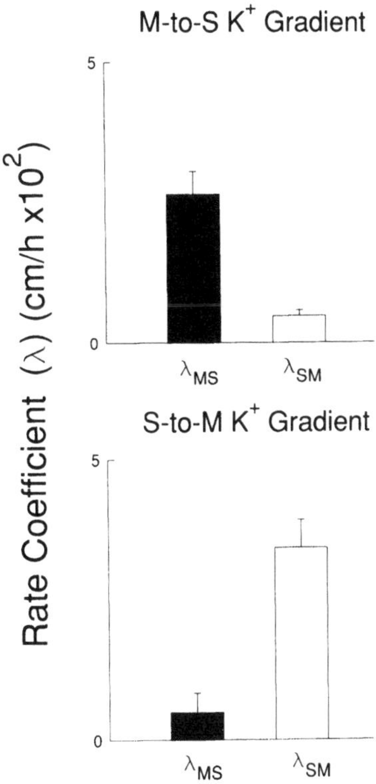

Fig. 2. Transmural rate coefficients (λ_{MS}, λ_{SM}) for $^{42}K^+$ flow across amphotericin permeabilized sheets of turtle colon in the presence of ouabain and a K^+ gradient. Transmembrane potential was clamped to 0 mV, and the applied K^+ gradient was 112:2.5. Rate coefficients were corrected for paracellular flow. Diffusional $^{42}K^+$ flow would result in $\lambda^*_{MS} = \lambda^*_{SM}$. Instead, the rate coefficients exhibited a marked asymmetry which was dependent on the orientation of the K^+ gradient, as expected for the positive coupling between tracer flow and unlabeled K^+ characteristic of single-file flow. Data taken from Germann et al. (1986a)

simultaneously by tracer and abundant K^+, indicated as states I and II in Fig. 1. If we focus on the movement of tracer K^+ (open circle) it is clear that the ionic repulsion due to abundant K^+ (solid circle) occupying the other site forces tracer K^+ to exit the channel in a preferred direction, i.e., away from the occupied site. On average the preferred direction of K^+ flow, therefore, is determined by the relative probabilities of finding the channel in state I or II. In the presence of a left-to-right abundant K^+ gradient, as indicated, state I is much more likely so that tracer, on average, is much more likely to move toward the right-hand bath. The anomalous flux ratio is a direct reflection of this doubly occupied state, which provides the mechanism for coupling the free energy inherent in the K^+ gradient to the flow of tracer K^+.

It was necessary to distinguish the swelling-induced K^+ conductance from at least two other basolateral K^+ conductances: a resting g_K, which appeared to be responsible for the recycling of K^+ and Rb^+ in the support of normal transport, and a calcium-activated g_K, which may play a rôle in the sequence of events which are set in motion by the binding of agonist to cholinergic receptors (Venglarik and Dawson 1986). This turned out to be relatively easy for several reasons. First, the swelling-induced g_K is relatively large, at least three to five times that of the resting g_K, so that in swollen cells it dominates the basolateral membrane. Second, the two blockers quinidine and lidocaine, although hardly specific in general, proved very useful here; neither had a marked effect on active Na^+ absorption, whereas Ba^{2+}, which inhibits both the resting- and swelling-activated conductances, caused a distinct inhibition of active Na^+ absorption. Thus, it appears that the swelling-activated g_K plays little, if any, role in the mechanism of active Na^+ absorption.

Cell swelling appears to be associated with the activation of a specific population of basolateral K^+ channels, a process which could, in principle, promote salt and water exit from the colonic cells, but many questions remain. One of these concerns the *anion* conductance of the basolateral membrane. At present we do not know whether a specific anion conductance is activated by cell swelling, as has been reported in some other cells (Cliff and Frizzell 1990). Another puzzle is the nature of the coupling between cell swelling and channel activation. It has been suggested that stretch-activated channels could directly transduce changes in cell volume into changes in channel activity if deformations of the plasma membrane alter channel gating (Sachs 1987). However, the 17-pS channel which appears to be responsible for the swelling-activated K^+ conductance was not activated by stretch (Richards and Dawson, unpubl.). Finally, it is not clear at this juncture whether the colonic cells actually exhibit volume regulation under conditions of normal transport.

4 Basolateral Na/H Exchangers Activated by Cell Shrinking

A shrinkage-activated Na/H antiporter (Post and Dawson 1992) is the second half of the putative basolateral volume regulation system. The antiporters identified in other cells have been shown to carry out the 1:1 exchange of Na^+ for either Na^+, Li^+, or H^+ and to be completely blocked by relatively high concentrations of amiloride (~1 mM) (Aronson 1985). Thus, we assayed for exchange activity in the basolateral membrane using the strategy diagramed in Fig. 3 in which gradients of Na^+, Li^+, or H^+ were imposed across amphotericin-permeabilized tissues to induce counterflow or *trans* acceleration of $^{22}Na^+$ flow; and amiloride, or one of its congeners, was used to inhibit counterflow. In cells incubated with Cl-free, gluconate

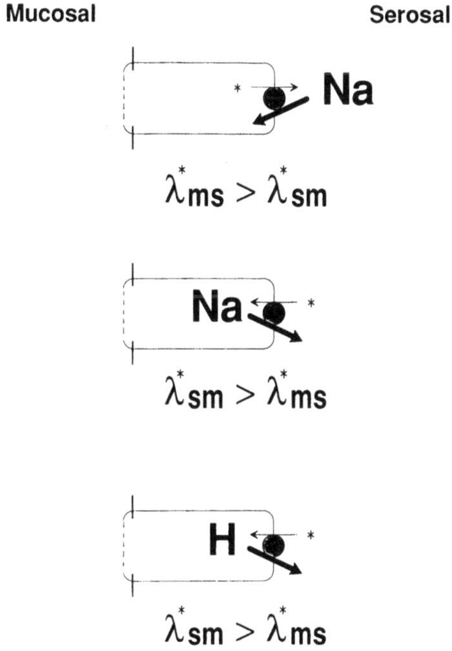

Fig. 3. Strategy employed to identify an antiporter by measuring counterflow of $^{22}Na^+$ driven by gradients of Na^+ or H^+. Each drawing shows a cell whose apical membrane has been rendered permeable to monovalent cations by amphotericin. In the presence of an S to M Na^+ gradient λ_{MS} is *trans* accelerated. Conversely, in the presence of an M to S Na^+ gradient λ_{SM} is *trans* accelerated. An M to S H^+ gradient was imposed and λ_{SM} was *trans* accelerated. Serosal amiloride reduced the rate coefficients to their pre-amphotericin permeabilization values, indicating that all of the basolateral $^{22}Na^+$ movement under these conditions resulted from the operation of the Na/H exchanger

Ringer's solution the results were quite dramatic, as summarized in Fig. 4 (top and middle panels). The imposition of a Na^+ gradient induced $^{22}Na^+$ counterflow, which was altered predictably by changing the orientation of the gradient, and was completely blocked by amiloride. Gradients of Li^+ or H^+ also produced amiloride-sensitive, $^{22}Na^+$ counterflow. All of these results were consistent with the presence in the basolateral membrane of a high level of Na/H antiport activity. Here again, the strictly anomalous behavior of the tracer rate coefficients in the presence of a gradient of the transported ion was a key element of the functional "signature", and allowed us to identify cation exchange unequivocally. Clearly, the relative homogeneity of the tissue and the high level of expression were major advantages in this endeavor.

In the course of these experiments we made an observation which was, to say the least, quite unexpected; as shown in Fig. 4 (bottom panel) the imposition of an M to S Na^+ gradient gave rise to an outward Na^+ current across the basolateral membrane which was blocked by serosal amiloride!

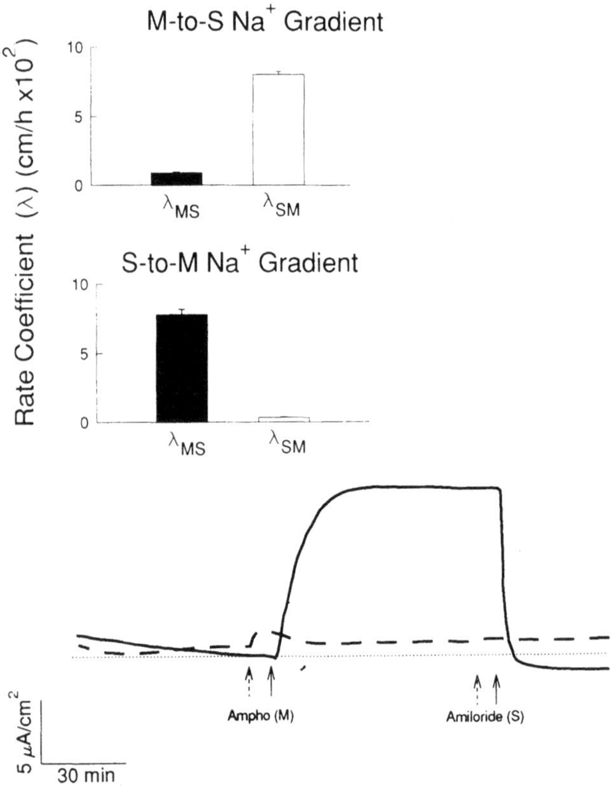

Fig. 4. Amiloride-sensitive unidirectional rate coefficients for transmural $^{22}Na^+$ movement and I_{SC} traces in the presence of M-to-S or S-to-M Na^+ gradients. *Top and middle* The mucosal bath (*top panel*) contained (in mM): 110 Na^+, 2 K^+, 1 Ca^{2+}, 115 gluconate, 20 TAPS, pH 6.5; serosal bath contained (in mM): 110 K^+, 2 Na^+, 1 Ca^{2+}, 115 gluconate, 20 TAPS, pH 6.5. Bath solutions were reversed for the *middle panel* (i.e., an S-to-M Na^+ gradient was imposed). Rate coefficients were determined in paired tissues in the presence of 10 μM mucosal amphotericin B and 100 μM serosal ouabain. Each *bar* represents the difference between the mean (±SE) of the rate coefficients (determined for at least three consecutive 10-min periods) determined first in the absence and then in the presence of 1 mM serosal amiloride. Shown are amiloride-sensitive rate coefficients. In the presence of 10 μM mucosal amphotericin-B and an M-to-S Na gradient the amiloride-sensitive portion of λ_{SM} was markedly accelerated (*top*); whereas an S-to-M Na gradient had the opposite effect, the amiloride-sensitive portion of λ_{MS} was accelerated (*middle*). *Bottom* I_{sc} in the presence of M-to-S (*solid trace*) or S-to-M (*dashed trace*) Na^+ gradients (upward deflection of I_{sc} trace indicates positive charge movement from M to S). Amphotericin permeabilization revealed a Na^+ selective basolateral membrane in the presence of elevated cytosolic Na^+ and a K^+ selective basolateral membrane in the presence of elevated cytosolic K^+. Na^+ conductance was eliminated by 1 mM amiloride in the serosal bath

Table 1. $K_{1/2}$ (μM) for inhibition of basolateral Na/Na antiport, basolateral Na current (I_{Na}^{bl}), and apical Na channels (I_{Na}^{ap}) by amiloride analogs

Analog	Na/Na[a]	I_{Na}^{bl}[a]	I_{Na}^{ap}[b]
Ethylisopropylamiloride	0.35	0.65	*N/D*[c]
Dimethylamiloride	0.70	1.0	300
Hexamethyleneamiloride	1.5	1.5	250
Amiloride	10	10	0.12
Phenamil	200	150	0.04

[a] Na/Na exchange and basolateral Na current were measured under the same conditions as in Fig. 3. The experimental paradigm was also similar to that depicted in Fig. 3, except that an additional period for determining the sensitivity of λ_{SM} to the inhibitor was included.
[b] The effect of amiloride analogs on the apical Na channel was determined by adding the test dose of analog to the mucosal bath. The mucosal bath contained 2 mM Na and was buffered to pH 6.5 to mimic the conditions used to determine the effect of amiloride analogs on λ_{MS} and I_{Na}^{bl}.
[c] Not determined.

We did not expect conductive Na^+ flow to be associated with obligatory cation exchange, but the Na^+ selectivity of this current and its sensitivity to amiloride were difficult to ignore. Fortunately, previous work on Na^+ channels and exchangers had produced an arsenal of amiloride analogs (Kleyman and Cragoe 1988) which we could use to test the hypothesis that the Na-selective conductance might be mediated by the Na/H antiporter. Table 1 lists the $K_{1/2}$ values for block of the basolateral Na^+ current by each of a series of amiloride analogs. Also tabulated are the $K_{1/2}$ values for blockade of basolateral Na/Na exchange and *apical* Na^+ conductance. The analogs tested range from benzamil, highly specific for blockade of apical Na^+ channels, to ethylisopropyl-amiloride, highly specific for inhibition of Na/H exchange. In each case, the efficacy of the analog for inhibition of the basolateral Na^+ current paralleled its efficacy in blocking Na/Na exchange, and was clearly different from that for blockade of the apical Na^+ channel. This result suggested, from the standpoint of blocker specificity, that the basolateral Na^+ conductance was more like a Na/H antiporter than an apical Na^+ channel.

Another striking similarity between the basolateral Na^+ conductance, Na/H, and Na/Na exchange was found in the regulation of these processes. As illustrated in Fig. 5 cell swelling attenuated Na/H and Na/Na exchange and inactivated the Na^+ conductance (Post and Dawson 1992), consistent with the notion that conductance and exchange could be mediated by the same or closely related membrane proteins. In subsequent studies we have extended the pharmacological profile for the basolateral Na^+ current and

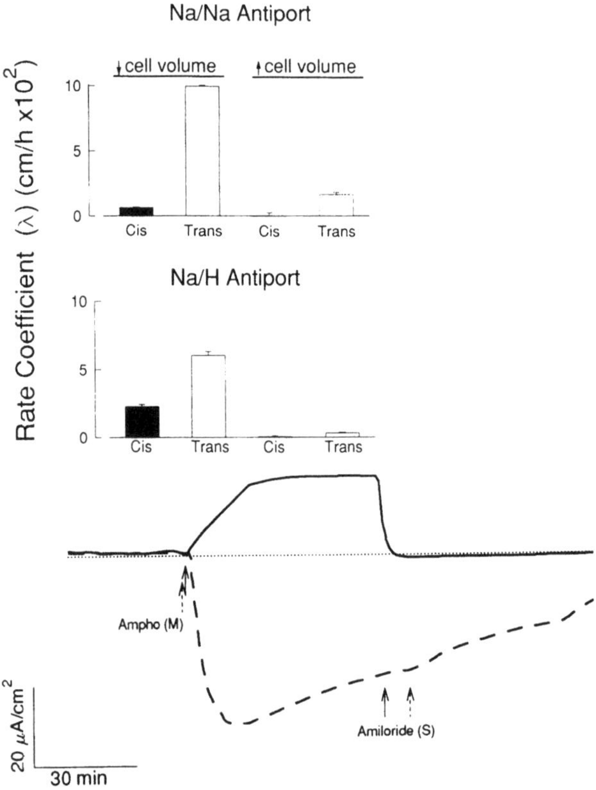

Fig. 5. Effect of cell volume changes on Na^+/Na^+ antiport (*top*) and Na^+/H^+ antiport (*middle*) and basolateral Na^+ current (*bottom*). Transmural rate coefficients were determined using two sets of paired tissues, one exposed to conditions which promote cell shrinkage (bath solutions as in Fig. 4) and the other exposed to conditions which promote cell swelling (gluconate salts were replaced by chloride salts). *Top panel* Robust Na^+/Na^+ exchange in the shrunken cells and markedly attenuated exchange in the swollen cells. *Middle panel* shows that alkalinizing the serosal bath to pH 8.5 induced a marked asymmetry in the rate coefficients under conditions of cell shrinkage, but the rate coefficients for swollen cells were greatly attenuated. *Bottom panel* Robust Na^+ currents in shrunken cells (upward deflection of *solid current trace*) that were eliminated by the serosal addition of amiloride (1 mM). Swollen cells exhibited K^+ currents (*dashed trace*) that were eliminated by the serosal addition of Ba^{2+} (10 mM) (not shown)

basolateral Na^+ exchange, and the results have been 100% consistent: any compound that blocks basolateral Na^+ current also blocks Na/Na exchange. Examples include quinidine (2 mM), diphenylamine-2-carboxylic acid (DPC, 1 mM), Ni^+ (1 mM), Co^{2+} (~1 mM), and Gd^{3+} (10 mM) a trivalent ion used as a blocker of stretch-activated, nonselective cation channels.

The ultimate proof of the hypothesis that the basolateral exchanger exhibits a conductive operating mode will rest on efforts to clone the Na/H exchanger and express it in a heterologous system in which conductive

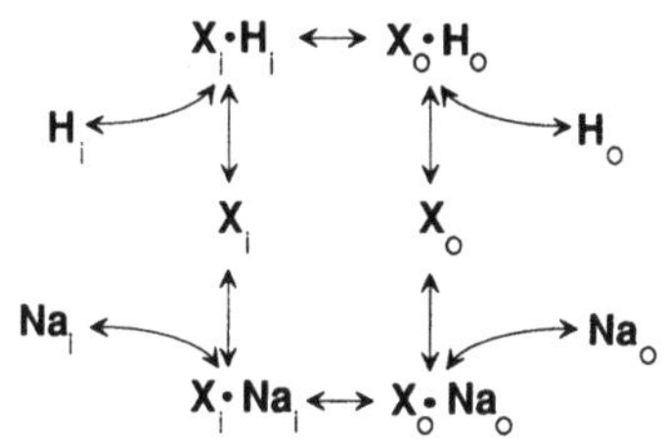

Fig. 6. Kinetic model for Na/H antiport based on a ping-pong mechanism. *X* represents the antiporter and the subscripts *o* and *i* denote the outward- and inward-facing conformations of the protein

behavior will be readily detectable, like a *Xenopus* oocyte. Northern blot and sequence analysis (Harris et al. 1992), in fact, suggest that the reptilian colonic exchanger is highly homologous to the human form cloned by Sardet et al. (1989), so that the tools for the definitive test will hopefully be available soon. In the meantime the absence of such data frees us from the constraints of reality to speculate on the implications of a conductive operating mode of the exchanger for models of the catalytic cycle of Na/H exchange.

5 Models for Na/H Exchange

How does a "carrier" differ from a "channel"? This question, provoked by our observations on basolateral Na/H exchange and current, was illuminated by the thoughts of P. Läuger (1987) who proposed that carrier- and channel-mediated transport could be viewed as points on a continuum. A channel was viewed as a linear sequence of binding sites which, when the channel was in the open or conducting conformation, were accessible from either side of the membrane. A carrier, in contrast, was viewed as having a binding site which was, in effect, translocated from one side of the membrane to the other along with the bound substrate so that at any instant the binding site was accessible from only one side of the membrane. The binding site would be either "inward-facing" or "outward-facing" depending on from which side of the membrane the last translocated substrate molecule had originated. Figure 6 (Fröhlich and Gunn 1986; Stein 1986) shows the simplest diagram of the reactions for a ping-pong catalytic cycle of a Na/H exchanger, where the subscripts "o" and "i" refer to the inward- and outward-facing form of the protein, respectively. Na^+ binds to the outward-facing conformation of the exchanger, X_o, forming $X_o \cdot Na_o$. The transition $X_o \cdot Na_o \leftrightharpoons X_i \cdot Na_i$ is envisioned as a conformational change associated with the inward translocation of one Na^+ ion, thereby generating one Na^+ in the cell and the empty, inward-facing form of the antiporter, X_i. An obligatory exchange mechanism is created by disallowing the conformational change $X_i \leftrightarrows X_o$, the translocation of the empty site. The antiporter may return to the outward-facing conformation only by binding another Na^+ or a H^+, the latter event leading to a net exchange of extracellular Na^+ for intracellular H^+ and the former to a "futile" exchange of extracellular Na^+ for intracellular Na^+. The proposed catalytic cycle provides the basis for an energy conversion process which resides in the use of free energy "invested" in the Na^+ gradient to drive protons out of the cell.

This simple ping-pong model leads to several important conclusions about cation exchange. First, the translocation of substrate is strictly coupled to the cycle of conformational changes of the exchange protein, and this probably limits the *rate* of the process. In contrast, translocation through a channel is due to diffusion or hopping from site to site. A second consequence of the link between substrate translocation and conformational change is that the conformational state of the exchanger is a function of the applied ion gradient. The imposition of a steep, inwardly directed Na^+ gradient, for example, by promoting the transition $X_o \cdot Na_o \rightarrow X_i \cdot Na_i$, leads to an increase in the probability of finding the antiporter in the inward-facing conformation (i.e., X_i). This effect of the ion gradient on the state of the protein is presumably the basis for *trans* acceleration, i.e., the rate of $^{22}Na^+$ (or H^+) flow out of the cell is enhanced as a result of the increased availability of the empty, inward-facing conformation of the antiporter.

Several aspects of the behavior of the exchanger are only apparent from more quantitative considerations, and although space limitation precludes a detailed analysis, we will mention two. First, it is likely that even at a fairly acidic cytosolic pH, Na/H and Na/Na exchange are *both* ongoing processes. Thus, in addition to the *net* fluxes of Na^+ and H^+, which would be detected in an experiment involving, for example, the measurement of a Na^+-dependent change in intracellular pH, there are also a significant number of "futile", Na/Na exchange events. Inasmuch as cytosolic Na^+ and H^+ are competing for the inward-facing conformation of the exchanger the relative proportion of these two events is an important mechanistic question. Given the fact that even a fairly acidic cytosolic pH of 6 represents a proton concentration which is 10000 fold *lower* than the typical value for intracellular Na^+ ($\sim$10 mM), it is apparent that, if the exchanger is to effect any significant H^+ flux, the affinity of the inward-facing conformation of the antiporter for protons must be many-fold higher than its affinity for Na^+. A related question is whether the exchanger has an inherent asymmetry which would promote operation in the "physiologically relevant" mode, i.e., Na^+ influx and H^+ efflux. For example, experiments in some (Aronson et al. 1982) but not all (Knickelbein et al. 1990) systems provide evidence for an intracellular "activating site" which effectively "turns on" the Na/H antiporter at acidic cytosolic pH.

Consideration of the exchange cycle brings us to the question of a mechanism for the postulated conductive behavior of the exchanger. We

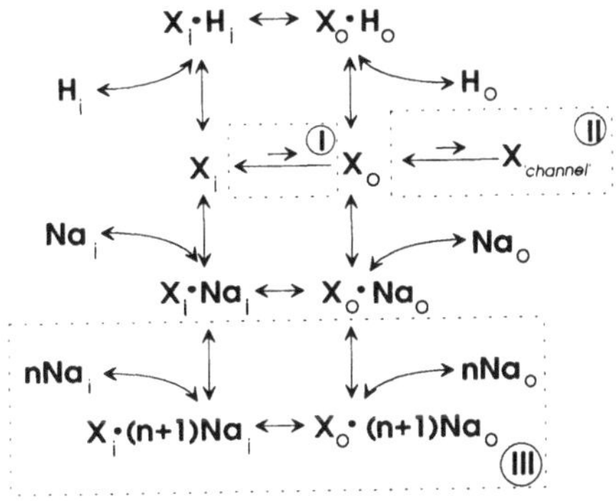

Fig. 7. Speculative model for the generation of a Na^+ conductance by the Na-H exchanger. The ping-pong kinetic model (Fig. 6) was modified to include three conductive mechanisms: *I* slip, *II* electrodiffusion via a channel-like conformation, and *III* electrogenic stoichiometry

have considered three possible mechanisms (Fig. 7): (I) "slip", (II) electrodiffusion via a channel-like conformation of the antiporter, and (III) a nonunity exchange stoichiometry. Slip has been considered previously in relation to the Cl-HCO_3 exchanger and is postulated to result from relaxation of the constraint that the empty site cannot translocate (Kaplan et al. 1983; Knauf et al. 1983). Comparison of Na^+ fluxes and Na^+ currents suggested that the turnover rate for exchange is about 20 times that for conduction, so that if once in 20 cycles the unbound form of the protein could recycle this would account for the current. A second possibility is simply a conducting conformation of the protein that is induced by elevated cytosolic Na^+ (i.e., is associated with the increased abundance of the X_o conformation of the exchanger). A third mechanism for conductive Na^+ flow via the antiporter would be a nonunity stoichiometry for the exchange cycle; $2Na^+/1H^+$, for example. Activation of this mode would require the binding of multiple Na^+ ions to one form of the protein and their translocation. This mechanism leads to the prediction that an imposed proton gradient could drive a conductive counterflow of Na^+. In fact, recent experiments (Post and Dawson, unpubl. observ.) suggest that steep, outwardly directed proton gradients can in fact produce an amiloride-sensitive inward Na^+ current in the absence of a Na^+ gradient.

6 Carrier to Channel and Back Again

In spite of the somewhat different pharmacology the similarities between the conductance that we tentatively attribute to the basolateral exchanger and the apical Na^+ channel are striking. Both exhibit high selectively, only Na^+

and Li^+ conduct, and blockade by analogs of amiloride. Structure-function analysis of cloned exchangers may reveal that these two amiloride binding proteins are more closely related than we might have initially thought. In the meantime it will be of interest to determine if the basolateral, amiloride-sensitive, Na^+ conductance is found in other epithelial cells and to determine what role the basolateral antiporter plays in the process of epithelial salt transport.

7 Putting the Parts Together

Perhaps the most striking feature of these two volume-sensitive transport elements is the high level of their expression in the basolateral membrane of the colonic cells. It would appear that the cell has an enormous capacity to respond to osmotic perturbations. Are these transporters involved in the changes in transcellular transport which serve to regulate body fluid composition or are they "fail-safe" mechanisms which are activated only under extreme conditions? Answering this question presents a formidable challenge in that it requires that we identify the cellular mechanisms which activate and inactivate the transporters and then put the parts together in an integrated regulatory scheme. Resolution of these and other issues will hopefully emerge as molecular biology and transport physiology collide in the study of epithelial salt transport processes.

Acknowledgments. This work was supported by NIH, Michigan Heart Association, and the Cystic Fibrosis Foundation.

References

Aronson PS (1985) Kinetic properties of the plasma membrane Na^+-H^+ exchanger. Annu Rev Physiol 47: 545–560

Aronson PS, Nee J, Suhm MA (1982) Modifier role of internal H^+ in activating the Na^+-H^+ exchanger in renal microvillus membrane vesicles. Nature 299: 161–163

Catterall WA (1988) Structure and function of voltage-sensitive ion channels. Science 242: 50–61

Chang D, Dawson DC (1988) Digitonin-permeabilized colonic cell layers. Demonstration of calcium-activated K^+ and Cl^- conductances. J Gen Physiol 92: 281–306

Chang D, Kushman NL,Dawson DC (1991) Intracellular pH regulates basolateral K^+ and Cl^- conductances in colonic epithelial cells by modulating Ca^{2+} activation. J Gen Physiol 98: 183–196

Cliff WH, Frizzell RA (1990) Separate Cl^- conductances activated by cAMP and Ca^{2+} in Cl^--secreting epithelial cells. Proc Natl Acad Sci USA 87: 4956–4960

Dawson DC (1987) Properties of epithelial potassium channels. In: Bronner F, Kleinzeller A (eds) Giebisch G (guest ed) Current topics in membranes and transport. Potassium transport: physiology and pathophysiology, vol 28. Academic Press, New York, pp 41–71

Dawson DC (1991) Principles of membrane transport. In: Schultz SG (section ed) Field M, Frizzell RA (volume eds) Rauner BB (executive ed) Handbook of physiology. Section 6. The gastrointestinal system. Vol IV. Intestinal absorption and secretion. American Physiological Society, Bethesda, MD, pp 1–44

Dawson DC, Chang D (1990) Turtle colon: keeping track of transporters in the apical and basolateral membranes. In: Fleischer S, Fleischer B (eds) Methods in enzymology, vol 192, Biomembranes: Part W. Cellular and subcellular transport: epithelial cells. Academic Press, New York, pp 734–745

Dawson DC, Richards NW (1990) Basolateral K conductance: role in regulation of NaCl absorption and secretion. Am J Physiol 259: C181–C195

Dawson DC, Van Driessche W, Helman SI (1988) Osmotically induced basolateral K^+ conductance in turtle colon: lidocaine-induced K^+ channel noise. Am J Physiol 254: C165–74

Dawson DC, Wilkinson DJ, Richards NW (1990) Basolateral K channel noise: signals from the dark side. In: Bronner F. (ed) Helman S, Van Driessche W (guest eds) Current topics in membrane and transport, vol 37, Channels and noise in epithelial tissues. Academic Press, New York, pp 191–212

Diamond JM (1982) Transcellular cross-talk between epithelial cell membranes. Nature 300: 683–685

Foskett JK (1990) $[Ca^{2+}]_i$ modulation of Cl^- content controls cell volume in single salivary acinar cells during fluid secretion. Am J Physiol 259: C990–C1004

Fröhlich O, Gunn RB (1986) Erythrocyte anion transport: the kinetics of a single-site obligatory exchange system. Biochem Biophysica Acta 864: 169–194

Germann WJ, Ernst SA, Dawson DC (1986a) Resting and osmotically induced basolateral K conductances in turtle colon. J Gen Physiol 88: 253–74

Germann WJ, Lowy ME, Ernst SA, Dawson DC (1986b) Differentiation of two distinct K conductances in the basolateral membrane of turtle colon. J Gen Physiol 88: 237–51

Grinstein S, Rothstein A (1986) Mechanisms of regulation of the Na+/H+ exchanger. J Membr Biol 90: 1–12

Grinstein S, Rothstein A, Sarkadi B, Gelfand EW (1984) Responses of lymphocytes to anisotonic media: volume-regulating behavior. Am J Physiol 246: C204–C215

Harris S, Richards NW, Logsdon CD, Pouyssegur J, Dawson DC (1992) Cloning of partial cDNAs homologous to the human NHE-1 antiporter from reptilian colon. J Gen Physiol (in press)

Harvey BJ, Ehrenfeld J (1988) Role of Na^+/H^+ exchange in the control of intracellular pH and cell membrane conductances in frog skin epithelium. J Gen Physiol 92: 793–810

Harvey BJ, Thomas SR, Ehrenfeld J (1988) Intracellular pH controls cell membrane Na^+ and K^+ conductances and transport in frog skin epithelium. J Gen Physiol 92: 767–791

Hodgkin AL, Huxley AF (1952) A quantitative description of membrane current and its application to conduction and excitation in the nerve. J Physiol (Lond) 117: 500–544

Hodgkin AL, Keynes RD (1955) The potassium permeability of a giant nerve fiber. J Physiol (Lond) 128: 61–88

Hoshi T, Zagotta WN, Aldrich RW (1990) Biophysical and molecular mechanisms of *Shaker* potassium channel inactivation. Science 250: 533–538

Kaplan J, Pring M, Passow H (1983) Band-3 protein-mediated anion conductance of the red cell membrane. Slippage vs ionic diffusion. FEBS Lett 156: 175–179

Kirk KL, Dawson DC (1983) Basolateral potassium channel in turtle colon: evidence for single-file ion flow. J Gen Physiol 82: 297–313

Kleyman T, Cragoe EJ Jr (1988) Amiloride and amiloride analogues as tools in the study of ion transport. J Membr Biol 105: 1–21

Knauf PA, Law FY, Marchant PJ (1983) Relationship of net chloride flow across human erythrocyte membrane to the anion exchange mechanism. J Gen Physiol 81: 95–126

Knickelbein RG, Aronson P, Dobbins JW (1990) Characterization of Na-H exchangers on villus cells in rabbit ileum. Am J Physiol 259: G802–G806

Läuger P (1987) Dynamics of ion transport systems in membranes. Physiol Rev 67: 1296–1326

Lechleiter J, Girard S, Peralta E, Clapham D (1991) Spiral calcium wave propagation and annihilation in *Xenopus laevis* oocytes. Science 252: 123–126
Palmer LG, Sackin H (1988) Regulation of renal ion channels. FASEB J 2: 3061–3065
Post MA, Dawson DC (1992) Basolateral Na/H antiporter: uncoupled Na transport produce an amiloride-sensitive conductance. Am J Physiol 262: C1089–C1094
Richards NW, Dawson DC (1986) Single potassium channels blocked by lidocaine and quinidine in isolated turtle colon epithelial cells. Am J Physiol 251: C85–C89
Richards NW, Wilkinson DJ, Logsdon CD, Dawson DC (1991) Expression of amiloride-sensitive Na^+ channels from turtle colon in *Xenopus* oocytes FASEB J 5: A689
Sachs F (1987) Baroreceptors at the cellular level. Fed Proc 46: 12–16
Sardet C, Franchi A, Pouysségur J (1989) Molecular cloning primary structure and expression of the human growth factor-activatable Na^+/H^+ antiporter. Cell 56: 271–280
Sariban-Sohraby S, Burg M, Turner RJ (1983) Apical sodium uptake in toad kidney epithelial cell line A6. Am J Physiol 245: C167–C171
Schultz SG (1981) Homocellular regulatory mechanisms in sodium-transporting epithelia: avoidance of extinction by "flush-through". Am J Physiol 241: F579–90
Schultz SG, Hudson RL (1986) How do sodium-absorbing cells do their job and survive? News Physiol Sci 1: 185–189
Tsien RY (1988) Fluorescence measurement and photochemical manipulation of cytosolic free calcium. Trends Neurosci 11: 419–424
Venglarik CJ, Dawson DC (1986) Cholinergic regulation of Na absorption by turtle colon: role of basolateral K conductance. Am J Physiol 251: C563–70
Watson JD, Crick FHC (1953) Molecular structure of nucleic acids: a structure for deoxyribose nucleic acid. Nature 171: 737–738
Wilkinson D, Dawson DC (1990) Cholinergic modulation of apical Na^+ channels in turtle colon: analysis of CDPC-induced fluctuations. Am J Physiol 259: C668–C674
Zagotta WN, Hoshi T, Aldrich RW (1990) Restoration of inactivation in mutants of *Shaker* potassium channels by a peptide derived from ShB. Science 250: 568–571

Chapter 4

Basic Characteristics and Hormonal Regulation of Ion Transport in Avian Hindguts

Erik Skadhauge[1]

Contents

1 Introduction

The transport of ions – particularly of NaCl – across the epithelia of the lower intestines of birds is of interest for two reasons. First, the epithelia of these gut segments respond in a number of avian species with marked changes in rate and mechanism of transport depending on the NaCl content of the diet, or following external injection of osmoregulatory hormones. This makes these epithelia valuable as transport models. Second, the fact that the cloaca and lower gut function as storage organs for ureteral urine and feces makes the mechanisms and regulation of ion transport across the walls very important in avian excretory physiology. The chicken coprodeum has not only the largest increase in transport of Na inducible by NaCl depletion encountered among vertebrates, it is also in the unique position of

[1]Department of Anatomy and Physiology, Section for Veterinary Physiology and Biochemistry, The Royal Veterinary and Agricultural University, 13 Bülowsvej, DK-1870 Frederiksberg C, Denmark

Advances in Comparative and Environmental Physiology, Vol. 16

being, so far, the only organ for which total organ adaptation has been quantified on the ultrastructural level. The morphological adaptation may then be related to the adaptation of apical and basolateral transport capacity on the single cell level and to transepithelial transport rates as measured in vitro and in vivo. Turnheim (1991) has in a recent review observed: "Since the emergence of cellular biology, we have been so centered on the individual cell that we may have lost sight of the fact that in higher animals cells do not occur as single entities but as a connected multitude". Here, the detailed aspects of adaptation of the "connected multitude" of cells have been studied on an ideal model epithelium. The studies of the chicken coprodeum may thus be seen as an application of the August Krogh principle.

This review will define "avian hindguts" as the organs into which (in most species) ureteral urine is regurgitated and stored, i.e., coprodeum, large intestine (colon/rectum), and the paired caeca (when present). These organs are provided with transporting epithelia (a single layer of columnar cells), in contrast to urodeum and proctodeum – the lower parts of the cloaca – which do not transport ions across their walls (the epithelium is a multi-layer squamous type), nor are they used to store the excreta. Coprodeum and colon form in most species a common chamber for storage of urine and feces, whereas the caeca receive only a fraction of urine and finer particles in solution from the ileal chyme (see Skadhauge 1981, Chap. 3). Coprodeum and colon are therefore responsible for the postrenal modification of ureteral urine, and serve together as the final regulator and integrator of excretion of salt and water. Resorption of NaCl, and secretion of K, take place in both organs and some resorption of water occurs, mainly in the colon (see surveys by Skadhauge 1981; Thomas 1982). The prolonged storage possible in the caeca allows, however, the breakdown of fibers and uric acid with resulting production and absorption of short chain fatty acids (SCFA). The breakdown of uric acid may be important in nitrogen conservation (Campbell and Braun 1986) and the capacity for NaCl and water absorption (Thomas and Skadhauge 1988, 1989b,c) makes, at least in the domestic fowl, the caeca of potential great importance in osmoregulation.

In 1988–89 reviews were published which described both the basic characteristics of ion transport in avian hindguts and the hormonal regulation (Clauss and Skadhauge 1988; Thomas and Skadhauge 1988) and symposia volumes appeared: Progress in Avian Osmoregulation (Hughes and Chadwick 1989) and Function of the Avian Cecum (Braun and Duke 1989). In addition, shorter reviews, mainly concerning hormonal regulation of NaCl transport in coprodeum and colon of the hen, were published by Skadhauge (1989a,b) and Skadhauge et al. (1989). For these reasons the present survey will concentrate on the latest progress on the subject. As in the past, particularly as far as mechanisms of transport are concerned, the majority of available information has been obtained on the domestic fowl.

2 Ion Transport Across Avian Coprodeum and Colon

2.1 Basic Characteristics of NaCl Transport

The absorption of Na and Cl across the epithelia of these organs is markedly influenced by the NaCl content of the diet. The ion transport patterns must therefore be described at both low and high NaCl intakes. Due to a meticulous study of plasma hormone levels and intestinal transport rates in vitro at six levels of NaCl intake (Árnason and Skadhauge 1991), the exact limits of "high" and "low" NaCl intake are now known in the domestic fowl (vide infra). The time for reaching steady state in plasma hormone levels and the rates of epithelial transport after abrupt changes in NaCl intake have previously been described (Thomas and Skadhauge 1982; Skadhauge et al. 1983; Rice et al. 1985) and reviewed (Skadhauge 1983; Skadhauge et al. 1985). Briefly, it takes a full week to reach the "low-Na" mode of Na transport, but apparently only 1–2 days to "resalinate" the bird. The temporal studies reveal, however, complicated patterns of interaction of target organ responses and hormone levels, which give further insight into the mechanisms controlling the actual rates of Na transport. These studies will therefore be included in this section.

In both coprodeum and colon from the low-NaCl hen a high rate of Na absorption occurs, which is an electrogenic Na transport, nearly equal to the short-circuit current (SCC), in full accordance with the classical Koefoed-Johnsen-Ussing model. The flux ratio for Cl is unity. The Na transport is not stimulated by glucose nor by amino acids in the mucosal medium, but mucosal amiloride (10^{-4} M) suppresses SCC fully, and often reverses the SCC slightly. This is due to K^+ and H^+ secretion, both of which have been verified experimentally (vide infra). A high-NaCl intake simply suppresses the Na absorption of the coprodeum, the flux ratio becomes unity, and SCC falls to near zero. In the colon, however, Na absorption continues, albeit at a somewhat lower rate, but now only in the presence of mucosal glucose and amino acids, as a cotransport with these non-electrolytes (see Lind et al. 1980a,b; Munck 1989). This symport is amiloride-independent. The pattern is probably general for seed-eating, Na^+-conserving birds, since roughly the same transport rates have been observed in the Australian parrot, the galah *Cacatua roseicapilla* (Skadhauge 1980; Skadhauge and Dawson 1980). This dietary/hormonal dependence is definitely different in other species such as the salt-gland-possessing, marine-adaptable bird, the domestic duck (Skadhauge et al. 1984) and the Australian emu (*Dromaius novaehollandiae*) (Skadhauge et al. 1991), in which NaCl loading apparently does not change the Na-transport pattern (vide infra).

The main question is the mechanism of induction of Na transport, which is zero on a high-NaCl diet, to a high level on low-NaCl intake. Secondly, what controls and limits the actual transport rate in the coprodeum, and what switches the transport from hexose-amino-acid cotransport to "pure" Na transport in the colon? The basic study of Bindslev (1979) clearly demonstrated, with two fully independent techniques, that the Na permeability of the apical membrane of the coprodeum undergoes a major increase from a high- to low-NaCl diet. The influx of Na across the mucosal membrane, as measured by the Schultz-Curran technique, was approximately $10\,\mu mol\,cm^{-2}\,h^{-1}$ in the low-NaCl birds, and only $0.9\,\mu mol\,cm^{-2}\,h^{-1}$ in birds on the high-NaCl diet. After ouabain treatment, 20:1 Na-dilution potentials at the mucosal side gave 17–18 mV in birds on a low- and less than 2 mV on a high-NaCl diet. The dilution potentials of birds on a low-NaCl diet were, fully reversibly, amiloride-suppressible. Although semiquantitative, Bindslev's experiments show that a major part of the adaptation to a low-NaCl diet is the opening of apical Na channels. Early experiments by C. Ellory and E. Skadhauge (briefly reported in Skadhauge 1980) indicated that the Na pump at the basolateral membrane does not undergo a major change with diet, since the ouabain-inhibitable Na/K-ATPase was only increased two-fold by low-NaCl adaptation. This observation was recently confirmed by Dantzer et al. (1988) who observed that the ouabain-inhibitable K-paranitrophenylphosphatase activity was also augmented less than two-fold (vide infra).

The change in apical permeability to Na was thoroughly tested by rapid current-voltage measurements of the coprodeum after depolarization of the basolateral membrane potential with a high K-concentration (Clauss et al. 1987). The specific Na permeability was defined as the difference before /after mucosal amiloride. At an amiloride-inhibitable Na current of 182 ± $18\,\mu A/cm^2$ the apical Na permeability was determined to be $(7.3 \pm 0.6) \times 10^{-2}$ cm/h on tissues from birds on a low-NaCl diet. Because the coprodeum exhibits no measurable Na transport (amiloride-sensitive SCC) on high-NaCl intake, the – presumably low – Na permeability in that physiological state could not be determined. Aldosterone treatment resulted, however, in one third of the low-NaCl permeability, and spironolactone medication reduced the low-NaCl value about 50%. This clearly indicates that the hormone opens apical channels. It is noteworthy that there is a linear relationship, for all experiments, between initial SCC, as measured before basolateral K-depolarization, and calculated apical Na permeability. This strongly supports the physiological validity of the measurements (Skadhauge 1989b). The main conclusion is, at this stage, that the adaptation of the coprodeum to low NaCl/high aldosterone involves first and foremost the opening of apical Na channels, enlarging the Na permeability by a factor of around 100, whereas the basolateral pumping capacity is only increased by a factor of about 2. The Na channel has not been fully characterized by patch-clamp analysis, but Christensen and Bindslev's (1982) frequency fluctuation study ("noise analysis") indicated a single-channel conductance of 4 pS and a maximal site density of $38/\mu m^2$. On embryonic coprodeum Hoffmann et al. (1990) observed a single-channel current of 2.2 pA, corresponding to a conductance of about 40 pS, but a channel density of only 0.16/μm^2. This suggests that age as well as diet play a role. Cuthbert et al. (1982) observed, by binding of the amiloride derivative benzamil, 7 sites/μm^2 on a low-NaCl and none on a high-NaCl diet. Recent findings by Heinz et al. (1991) suggest that the ontogenetic development of Na channels may be complex. These authors studied the coprodeum of the chicken embryo and observed electrogenic Na absorption with both amiloride-sensitive and -insensitive components. Part of the latter was activated by removal of mucosal Ca^{2+}.

There are basically three ways in which aldosterone may increase the apical Na permeability: (1) de novo synthesis of Na channels; (2) activation of preexisting quiescent channels; and (3) an increase in the open probability of already conducting channels (Garty and Asher 1991). Concerning these possibilities, the third is unlikely as the sole explanation for simple quantitative considerations. Insertion of preformed channel protein from cytosol into the membrane was suggested by amiloride-binding studies (Cuthbert et al. 1982). Garty and Asher (1991) and Asher et al. (1992) now bring forward strong evidence suggesting that aldosterone/NaCl depletion evokes modulations in the level of mRNA which codes either for the Na-channel or a posttranscriptional regulator of the channel. These authors injected *Xenopus* oocytes with poly A+ RNA isolated from both coprodeum and colon, taken from birds which were either "fed for 10 days a zero sodium diet" or "fed normal food (containing 0.15% w/w NaCl)", i.e., a high-NaCl diet. The 22 Na uptake was measured in the absence and presence of amiloride (0.4 μM). The RNA from the low-NaCl birds induced a high rate of Na uptake which was fully amiloride-suppressible, whereas the RNA from the high-NaCl-diet group induced none. In addition, an amiloride analogue, which is a poor channel blocker but a potent inhibitor of Na/H exchange, did not block the Na influx at all. This provides strong evidence that the protein expressed is the apical Na channel. The authors therefore concluded that a chronic high level of aldosterone brings about the induction of apical Na channels but they expressed the reservation that it is not certain that NaCl depletion augments transcription of channels only through an increased aldosterone level. The huge change in Na uptake with dietary adaptation obviously makes it possible to quantify the de novo synthesis of Na channels in relation to acute Na depletion and resalination.

With regard to chloride *absorption*, both classical (Choshniak et al. 1977; Lind et al. 1980a; Holtug and Skadhauge 1982) and recent in vitro studies of both coprodeum and colon (Clauss et al. 1988, 1991) have indicated flux ratios of unity in the short-circuited state, and "pure" electrogenic driving in open circuit. Under these conditions, when the Na pump works against the potential difference it creates, the actual Na-absorption rate is halved (see Skadhauge 1982). A recent study of the localization of Na current by extracellular recording with the vibrating microprobe has demonstrated that the amiloride-suppressible inward current (i.e. Na absorption) is distributed equally over the villi/folds of the coprodeum (Holtug et al.

1991a). Within the power of resolution of the probe, about 10 μm, equal to the width of the cells, special transporting cells could not be detected in open circuit. Over the crypts an outwardly directed current was detected, demonstrating ion migration (Cl inwards, Na outwards). This current disappeared by transepithelial short-circuiting. The electrogenic Cl absorption through the crypts accounted, however, only for 4% of the total Cl absorption. The major part of the Cl transport follows Na over the villi, presumably through paracellular shunts.

2.2 Chloride Secretion

In the above-mentioned experiments we observed after amiloride and theophylline, a return of inwardly directed current equally distributed over the villi/folds, having a magnitude of approximately one third of the Na current. An outwardly directed current was observed over the crypts, as before reflecting Na and Cl migration in the electric field, leading now to a net movement of Na following the Cl secretion. Were the observed Cl secretion to occur from the crypts we could calculate that not only should the direction of current be opposite to that observed, but the magnitude 10000-fold larger. We have therefore shown, beyond reasonable doubt, that Cl is secreted from the same population of villar cells which normally absorbs Na (Holtug et al. 1991a). Secretion of Cl is also inducible in the colon as reported previously (Munck et al. 1984). Recently, these studies have been extended to include dietary and hormonal actions, and the time effects of hormones, on electrolyte transport. Clauss et al. (1991) actually report the colonic data from the same group of birds from which coprodeal observations have already been published (Clauss et al. 1988). Our study confirms, first of all, the effects of dietary change, and external aldosterone injection, which switches the electrogenic Na-amino-acid/hexose cotransport system to the amiloride-sensitive Na channel. Secondly, electrogenic Cl secretion was induced by theophylline and inhibited by bumetanide; this drug is presumed to act on the basolateral NaK2Cl cotransport. The modulation of Cl secretion induced by a low-NaCl diet or aldosterone injection only varied by a factor of 2, from 7 to 14 $\mu mol\,cm^{-2}\,h^{-1}$. As in the coprodeum this rate of secretion is a fairly stable fraction of the normal rate of Na absorption, in the colon approximately 50%. In the colon a minor fraction of cells precipitates Ag, unchanged after dietary modification or hormonal injection. Furthermore, apical microvilli are always present (Clauss et al. 1991). It is noteworthy that the maximal stimulation of Na absorption after aldosterone injection is reached already after 24 h of treatment, in agreement with the absence of an apparent fall in aldosterone sensitivity after aldosterone treatment in the days after acute resalination (Clauss et al. 1984).

Andersen et al. (1990) subjected the interaction of Cl concentration and bumetanide sensitivity to closer scrutiny and observed a Hill coefficient of near 2 (1.93 ± 0.03). This lends further support to the notion that Cl secretion is mediated by a basolateral NaK2Cl cotransport process. Chloride secretion can also be induced by serotonin with cAMP as the probable intracellular mediator (Hansen and Bindslev 1989). Clauss et al. (1990) and Fischer et al. (1991) have recently addressed the question of the nature of the apical Cl channel by patch-clamp technique. The colonic enterocytes were isolated in Ca-free solution, and cell-attached, patch clamping instituted at 37°C. In spite of a seal rate of 80%, ion channels were only observed in 10% of the colonic patches, and not at all in the coprodeum. A total of nine channels were analyzed. They fell in two groups with average conductances of 9 and 35 pS at negative holding potentials. These data suggest that either the preparation process closes the channels or that most of the Cl conductance occurs through subpicosiemens channels. The low fraction of patches having detectable channels should therefore not be taken as proof of cluster occurrence of Cl channels (i.e. presence of "Cl cells"). Recent studies on hen ileŭm have demonstrated phorbol ester stimulation of active anion secretion (Chang et al. 1990), which involves arachidonic acid metabolism (Musch et al. 1990).

2.3 Potassium Transport

Both coprodeum and colon exhibit a small electrogenic secretion of K, usually around 0.5 $\mu mol\,cm^{-2}\,h^{-1}$ in the short-circuited state. This is, together with the Na/H antiport (vide infra), responsible for the negative current (lumen positive), which is left after amiloride treatment. In the colon the K-secretion was higher, 0.7 $\mu mol\,cm^{-2}\,h^{-1}$ on a high-NaCl diet with glucose and amino acids to stimulate Na transport, than at low-NaCl intake, 0.4 $\mu mol\,cm^{-2}\,h^{-1}$ (Lind et al. 1980a). Two recent studies have addressed the colonic regulation of K-secretion. Grubb and Bentley (1990) studied the unstripped epithelium from birds on a "normal" (i.e., high-NaCl) diet in the short-circuited state. The bathing media contained glucose but no amino acids. A net secretion of 0.05 $\mu mol\,cm^{-2}\,h^{-1}$ was observed. This was abolished by K-loading the birds, due to an increase in the mucosa to serosa movement of K by one order of magnitude. Munck and Munck (1990) observed on stripped epithelium from hens on a low-NaCl diet a secretion of only 0.2 $\mu mol\,cm^{-2}\,h^{-1}$. This was on high-NaCl-diet birds augmented from a control value of 0.08 ± 0.04 to 0.42 ± 0.07 with glucose in the bathing media and to 0.60 ± 0.07 $\mu mol\,cm^{-2}\,h^{-1}$ with the further addition of amino acids. Although the net Na absorption is higher in the low-NaCl colon than in even the maximally glucose/amino acid-stimulated high-NaCl colon (in which plasma aldosterone is suppressed), the K-secretion is, as reported above, smaller. Data reported by Munck and Munck (1990) lend support to the notion that the observed difference in the rate of K-secretion is caused by the relative K-impermeability of the apical membrane in the low-NaCl colon. Insofar as K-depletion indicates NaCl repletion, the study of Grubb and Bentley is at variance with that of Munck and Munck concerning the effect of alkaline metal ion balance on net K-transport across the colonic epithelium. Only studies with full hexose/amino acid stimulation of a NaCl-replete colon, and plasma aldosterone recordings during K-depletion can clarify the issue.

Both Andersen et al. (1990) and Munck and Munck (1990) have observed increased unidirectional fluxes of K in high-NaCl colon exposed to theophylline with reduction after bumetanide. This is in full accordance with the current model for Cl secretion (Dawson 1991), which requires a simultaneous increase in the K-conductance. It should be noted that Montero et al. (1988, 1991), on isolated epithelial cells from the chicken small intestine ("midintestine"), made observations which suggest that the K-efflux occurs through Ca-activated K-channels. However, 86 Rb was used as a marker for K and this may underestimate the true K-flux (Grubb et al. 1988).

2.4 Na/H Antiports

Two classical observations suggest the presence of these antiports in the lower gut of birds: (1) the occasional observation of a reversal of current after amiloride in both the coprodeum (Clauss et al. 1988) and colon (Clauss et al. 1991) to values of higher negativity than likely to be accounted for by secretion of K; and (2) the re-emergence of positive current (serosa positive) after the addition of acetazolamide to amiloride-treated colon (Lind et al. 1980a). The recent discovery of an acid microclimate in hen colon provides compelling evidence for the presence of a Na/H antiport. Holtug et al. (1991b) superfused the mucosal surface of a Krebs-phosphate buffer with a constant pH of 7 and observed with small pH electrodes a microclimate with a pH of 6.27 ± 0.05. Since the mucosal presence of 23 mM butyrate reduced net Na absorption by 80%, Holtug (1989) speculated that SCFAs enter the colonic cell in the protonated form inhibiting the apical Na channel by the resulting low intracellular pH. Laverty and Alberici (1991) and V.S. Elbrønd et al. (pers. comm.) observed acetazolamide-suppressible histochemical staining for carbonic anhydrase activity in colons of the European starling (*Sturnus vulgaris*) and the domestic fowl, respectively. Laverty and Albirici found scattered cells with intense apical brush border membrane staining as well as some cytoplasmic staining. Cells lining the coprodeum stained less intensively. Elbrønd observed neither a major difference between coprodeum and colon, nor between high- and low-NaCl diet. The staining was most pronounced in the crypts. On the cellular level the lateral membranes were most strongly stained. The physiological role of the carbonic anhydrase is not clear, but the enzyme assists presumably first and foremost the transport of SCFAs. The Na/H exchange has been studied in isolated enterocytes from the small intestine at three sites (Montrose et al. 1985) and "the distal half" (Semrad and Chang 1987). Montrose et al.

demonstrated a rapid exchange in intracellular pH by the Na/H exchange system in response to imposed Na gradients, and Semrad and Chang's observations indicate that an increase in intracellular Ca concentration is essential for the cAMP-induced inhibition of amiloride-sensitive Na/H exchange. It should be noted that Cl^-/base exchangers also are observed in the small intestine (Calonge et al. 1992).

3 Hormonal Regulation of Ion Transport

3.1 Regulation by Aldosterone

Previous work, both in vivo (see Skadhauge 1981) and in vitro (Thomas et al. 1980), have clearly shown aldosterone to be a major regulator of Na transport in the coprodeum and colon. This has also been established by measurements of changes in plasma concentrations of possible osmoregulatory hormones (aldosterone, corticosterone, prolactin, arginine vasotocin) and target organ responses during NaCl depletion and repletion (resalination) (Skadhauge et al. 1983; Rice et al. 1985; Thomas and Skadhauge 1982), and changes in plasma hormone concentrations in response to dehydration (Árnason et al. 1986) or other osmotic stimuli (Arad et al. 1986) at high- and low-NaCl intake. Briefly, aldosterone changes from the near-detection limit by radioimmunoassay at high- to around 300–400 pg/ml at low-NaCl intake, whereas corticosterone at a value of around 1 ng/ml undergoes no major change. Similar observations were made by Radke et al. (1984) and Rosenberg and Hurwitz (1987). Both prolactin and AVT increase proportionally to plasma osmolality (and therefore plasma Na and Cl concentrations), and reach higher concentrations at higher NaCl intakes. During NaCl depletion both target organ responses and plasma aldosterone concentrations reach new steady states over 4–6 days, whereas resalination reduces the plasma aldosterone concentration to near zero in 8 h and brings about in 24 h the reduction of coprodeal SCC to near zero and the switch-over of colonic Na transport from the amiloride-suppressible to the amino-acid-stimulated cotransport. These temporal changes in plasma aldosterone in response to diet were recently confirmed by Sandor et al. (1989).

This section addresses the first question: What is actually "high" and "low" NaCl intake as experienced by the hen? Árnason and Skadhauge (1991) subjected groups of hens to long-term adaptation to six levels of NaCl intake ranging from 0.2–25 $\mu mol\, kg^{-1}$ body weight day^{-1} at near constant intake of K. The plasma osmolality, and concentrations of Na, K, and Cl, as well as aldosterone, corticosterone, prolactin and AVT, were measured at the six levels. The Na absorption and Cl secretion of the coprodeum and colon were characterized in the Ussing chamber by the effects of amiloride and theophylline and by hexose and amino acids (colon) on SCC and the electrical potential difference (PD). The very wide range in NaCl intake was chosen in order to detect the defense mechanisms against NaCl lack and surplus (i.e. osmotic stress) respectively, and to demonstrate the level of NaCl which presents the least osmoregulatory stress, i.e. neither a NaCl deficit nor a NaCl load is detected. This will also quantify the ranges of "high" and "low" NaCl intakes. The six levels, labelled 1–6 from low to high, were designed to present a doubling of Na intake at each step. At the low levels the most conspicuous change was a concomitant change in plasma aldosterone concentration and amiloride-suppressible SCC of both coprodeum and colon. The amiloride-suppressible SCC was nearly absent from level 4 and upwards in both organs. From this level plasma osmolality and Na and Cl concentrations rose proportionally to prolactin and AVT concentrations. Concomitantly, the hexose/amino-acid stimulation of the colon emerged, but with no change from levels 5–6. This suggests that these hormones have no direct effect on colonic Na transport. Corticosterone underwent a slight reduction with increased NaCl intake as did plasma-K. The theophylline-stimulated Cl secretion was independent of NaCl intake in the colon, but disappeared in the coprodeum from level 4 and upwards. The study confirms that aldosterone provides the major hormonal defense against NaCl depletion, and that both prolactin and AVT rise in proportion to the osmotic stress but with no apparent correlation to intestinal transport. The level of NaCl intake which is osmoregulatory neutral as based on both hormonal and intestinal parameters, is in the range of 2.3–5.6 mmol Na kg^{-1} body weight day^{-1} (between levels 3 and 4), or as calculated as Na content in food: 35–70 mmol/kg. The intakes of NaCl previously denoted by our group as "high" and "low" are now proven to be so. The so-called normal Na intake reported by other groups is clearly high in physiological terms.

3.2 Hormone Injection Studies

Early injection studies made it clear that aldosterone augments Na absorption, as studied by in vivo perfusion (Thomas and Skadhauge 1979; Thomas et al. 1979) and in vitro in the Ussing chamber (Thomas et al. 1980). These observations were amply confirmed by the in vitro study by Clauss et al. (1988), who found that 24 h of aldosterone treatment induced an amiloride-suppressible increase of $2.1 \pm 0.4\,\mu mol\,cm^{-2}\,h^{-1}$. This is, however, only a fraction of the low-NaCl diet Na absorption of $14.2\,\mu mol\,cm^{-2}\,h^{-1}$. In colon Clauss et al. (1991) observed in the same group of birds a delta-amiloride value of $23.3\,\mu mol\,cm^{-2}\,h^{-1}$, nearly equal to the low-NaCl value of $19.2\,\mu mol\,cm^{-2}\,h^{-1}$. A possible explanation for this apparent difference in the ability of aldosterone to mimic low-NaCl adaptation is undoubtedly, in part, the long time necessary for the development of apical microvilli in the coprodeum (vide infra). In the presence of glucose, but without amino acids, the colon from high-NaCl birds had an SCC of only $3.3\,\mu mol\,cm^{-2}\,h^{-1}$, which rose to $8.4\,\mu mol\,cm^{-2}\,h^{-1}$ after the addition of amino acids. This demonstrates two aspects of the coprodeum/colon difference: First, that the amiloride-suppressible Na channel has a higher permeability than the cotransporters; second, that the amino acid stimulation, as already noted by Lind et al. (1980a) and confirmed by Arnason and Skadhauge (1991, see their Fig. 4), induces a higher rate of cotransport than the hexoses. Whether this indicates the presence of two different kinds of cotransporters or the activation of different cell types remains to be established.

The acute resalination/aldosterone injection study (Clauss et al. 1991) shows a time lag between the prompt (over 24 h) induction of maximal amiloride-suppressible Na transport by aldosterone, but a slower disappearance of the amino acid stimulation. Following resalination it took 2–3 days for the amino acid stimulation to reach the high-NaCl level. These delays were already observed by Thomas and Skadhauge (1982). Since a change in Na transport in the colon cannot be related to growth of microvilli (Clauss et al. 1991) nor to a change in basolateral enzyme activity (Skadhauge 1980), it might be speculated that it is the synthesis and insertion of the amino acid cotransporter which change at a slower rate. The effect of a low-Na diet on colonic Na absorption was confirmed by Grubb and Bentley (1987) who switched their hens from feed containing 87 µmol Na/kg (which is between Árnason and Skadhauge's levels 4 and 5) to 4.3 µmol/kg (between 1 and 2). A comparison of studies must be made with caution since these workers used unstripped colon and did not stimulate with amino acids. An amiloride concentration of 10^{-5} M was used which, if the colon has the same sensitivity as coprodeum, may be borderline with regard to the complete suppression of the amiloride-sensitive current (Bindslev et al. 1982).

Since coprodeum and colon have partly overlapping mineralo- and glucocorticoid receptors (vide infra), it is not surprising that both corticosterone and dexamethasone augment Na transport across these epithelia. Clauss et al. (1987), in their instantaneous current-voltage study, observed that 12 h of dexamethasone treatment to high-NaCl birds induced an apical Na permeability of 20% of the low-NaCl value. Grubb and Bentley (1989) found in a comprehensive study that a single injection of dexamethasone (650 µg/kg) quadrupled the amiloride-suppressible SCC. Three or 7 days of treatment resulted in a lower absolute value of SCC, but the delta-amiloride value was higher, 75 and 73 $\mu A/cm^2$, respectively. This reflects most likely an increased H^+-secretion, as the K- and Cl fluxes, although increased, did not change the net movement of charge.

In spite of the obvious effects of aldosterone on Na transport demonstrated after in vivo injection, the hormone does not seem able to augment the SCC after in vitro addition to the bathing media. The expected time lag before any transport effect was likely to be measureable is, however, close to the maximal survival time for an in vitro preparation from a warm-blooded animal. Recently, the ability to improve in vitro survival (i.e., to sustain SCC) of the "cocktail" of stimulators of intestinal metabolism (Windmueller and Spaeth 1978) was tested by Elbrønd and Skadhauge

(1992), and the aldosterone effect in the presence of these stimulators was tested by Hoffmann et al. (1990). The bathing medium was supplemented with 10 mM β-hydroxybutyrate, 0.5 mM glutamine, and 10 mM mannose (BGM). Elbrønd and Skadhauge measured the ability of various media to sustain SCC over 8 h. The effect of hourly replacements of the media was also tested. This replacement resulted in significantly higher SCC from 2–8 h but neither HCO_3 medium nor the addition of BGM maintained SCC as well as the standard Krebs-phosphate buffer used by our group since our first in vitro publication (Choshniak et al. 1977). This standard medium alone (with replacement) resulted in an SCC of low-NaCl coprodea at 4 h of $156 \pm 21\,\mu A/cm^2$, and 8 h $77 \pm 14\,\mu A/cm^2$. This current was amiloride-sensitive. Aldosterone, even in high concentrations, failed to produce any increase in SCC (from zero) on coprodea from birds exposed to a chronic high-NaCl diet and after 48 and 72 h of resalination. Unexpectedly, control tissue from 24-h resalinated birds rose maximally after 6 h to more than $200\,\mu A/cm^2$, fully amiloride-sensitive. It is speculated that this might be due to the washout of a transport inhibitor (endogenous digitalis-like compound). It is encouraging that the coprodeum can sustain a relatively high current over 8 h and further studies in the presence of BGM should be carried out. Hoffmann et al. (1990 and unpubl.) mounted embryonic coprodeum in Ringer's with BGM but did not find a significant increase in SCC after aldosterone (10^{-7}). The SCC, however, was augmented five- to six-fold by simultaneous treatment with thyroxine (10^{-6} M), which in itself produced a smaller rise in SCC. The possible permissive effects of corticosterone should also be tested, since it is possible that "full" aldosterone effects require not only the occupation of mineralocorticoid receptors (MR), but also glucocorticoid receptors (GR). In addition, compounds affecting the 11 β-hydroxysteroid dehydrogenase activity should be tested.

3.3 Studies of Intestinal Steroid Receptors

The intestinal mineralocorticoid (MR) and glucocorticoid (GR) receptor systems have been characterized in both the domestic duck (colon) (DiBattista et al. 1985, 1989) and the hen (coprodeum and colon) maintained on either a low-Na or high-Na diet (Sandor et al. 1989). These organs possess cytosolic and nuclear MR and GR of high affinity and low capacity. The cytosolic equilibrium dissociation constants were in the nanomolar range, and the concentration of receptors was in the 10–500 fmol/mg protein range. The cytosolic MR-binding parameters showed a "downregulation" on a low-Na diet in both species, whereas the cytosolic GR-binding parameters were little influenced by diet. The downregulation on high aldosterone levels was not observed in hen nuclear receptors. The temporal changes in RNA polymerase activity in hen coprodeum paralleled, during both Na depletion and resalination, the MR downregulation. Similarly, the RNA polymerase activity was inversely proportional to the levels of plasma aldosterone. The affinity of the steroids to the MR and GR did not change with Na depletion or resalination. A marked difference was observed in the affinities of the steroids for duck and hen colon respectively, with the hen having the lowest equilibrium dissociation constants. The average values of the maximum binding sites of colonic and coprodeal cytosol were 64 and 47 fmol/mg protein on a low-Na diet, but 221 and 125 fmol/mg protein on high-Na intake respectively. This latter value is fully confirmed by Rafestin-Oblin et al. (1989, 1991) who reported 120–160 fmol/mg protein in the "distal intestine" (coprodeum), but only 40–60 in the "proximal intestine" (colon). Since the latter value is not far from the low-Na value of Sandor et al., differences in actual NaCl intake may partly explain the discrepancy. It is noteworthy that there is some overlapping in binding of both MR and GR to aldosterone, corticosterone and other steroid ligands such as dexamethasone (DiBattista et al. 1985; Sandor et al. 1989; Rafestin-Oblin et al. 1991). The physiological implications of these studies are first, that they confirm the epithelia of the lower intestines as target organs for steroid action. Second, the duck/hen difference tallies well with the marine/terrestrial background of the two species. Third, the downregulation of MR rules "receptor recruitment" out as a possible explanation for the decrease in the apparent aldosterone effect during abrupt resalination (Clauss et al. 1984, 1988). The concomitant downregulation of RNA polymerase II activity of hen coprodeal nuclei during Na depletion and upregulation during resalination seem, however, to go against the de novo synthesis of Na channels, since transcriptional activity involving mRNA synthesis depends upon the polymerase action. This apparent paradox may perhaps first be resolved, when it becomes possible to follow changes in intracellular Na activity by independent quantitative measurements.

3.4 Effects of Other Hormones

This section must necessarily be short, since no hormones other than adrenocortial steroids have been systematically tested for their effect on ion transport in avian coprodeum or colon. Potential candidates are, however, besides prolactin and thyroxine (Hoffmann et al. 1990), gastrointestinal hormones, AVT, and the renin/angiotensin system. Birds possess the latter as surveyed by Wilson (1989). Concerning VIP a slight effect of Cl secretion was reported by Munck et al. (1984) on hen colon but no physiological role can presently be assigned to GI hormones on the lower gut. The atriopeptins or atrial natruretic factor (ANF) are produced by the chicken heart and affect the kidney (see Gregg and Wideman 1986) and ANF affects aldosterone secretion in chickens (Rosenberg et al. 1989) and ducks (Gray et al. 1991), but a regulatory role in relation to intestinal ion transport cannot (yet?) be defined. This field is open for further investigation; interesting differences are to be expected between terrestrial and marine birds.

3.5 Correlation of Structure and Function During Long-Term Aldosterone Exposure

A major inspiration for several of the experiments reported in this section was the discrepancy between the modest SCC after 24 h of aldosterone injection compared to the SCC induced by low-NaCl adaptation of coprodeum (Thomas et al. 1980). Two a priori explanations are at hand: other hormones or non-hormonal consequences of NaCl depletion are essential, or it takes the epithelium longer than 24 h to adapt fully to a high plasma aldosterone level. One experiment was crucial to elucidate both possibilities: Árnason and Skadhauge (reported by Skadhauge 1983) and Clauss et al. (1984) observed that 24 h of external aldosterone injection given simultaneously with resalination fully maintained SCC and PD at the high low-NaCl values. Otherwise, both SCC and PD would have been reduced to near zero. This demonstrates first, that aldosterone alone is sufficient to sustain the highest possible rate of Na absorption in spite of an expanded extracellular volume, etc. Second, the experiment raises the question of how soon this effect is reduced to the much lower stimulation (ca. 20%) of Na transport by aldosterone injection to birds on a chronic high-NaCl diet. Clauss et al. (1984, 1988) observed that the decline of the aldosterone effect on Na transport as well as on inducible Cl secretion took place stepwise over 3–5 days. Since this is a time scale to be expected (but not confirmed, vide infra!) for the exchange of the intestinal cell population, the possible correlation with the anatomical development of the epithelium was explored. Clauss et al. (1988) had already demonstrated a difference in development of the apical microvilli, which were observed to be nearly absent in high-NaCl birds, but densely packed microvilli were observed in low-NaCl birds. Exactly the same change was observed by Goldstein et al. (1990) in the savannah sparrow (*P.S. beldingi*) in a NaCl-exposed versus a NaCl-deprived subpopulation. Concerning the effect of long-term external aldosterone injection simultaneously with resalination, Árnason and Skadhauge (1985; see also Skadhauge 1989b) demonstrated that the SCC was maintained over 5 days at around 70% of the low NaCl value. This is a somewhat intermediate result that barely provides proof that aldosterone alone is sufficient to maintain the highest rate of Na transport. The obvious experiment to perform would be continued aldosterone infusion over several days to high-NaCl birds. This experiment was not carried out since a possible K-depletion during the expected "escape" of Na retention might blur the issue.

This section will deal with three types of study: (1) histological changes in the hen coprodeum in relation to dietary and hormonal manipulations; (2) cell turnover; and (3) possible evidence for special transporting cells. Eldrup et al. (1979, 1980) were the first to study in detail the coprodeum of hens maintained on a low- and high-NaCl diet by toluidine-stained thin section light microscopy, by transmission electron microscopy and by freeze fracture (see also Skadhauge 1981, Fig. 6.3). Eldrup et al. observed that a low-NaCl diet is associated with the augmentation of the number of dark mitochondrion-rich cells and the induction of rod-shaped particles in the luminal membrane. As these spanned the membrane, the authors suggested that they might function as Na channels. The presence of a small population of dark cells by toluidine staining was confirmed by Clauss et al. (1988), but no change in mitochondria content in comparison to the neighbouring light cells nor a variation with diet was observed. The number of microvilli of the dark cell, however, seems to be slightly higher than that of the light cells. Although Clauss et al. did observe so-called Cl cells by Ag precipitation in hens on a low-NaCl diet, no convincing correlations emerged, which might suggest a role for these cells in inducible Cl secretion, nor were special transporting cells found by electrophysiological methods (vide supra). The exact reason for the Ag precipitation, and whether it expresses a physiological function, remains unknown. Local trapping of ions with low solubility products (not only Cl, but also CO_3 and PO_4) is a possibility.

Elbrønd et al. (1991) used the precise quantitative sampling scheme developed by Mayhew et al. (1990) to estimate the total microvillus surface area of the coprodeum of hens on high- and low-NaCl diets. The same scheme permits estimations of the cell number and mean cell sizes. Elbrønd et al. observed an increase in total microvillus membrane surface area (of the entire organ) from $87 \pm 9.7\,cm^2$ in hens on high- to $202 \pm 38\,cm^2$ on a low-NaCl diet. No differences were found in mean cell height or mean cell volume. There were two main reasons for the observed enlargement in apical surface area: (1) the extent to which mucosal folding increased the serosal surface area (high-NaCl 2.6-fold, low-NaCl 3.2-fold); (2) the average microvillus enlargement of the "simple" apical surface area (high-NaCl 2.1-fold, low-NaCl 3.6-fold). The changes result in a total number of cells per organ of 270 million on high-NaCl and 420 million on low-NaCl diet. When the folding factor and the microvillus factor are both taken into account (the serosal area versus the apical area), i.e., the difference directly affecting measurements in the Ussing chamber, the high-NaCl apical area is $2.6 \times 2.1 = 5.6$-fold enlarged and the low-NaCl is $3.2 \times 3.6 = 11.7$-fold enlarged, i.e., the difference is doubled (factor = 2.1). In order to determine the adaptation which causes these differences, the observations can be profitably related to the rate of cell turnover (V.S. Elbrønd, in prep.). Elbrønd used bromodeoxyuridine incorporation to mark cell turnover and cell movement from the crypts of coprodea from hens on high-NaCl and low-NaCl diets respectively. The bromodeoxyuridine is incoporated in DNA in the synthesis phase of the mitosis. It is localized by specific immunohistochemical techniques. Elbrønd observed birds on the day of injection, and after 2, 4, 6, and 8 days. A larger number of proliferations were probably present in birds on a low-NaCl diet, but the site of proliferation and the kinetics seemed to be the same for both diets. Up to and including day 4 only the bottoms of the crypts were labelled. At day 6 the crypt walls were marked, and only at day 8 labelling had extended slightly beyond the crypt opening, i.e., to the sides of the villi/folds. The immediate conclusion to be drawn from Elbrønd's experiments is that the rate of cell turnover in the coprodeum is much lower than that observed in other warm-blooded intestinal epithelia.

The direct effect of external aldosterone injections to birds on a high-NaCl diet was investigated by Dantzer et al. (1988) and V.S. Elbrønd (in prep.), who injected the hormone over 2, 4, and 6 days. Three daily injections of 128 μg/kg were given. The injections caused neither a weight increase nor change in plasma electrolyte concentrations, suggesting that the possible K-depletion might not be as serious in birds as feared from analogous mammalian injections. This long-term aldosterone treatment resulted in both a major development of the apical microvillus population and a substantial increase in amiloride-sensitive SCC, higher than that inducible by 24 h of aldosterone treatment. Taken together, the experiments quoted so far provide strong circumstantial evidence for two stages in the chronic adaptation to low-NaCl diet/high aldosterone: (1) over 5 days, aldosterone induces the development of apical microvilli of the existing cell population and an increase in the basolateral ATPase activity. This permits an increase in total SCC by 400% in comparison with the SCC induced by opening of the available apical Na channels after 24 h of aldosterone treatment to chronically high-NaCl birds. (2) Besides this adjustment, there is late growth (over 2–3 weeks) of a new cell population with enlargement of the mucosal folding and the total cell number. This provides, ceteris paribus, a further possible 23% increase in SCC of the long-term adapted low-NaCl birds. Whether this adaptation is partially or totally attributable to aldosterone remains to be established.

The basolateral adaptation was explored in a further study (Mayhew et al. 1992) using the quantitative sampling technique (Mayhew et al. 1990). We found no change in either the lateral or basal membrane surface area with diet, the lateral area being nearly 1000 μm^2/cell and the basal area approximately 17 μm^2/cell. There is thus no structural basis for enlarged *basolateral* Na-pumping activity at the single cell level. It should be noted that the observed (vide supra) doubling of ATPase activity of epithelial homogenates may be due partly to the increased *mucosal* folding in birds on a low-NaCl diet. The total membrane area increased, as to be expected, from high-NaCl to low-NaCl not only as reported above regarding the apical membrane, but also the lateral area increased from 2530–4040 cm^2, and the basal area from 45–68 cm^2 due to the general increase in the number of cells. Of interest is also the fraction of different cell populations in the birds on the two diets. There was no change in either the goblet cell fraction (19%) or dark cell fraction (3%) and only the principal light cells, believed to be a uniform population of transporting cells, increased relatively since their absolute number and volume increased in birds on a low-NaCl diet. It has not been possible to identify the adaptation of special cells, which could be likely candidates for high rates of Na absorption/Cl secretion, either by

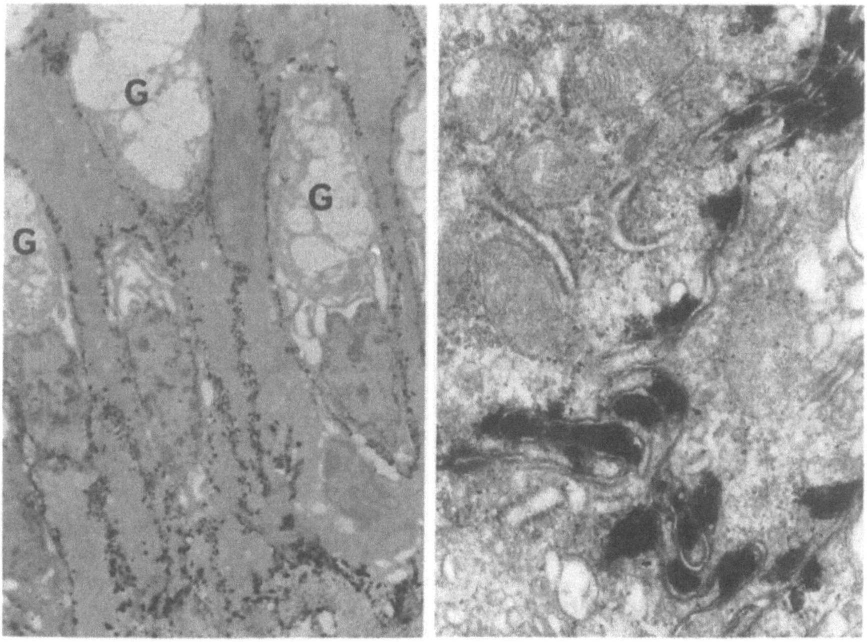

Fig. 1. Localization of Na-K-ATPase activity in the epithelium of chicken coprodeum. *Left* (×3.450) a precipitation which indicates the presence of the enzyme along the basolateral membranes of the principal (transporting) cells. *G* denotes goblet cells. A higher magnification from the basolateral area (×44 000) is shown to the *right* to demonstrate that the reaction products are membrane-bound. (Møller et al. 1989 and unpubl.)

LM or EM. The absence of special cells was confirmed by histochemical and immunochemical techniques.

Møller et al. (1989 and unpubl.) used ouabain-sensitive K-dependent p-nitrophenylphosphatase (K-pNPPase) activity to localize the Na-K-ATPase complex in coprodeum by transmission EM. The major finding was that the "Na pump" was localized in the basolateral membranes of the principal cells. The "local" amount of enzyme was identical in birds on a low-NaCl and high-NaCl diet, with an apparently equal distribution among cells in both dietary regimes (Fig. 1). No evidence for the presence of special cells with a high transport *capacity* for Na was thus found by this histochemical technique. Likewise, Smith et al. (1991 and unpubl.) used immunochemical methods to identify the amiloride-sensitive, apical Na channel. Polyclonal antibodies generated against purified, bovine, renal amiloride-sensitive Na channels were used. The distribution of the channel protein in coprodeum and colon from low-NaCl-diet birds was determined at the light microscopic level using indirect immunofluorescence and immunoperoxidase techniques. In both coprodeum and colon the

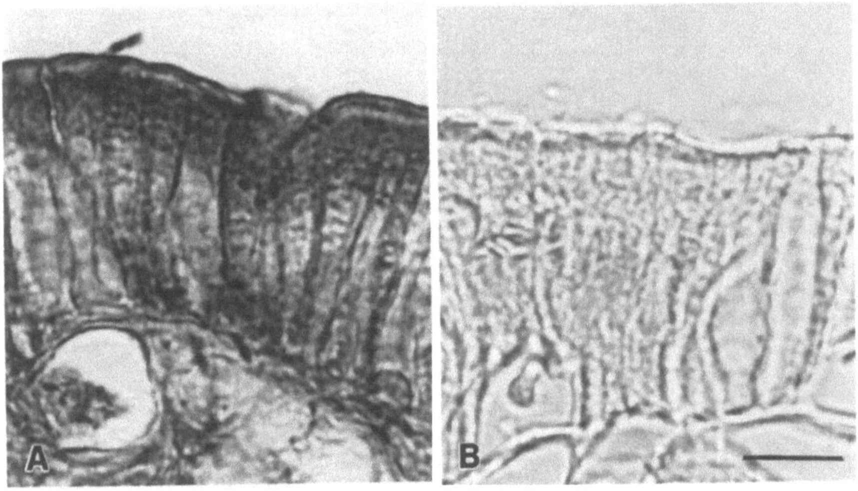

Fig. 2A,B. Immunoperoxidase localization of amiloride-sensitive Na^+ channels within the hen colonic mucosa. High magnification photomicrographs of paraffin sections which were stained with rabbit anti-Na^+ channel antibody (**A**) or non-immune rabbit IgG (**B**) followed by HRP-conjugated secondary antibody. As shown in **A** immunoreactivity is predominantly associated with the apical membrane and apical cytoplasma of the principal cells. Staining is also apparent within the basolateral cytoplasm of these cells. Control sections (**B**) do not exhibit a reaction product. The technique is described in Smith and Benos (1991). Scale: 10 μm

immunofluorescence and immunoperoxidase localization was predominantly associated with the apical membrane and apical cytoplasm of the principal cells. Staining was also apparent within the basolateral cytoplasm. Goblet cells did not exhibit any reaction products or immunofluorescence. No reaction appeared in the crypts. The cytoplasm was mainly stained near the apical and basolateral membranes. The staining was uniform along the villar folds of both the coprodeum and colon (Fig. 2). Therefore this study also supports the concept that a uniform population of principal cells is responsible for Na influx across the apical membrane. The cytoplasmic staining may represent vesicles containing amiloride-sensitive Na channels, as originally suggested from the binding of the amiloride derivative benzamil to homogenates of the coprodeal epithelium (Cuthbert et al. 1982).

4 Ion Transport in the Avian Caecum

Since this subject has been reviewed in a series of excellent papers in 1989 (Braun and Duke) little needs to be added here. Briefly, the anatomical development in various species was well characterized both macro- and microscopically, motility and movements of solutes and particles into the caeca were described, fermentation rates of carbohydrates and breakdown of uric acid were presented and absorption of ions, water and nutrients measured in relation to dietary and hormonal manipulations (aldosterone). The major conclusion is that in some species these organs function as a "supplementary rumen", and they may also be important in NaCl and water conservation. A major problem is the great adaptability of birds in this respect since there is no change in the digestion of the bird a few weeks after surgical removal of the caeca. This may, however, only occur during generous and optimal feeding, and the importance of the organs may only show up, as already pointed out by Thomas (1982), when a given species is presented with poor feed possibly in association with lack of both Na and water.

Concerning the regulation of ion absorption, Grubb (1991) demonstrated that the unstripped caecum of young chickens maintains Na transport well in vitro and it can be doubled by aldosterone. Either glucose, acetate or butyrate stimulated Na transport, glucose acting as an energy substrate since Na transport was not stimulated with a non-metabolizable glucose derivative on the mucosal side. Propriọnate attenuated Na transport, and electrogenic Cl secretion was observed in the short-circuited state. Moretó et al. (1991) observed Na-dependent L-proline and α-methyl-D-glucoside absorption by chicken proximal caecum, further testifying to the importance of Na-linked cotransport in nutrient uptake of these epithelia. The caecum may also transport dipeptides (Calonge et al. 1990). It should finally be noted that a thorough comparison by LM and EM of caecum in Gambel's quail (*Callipepla gambelii*) and the domestic fowl has been made by Strong et al. (1990).

5 An Update of Ion Transport in Hindguts of Species Other than the Fowl

The literature of the last decade has been excellently summarized by Goldstein in his chapter: Transport of water and electrolytes by the lower intestine and its contribution to avian osmoregulation (Hughes and Chadwick 1989) and little can be added at present. In brief, granivorous, terrestrial birds seem to have similar patterns of ionic regulation in the lower gut as the domestic fowl, whereas marine, salt-gland-possessing birds such as the domestic duck seem to regulate ion transport across the lower intestines to a much smaller degree. One of the large, flightless, ratite birds shows special adaptations. The Australian emu, *Dromaius novaehollandiae*, has a poor renal concentrating ability but compensates with a high rate of solute-linked water flow in coprodeum/colon. It was recently demonstrated (Dawson et al. 1991) that the emu kidney reabsorbs NaCl, and surprisingly also water, nearly as efficiently, as do mammals in spite of the limited concentrating ability. Skadhauge et al. (1991) studied emus in the Australian semi-desert in the dry season. Even under these harsh conditions the birds did not appear dehydrated and the electrolyte composition of the fluid fraction of the gut content permitted a high rate of solute-linked water absorption. These birds were found to ingest plant material of high-NaCl content, but a preliminary examination of coprodeum and colon in the Ussing chamber revealed a high SCC, which was amino acid-insensitive and amiloride-suppressible, regardless of the NaCl content of the diet, as studied on a group of captive emus. The unusually high rate of Na absorption (and solute-linked water flow) of the emu's coprodeum-colon observed previously (Dawson et al. 1985) was due to a high degree of mucosal folding, augmenting the serosal area by a factor of 5. Concerning the precise origin of solutes and water presented for possible resorption in the avian lower gut, Goldstein and Braun (1988) reported in the house sparrow (*Passer domesticus*) that due to the differences in osmolality and flow rate the hydrated birds principally reabsorb water of urinary origin in the lower guts, whereas in dehydrated birds it is water from ileal contents.

6 Conclusions

NaCl depletion induces, in the hen coprodeum, rheogenic Na absorption, which goes from near zero to 300–400 $\mu A/cm^2$ (11–15 $\mu mol\, cm^{-2} h^{-1}$). This change, which is the largest observed in vertebrate epithelia, is mainly regulated by aldosterone. The hormone has acute (4–24 h), chronic (3–5 days), and very late effects. The acute action is represented by an enlargement of amiloride-sensitive short-circuit current by a factor of 50–100 by the opening of 4-pS apical Na channels; the chronic effect is represented by a doubling of both the apical membrane area through the development of microvilli, and the basolateral pumping activity (ATPase). This occurs in existing cells since cell turnover is even slower than the chronic effect. The net effect is a quadrupling of aldosterone-inducible, short-circuit current. The very late effect is the growth of new cells, by more than 50%, mainly by increased folding of the epithelium. The Na transport takes place evenly across a homogeneous layer of principal cells at the villi/folds as shown by extracellular recording with the vibrating microprobe, by histoimmunological localization of the amiloride-sensitive Na channel, and by histochemical localization of K-paranitrophenylphosphatase activity. Cyclic AMP induces, particularly as observed after amiloride treatment, rheogenic Cl secretion of approximately one third of the initial Na-absorption rate. This transport also occurs across the principal Na-absorbing villus cells, most likely by opening of subpicosiemens Cl channels. The flux ratio is in the short-circuited state unity for both Cl under normal conditions and for Na during induced Cl secretion. The co-ion migration of both ions leads under open circuit conditions to near-neutral NaCl transport. Calculations based on the vibrating microprobe data demonstrate that most co-ion migration takes place across the villi/folds with a shunt of only 4% going through the crypts.

The epithelium of hen colon from birds on a low-NaCl diet has exactly the same pattern of Na absorption and induced Cl secretion as coprodeum but without chronic adaptation of either Na absorption or Cl secretion after resalination/aldosterone injection. This may be due to the presence of apical microvilli in birds on both high- and low-NaCl intake and the fact that a high-NaCl diet does not suppress Na absorption. It switches gradually, depending on the level of NaCl intake, from amiloride-suppressible Na absorption to rheogenic Na-amino acid/hexose cotransport. The molecular mechanism for this interaction has not been studied. Both coprodeum and colon secrete K and protons, most likely stimulated by aldosterone. The rate of K-secretion is low, rarely exceeding 1 $\mu mol\,cm^{-2}\,h^{-1}$. Unexpectedly, the rate of secretion is lower in birds on low-NaCl than on a high-NaCl diet; circumstantial evidence points to a reduced permeability of the apical membrane. The proton secretion, although not quantified, has been proven by the observation of an acid microclimate in the colon.

The predominant regulation of Na transport by aldosterone has been documented by three types of study: (1) external injection to birds receiving a high-NaCl diet and subsequent measurements in vivo and in vitro. (2) Correlation of plasma level of aldosterone and Na transport during Na depletion and repletion (resalination). (3) Finally, more recently, this correlation was observed at several stable levels of NaCl intake. Na depletion leads to parallel rises in plasma aldosterone and Na transport, and Na loading leads to osmotic stress with rising levels of prolactin and AVT. Corticosterone undergoes no major change. A neutral NaCl level, sensed by the fowl as neither surplus nor lack of NaCl, is found between 2.3 and 5.6 mmol/kg body weight day^{-1} corresponding to a concentration in feed of 35–70 µmol NaCl/kg.

Although the epithelia of the lower gut maintain Na transport for 5–8 h in vitro, addition of aldosterone to the incubating media results in only slight increases in Na transport in the coprodeum, whereas a doubling of SCC was observed in the caecum. The reason for this difference is unknown. An apparent attenuation of the aldosterone effect in vivo and in vitro cannot be attributed to receptor recruitment during long-term exposure to aldosterone since this, on the contrary, leads to down-regulation of the mineralocorticoid receptor of the cytosol of both coprodeum and colon. This is a high-affinity, low-capacity receptor with a maximal concentration of 0.5 pmol/mg protein. The aldosterone-induced opening of apical Na channels has been studied by amiloride binding. These studies suggest insertion of preformed channel material from the cytosol into the membrane, as confirmed by histoimmunological localization. De novo synthesis of Na channels is induced by aldosterone since RNA from low-NaCl but not high-NaCl coprodeum expresses amiloride-suppressible Na uptake after injection into *Xenopus* oocytes.

The chronic and late effects of aldosterone/low-NaCl diet on the development of the hen coprodeum have been measured by quantitative stereological methods. This epithelium is therefore the first in which the cellular adaptation of Na transport has been characterized for the entire organ. The role of the hindguts in avian osmoregulation has been quantified for a number of granivorous, terrestrial birds (including the role of the caeca), for some seabirds, particularly the domestic duck, and for a member of the ratite group, the emu.

Acknowledgments. The author thanks Drs. W. Clauss, V. Dantzer, and V.S. Elbrønd for making unpublished material available for this review, and for valuable comments to the manuscript. The excellent secretarial assistance of Miss Jytte Ebbesen-Andersen is greatly appreciated. The authors experimental investigations were supported by the Danish Agricultural and Veterinary Research Council, the European Community, NATO, the Novo Foundation, the Leo Research Foundation, the Lundbeck Foundation and the Velux (1981) Foundation.

References

Andersen V, Munck BG, Munck LK, Skadhauge E (1990) In vitro studies of theophylline-induced changes in Na, K and Cl transport in hen (*Gallus domesticus*) colon suggesting bidirectional, basolateral NaK2Cl cotransport. Comp Biochem Physiol 96A: 187–193

Arad Z, Chadwick A, Rice GE, Skadhauge E (1986) Osmotic stimuli and NaCl-intake in the fowl; release of arginine vasotocin and prolactin. J Comp Physiol B 156: 399–406

Árnason SS, Skadhauge E (1985) Effects of prolonged injections of aldosterone during resalination on coprodeal short-circuit current (SCC) of the domestic hen. Acta Physiol Scand 123: 50A

Árnason SS, Skadhauge E (1991) Steady-state sodium absorption and chloride secretion of colon and coprodeum, and plasma levels of osmoregulatory hormones in hens in relation to sodium intake. J Comp Physiol B 161: 1–14

Árnason SS, Rice GE, Chadwick A, Skadhauge E (1986) Plasma levels of arginine vasotocin, prolactin, aldosterone and corticosterone during prolonged dehydration in the domestic fowl. J Comp Physiol B 156: 383–397

Asher C, Singer D, Eren R, Yeger O, Dascal N, Garty H (1992) NaCl-dependent expression of amiloride-blockable Na^+ channel in *Xenopus* oocytes. Am J Physiol 262: G244–G248

Bindslev N (1979) Sodium transport in the hen lower intestine. Induction of sodium sites in the brush border by a low sodium diet. J Physiol (Lond) 288: 449–466

Bindslev N, Cuthbert AW, Edwardson JM, Skadhauge E (1982) Kinetics of amiloride action in the hen coprodeum in vitro. Pflügers Arch Eur J Physiol 392: 340–346

Braun EJ, Duke GE (1989) Function of the avian cecum. J Exp Zool Suppl 3: 1–130

Calonge ML, Ilnudáin A, Bolufer J (1990) Glycylsarcosine transport by epithelial cells isolated from chicken proximal caecum and rectum. Am J Physiol 258: G660–G664

Calonge ML, Molina MT, Ilundáin A (1992) Cl^-/base exchange and cellular pH regulation in enterocytes isolated from chick small intestine. Biochim Biophys Acta 1103: 45–50

Campbell CE, Braun EJ (1986) Cecal degradation of uric acid in Gambel quail. Am J Physiol 20: R59–R62

Chang EB, Wang N-S, Rao MC (1985) Phorbol ester stimulation of active anion secretion in intestine. Am J Physiol 249: C356–C361

Choshniak I, Munck BG, Skadhauge E (1977) Sodium chloride transport across the chicken coprodeum. Basic characteristics and dependence on sodium chloride intake. J Physiol (Lond) 271: 489–504

Christensen O, Bindslev N (1982) Fluctuation analysis of short-circuit current in a warm-blooded sodium-retaining epithelium: site current, density, and interaction with triamterene. J Membr Biol 65: 19–30

Clauss W, Skadhauge E (1988) Modulation of Na and Cl transport by mineralocorticoids. Comp Biochem Physiol 90A: 583–589

Clauss W, Árnason SS, Munck BG, Skadhauge E (1984) Aldosterone-induced sodium transport in lower intestine. Effect of varying NaCl-intake. Pflügers Arch Eur J Physiol 401: 354–360

Clauss W, Dürr JE, Guth D, Skadhauge E (1987) Effects of adrenal steroids on Na-transport in the lower intestine (coprodeum) of the hen. J Membr Biol 96: 141–152

Clauss W, Dantzer V, Skadhauge E (1988) A low salt diet facilitates Cl secretion in hen lower intestine. J Membr Biol 102: 83–96

Clauss W, Skadhauge E, Krattenmacher R, Hoffmann B, Fischer H (1990) Elektrogene Chlorid-Sekretion im Colon vom Huhn (*Gallus domesticus*). Verh Dtsch Zool Ges 83: 528–529

Clauss W, Dantzer V, Skadhauge E (1991) Aldosterone modulates electrogenic Cl-secretion in the colon of the hen (*Gallus domesticus*). Am J Physiol 261: R1533–R1541

Cuthbert AW, Edwardson JM, Bindslev N, Skadhauge E (1982) Identification of potential components of the transport mechanism for Na^+ in the hen colon and coprodeum. Pflügers Arch Eur J Physiol 392: 347–351

Dantzer V, Møller O, Skadhauge E (1988) Morphological and enzymic adaptation to aldosterone of the epithelium of the caudal hindgut of the hen. J Physiol (Lond) 396: 30P

Dawson DC (1991) Ion channels and colonic salt transport. Annu Rev Physiol 53: 321–339

Dawson TJ, Herd RM, Skadhauge E (1985) Osmotic and ionic regulation during dehydration in a large bird, the emu (*Dromaius novaehollandiae*): an important role for the cloaca-rectum. Q J Exp Physiol 70: 423–436

Dawson TJ, Maloney SK, Skadhauge E (1991) The role of the kidney in electrolyte and nitrogen excretion in a large flightless bird, the emu, during different osmotic regimes, including dehydration and nesting. J Comp Physiol B 161: 165–171

DiBattista AJ, Mehdi AZ, Sandor T (1985) A profile of the intestinal mucosal corticosteroid receptors in the domestic duck. Gen Comp Endocrinol 59: 31–49

DiBattista AJ, Mehdi AZ, Sandor T (1989) Steroid C-20 oxidoreductase activity of the duck intestinal mucosa: The interrelations of the enzymatic activity with steroid binding. Gen Comp Endocrinol 74: 136–147

Eldrup E, Møllgård K, Bindslev N (1979) Possible sodium channels in the luminal membrane of the hen lower intstine visualized by freeze fracture. INSERM 85: 253–262

Eldrup E, Møllgård K, Bindslev N (1980) Possible epithelial sodium channels visualized by freeze-fracture. Biochim Biophys Acta 596: 152–157

Elbrønd VS, Skadhauge E (1992) Na-transport during long-term incubation of the hen lower intestine. Comp Biochem Physiol 101A: 203–208

Elbrønd VS, Dantzer V, Mayhew TM, Skadhauge E (1991) Avian lower intestine adapts to dietary salt (NaCl) depletion by inncreasing transepithelial sodium transport and microvillous membrane surface area. Exp Physiol 76: 733–744

Fischer H, Kromer W, Clauss W (1991) Two types of chloride channels in hen colon epithelial cells identified by patch-clamp experiments. J Comp Physiol B 161: 333–338
Garth H, Asher C (1991) Does aldosterone induce de novo synthesis of Na^+ channels? In: Bonvalet JP, Farman N, Lombès M, Rafestin-Oblin ME (eds) Aldosterone: fundamental aspects, vol 125. Colloque INSERM, John Libbey Eurotext, London, pp 273–283
Goldstein DL (1989) Transport of water and electrolytes by the lower intestine and its contribution to avian osmoregulation. In: Hughes MR, Chadwick A (eds) Progress in avian osmoregulation. Leeds Philosophical and Literary Soc, Leeds, pp 271–294
Goldstein DL, Braun EJ (1988) Contributions of the kidneys and intestines to water conservation and plasma levels of antidiuretic hormone, during dehydration in house sparrows (*Passer domesticus*). J Comp Physiol B 158: 353–361
Goldstein DL, Williams JB, Braun EJ (1990) Osmoregulation in the field by salt-marsh savannah sparrows *Passerculus sandwichensis beldingi*. Physiol Zool 63: 669–682
Gray DA, Schütz H, Gerstberger R (1991) Interaction of atrial natriuretic factor and osmoregulatory hormones in the Pekin duck. Gen Comp Endocrinol 81: 246–255
Gregg CM, Wideman RF (1986) Effects of atriopeptin and chicken heart extract in *Gallus domesticus*. Am J Physiol 251: R543–R551
Grubb BR (1991) Avian cecum: role of glucose and volatile fatty acids in transepithelial ion transport. Am J Physiol 260: G703–G710
Grubb BR, Bentley PJ (1987) Aldosterone-induced, amiloride-inbibitable short-circuit current in the avian ileum. Am J Physiol 253: G211–G216
Grubb BR, Bentley PJ (1988) Relationship of transmural electrical parameters to the luminal Na concentration in the colon of the fowl (*Gallus domesticus*). J Comp Physiol B 158: 19–24
Grubb BR, Bentley PJ (1989) Avian colonic ion transport: effects of corticosterone and dexamethasone. J Comp Physiol B 159: 131–138
Grubb BR, Bentley PJ (1990) Potassium transport across the intestines of the fowl *Gallus domesticus*. J Comp Physiol B 160: 17–22
Grubb BR, Driscoll SM, Bentley PJ (1987) Electrical PD, short-circuit current and fluxes of Na and Cl across avian intestine. J Comp Physiol B 157: 181–186
Grubb BR, Brown HV, Bentley PJ (1988) Anomalous behaviour of ^{86}Rb as a tracer for transintestinal potassium transport in the fowl, *Gallus domesticus*. J Exp Biol 135: 487–490
Hansen MB, Bindslev N (1989) Serotonin-induced chloride secretion in hen colon. Possible second messengers. Comp Biochem Physiol 94A: 315–321
Heinz M, Krattenmacher R, Hoffmann B, Clauss W (1991) Different modes of electrogenic Na^+ absorption in the coprodeum of the chicken embryo: role of extracellular Ca^{2+}. J Comp Physiol B 161: 363–370
Hoffmann B, Krattenmacher R, Heintz M, Habura B, Clauss W (1990) Aldosteron und Thyroxin stimulieren den elektrogenen Natrium-Transport im embryonalen Coprodeum des Huhns (*Gallus domesticus*). Verh Dtsch Zool Ges 83: 534–535
Holtug K (1989) Mechanisms of absorption of short chain fatty acids – coupling to intracellular pH regulation. Acta Vet Scand Suppl 86: 126–133
Holtug K, Skadhauge E (1982) NaCl transport across hen colon. Pflügers Arch Eur J Physiol 394: 222–225
Holtug K, Shipley A, Dantzer V, Sten-Knudsen O, Skadhauge E (1991a) Localization of sodium absorption and chloride secretion in an intestinal epithelium. J Membr Biol 122: 215–229
Holtug K, Skadhauge E, McEwan GTA (1991b) Effect of propionate on colon acid-microclimate. Scand J Gastroenterol 26 (183): 77 (Suppl)
Hughes MR, Chadwick A (1989) Progress in avian osmoregulation. Leeds Philosophical and Literary Soc, Leeds. 346pp
Laverty G, Alberici M (1991) Carbonic anhydrase activity in kidney and lower intestine of the European starling. J Morphol 207: 273–381
Lind J, Munck BG, Olsen O, Skadhauge E (1980a) Effects of sugars, amino acids and inhibitors on electrolyte transport across hen colon at different sodium chloride intakes. J Physiol (Lond) 305: 315–325

Lind J, Munck BG, Olsen O (1980b) Effects of dietary intake of sodium chloride on sugar and amino acid transport across isolated hen colon. J Physiol (Lond) 305: 327–336
Mayhew TM, Dantzer V, Sødring-Elbrønd V, Skadhauge E (1990) A sampling scheme for studying the functional morphology of the coprodaeum in hens on high- and low-NaCl diets. J Anat 170: 246–248
Mayhew TM, Elbrønd VS, Dantzer V, Skadhauge E, Møller O (1992) Structural and enzymatic studies on the plasma membrane domains and sodium pump enzymes of absorptive epithelial cells in the avian lower intestine. Cell Tissue Res 270: 577–585
Møller O, Dantzer V, Skadhauge E (1989) Salt transport related changes in the hen lower intestinal tract. A morphological and histochemical study. Biol Cell 66: 200–201
Montero MC, Bolufer J, Ilundain A (1988) Potassium transport in epithelial cells isolated from small intestine of the chicken. Pflügers Arch Eur J Physiol 412: 422–426
Montero MC, Bolufer J, Ilundáin A (1991) Influence of external K^+ on potassium efflux in isolated chicken enterocytes. Comp Biochem Physiol 99A: 31–36
Montrose MH, Bebernitz G, Kimmich GA (1985) Evaluation of ion gradient-dependent H^+ transport systems in isolated enterocytes from the chick. J Membr Biol 88: 55–66
Moretó M, Amat C, Puchal A, Buddington RK, Planas JM (1991) Transport of L-proline and α-methyl-D-glucoside by chicken proximal cecum during development. Am J Physiol 260: G457–G463
Munck BG (1989) Amino acid transport across the hen colon: interactions between leucine and lysine. Am J Physiol 256: G532–G539
Munck BG, Andersen V, Voldsgård P (1984) Chloride transport across the isolated hen colon. In: Skadhauge E, Heintze K (eds) Intestinal absorption and secretion. MTP Press, Boston, pp 373–385
Munck LK, Munck BG (1990) Intestinal transport of potassium. Effects of changing apical and basolateral influx of sodium in the isolated mucosa of the hen (*Gallus domesticus*) colon. Comp Biochem Physiol 96A: 181–186
Musch MW, Nahkla AM, Chang EB (1990) Phorbol ester-stimulated secretion in chicken ileum: role of arachidonic acid metabolism. Gastroenterology 99: 393–400
Radke WJ, Albasi CM, Harvey S (1984) Dietary sodium and adrenocorticoid activity in ducks (*Anas platyrhyncos*) and chickens (*Gallus domesticus*). Gen Comp Endocrinol 56: 121–129
Rafestin-Oblin ME, Couette B, Lombès M, Baulieu EE (1989) Mineralocorticosteroid receptor of the chick intestine. Oligomeric structure and transformation. J Biol Chem 264: 9304–9309
Rafestin-Oblin ME, Couette B, Lombès M, Baulieu EE (1991) Biochemical studies of the mineralocorticosteroid receptor: oligomeric structure and anti-hormone action. In: Bonvalet JP, Farman N, Lombès M, Rafestin-Oblin ME (eds) Aldosterone: fundamental aspects, vol 125. Colloque INSERM, John Libbey Eurotext, London, pp 55–64
Rice GE, Skadhauge E (1982) Caecal water and electrolyte absorption and the effects of acetate and glucose, in dehydrated, low-NaCl diet hens. J Comp Physiol B 147: 61–64
Rice GE, Árnason SS, Arad Z, Skadhauge E (1985) Plasma concentrations of arginine vasotocin, prolactin, aldosterone and corticosterone in relation to oviposition and dietary NaCl in the domestic fowl. Comp Biochem Physiol 81A: 769–777
Rosenberg J, Hurwitz S (1987) Concentration of adrenocortical hormones in relation to cation homeostasis in birds. Am J Physiol 253: R20–R24
Rosenberg J, Pines M, Hurwitz S (1989) Inhibition of aldosterone secretion by atrial natriuretic peptide in chicken adrenocortical cells. Biochem Biophys Acta 1014: 189–194
Sandor T, Skadhauge E, DiBattista JA, Mehdi AZ (1989) Interrelations of the intestinal glucocorticoid and mineralocorticoid receptor systems with salt homeostasis. In: Hughes MR, Chadwick A (eds) Progress in avian osmoregulation. Leeds Philosophical and Literary Soc, Leeds, pp 305–332
Semrad CE, Chang EB (1987) Calcium-mediated cyclic AMP inhibition of Na-H exchange in small intestine. Am J Physiol 252: C315–C322
Skadhauge E (1980) Intestinal osmoregulation. In: Epple A, Stetson MH (eds) Avian endocrinology. Academic Press, New York, pp 481–498

Skadhauge E (1981) Osmoregulation in birds. Springer, Berlin Heidelberg New York, 203pp

Skadhauge E (1982) A quantitative survey of salt and water excretion. Comp Biochem Physiol 71A: 481–483

Skadhauge E (1983) Temporal adaptation and hormonal regulation of sodium transport in the avian intestine. In: Gilles-Baillien M, Gilles R (eds) Intestinal transport: fundamental and comparative aspects. Springer, Berlin Heidelberg New York, pp 284–294

Skadhauge E (1989a) Regulation of NaCl transport across avian hindgut. In: Smith MW, Sepulveda FV (eds) Adaptation and development of gastrointestinal function. Manchester University Press, Manchester, pp 41–48

Skadhauge E (1989b) Hormonal regulation of sodium absorption and chloride secretion across the lower intestine of birds. Zool Sci 6: 437–444

Skadhauge E, Dawson TJ (1980) In vitro studies of sodium transport across the lower intestine of a desert parrot. Am J Physiol 239: R285–RR290

Skadhauge E, Thomas DH, Chadwick A, Jallageas M (1983) Time course of adaptation to low and high NaCl diets in the domestic fowl: effects on electrolyte excretion and on plasma hormone levels (aldosterone, corticosterone and prolactin). Pflügers Arch Eur J Physiol 396: 301–307

Skadhauge E, Munck BG, Rice GE (1984) Regulation of NaCl and water absorption in duck intestine. In: Pequeux A, Gilles R, Bolis L (eds) Osmoregulation in estuarine and marine animals. Lecture Notes on Coastal and Estuarine Studies 9. Springer, Berlin Heidelberg, New York, pp 132–142

Skadhauge E, Clauss W, Árnason SS, Thomas DH (1985) Mineralocorticoid regulation of lower intestinal ion transport. In: Gilles R, Gilles-Baillien M (eds) Transport processes, iono- and osmoregulation. Proceedings in Life Sciences. Springer, Berlin Heidelberg New York, pp 118–133

Skadhauge E, Clauss W, Dantzer V (1989) Regulation of electrogenic Na-absorption and induced Cl-secretion in an intestinal epithelium: delayed effects of aldosterone. Acta Physiol Scand Suppl 136: 69–73

Skadhauge E, Maloney SK, Dawson TJ (1991) Osmotic adaptation of the emu (*Dromaius novaehollandiae*). J Comp Physiol B 161: 173–178

Smith PR, Benos DJ (1991) Epithelial Na^+ channels. Annu Rev Physiol 53: 509–530

Smith PR, Saccomani G, Bradford AL, Dantzer V, Benos DJ, Skadhauge E (1991) Immunochemical identification of amiloride-sensitive sodium channels from an intestinal epithelium. In: Bonvalet JP, Farman N, Lombès M, Rafestin-Oblin ME (eds) Aldosterone: fundamental aspects, vol 125. Colloque INSERM, John Libbey Eurotext, London, p 319

Strong TR, Reimer PR, Braun EJ (1990) Morphometry of the galliform cecum: a comparison between Gambel's quail and the domestic fowl. Cell Tissue Res 259: 511–518

Thomas DH (1982) Salt and water excretion by birds: the lower intestine as an integrator of renal and intestinal excretion. Comp Biochem Physiol 71A: 527–535

Thomas DH, Skadhauge E (1979) Chronic aldosterone therapy and the control of transepithelial transport of ions and water by the colon and coprodeum of the domestic fowl (*Gallus domesticus*) in vivo. J Endocrinol 83: 239–250

Thomas DH, Skadhauge E (1982) Time course of adaptation to low and high NaCl diets in the domestic fowl: effects on electrical behaviour of isolated epithelia from the lower intestine. Pflügers Arch Eur J Physiol 395: 165–170

Thomas DH, Skadhauge E (1988) Transport function and control in bird caeca. Comp Biochem Physiol 90A: 591–596

Thomas DH, Skadhauge E (1989a) Functional role of the flow of urine and digesta in the avian lower intestine. Acta Vet Scand Suppl 86: 212–218

Thomas DH, Skadhauge E (1989b) Function and regulation of the avian caecal bulb: influence of dietary NaCl and aldosterone on water and electrolyte fluxes in the hen (*Gallus domesticus*) perfused in vivo. J Comp Physiol B 159: 51–60

Thomas DH, Skadhauge E (1989c) Water and electrolyte transport by the avian ceca. J Exp Zool Suppl 3: 95–102
Thomas DH, Skadhauge E, Read MW (1979) Acute effects of aldosterone on water and electrolyte transport in vivo by the colon and coprodeum of the domestic fowl (*Gallus domesticus*). J Endocrinol 83: 229–237
Thomas DH, Jallageas M, Munck BG, Skadhauge E (1980) Aldosterone effects on electrolyte transport of the lower intestine (coprodeum and colon) of the fowl (*Gallus domesticus*) in vitro. Gen Comp Endocrinol 40: 44–51
Turnheim K (1991) Intrinsic regulation of apical sodium entry in epithelia. Physiol Rev 71: 429–445
Wilson JX (1989) The renin-angiotensin system in birds. In: Hughes MR, Chadwick A (eds) Progress in avian osmoregulation. Leeds Philosophical and Literary Soc, Leeds, pp 61–79
Windmueller HG, Spaeth AE (1978) Identification of ketone bodies and glutamine as the major respiratory fuels in vivo for postabsorptive rat small intestine. J Biol Chem 253: 69–76

Chapter 5

Segmental Heterogeneity and Regulation of Electrolyte Transport in the Rabbit Large Intestine

Wolfgang Clauss[1]

Contents

1 Introduction

Ion transport across rabbit colon has been a subject of investigation for more than 20 years. Based on fundamental principles of ion transport found in frog skin (Ussing and Zerahn 1951), the pioneering studies of Schultz and associates (Frizzell et al. 1976; Schultz et al. 1977; Turnheim et al. 1977, 1978, 1987; Frizzell and Schultz 1978; Frizzell and Turnheim 1978) elaborated the first detailed cellular models of sodium, potassium and chloride transport in rabbit colon. At that time, only the descending colon was investigated, and investigators aimed to find general models for ion transport in the mammalian colon (Schultz 1981, 1984). After some more years of research on this and other mammalian colonic epithelia, remarkable species differences in colonic ion transport became evident, and in many species under investigation, a pronounced segmental heterogeneity of transport functions was found (Clauss et al. 1985a; Halm and Frizzell 1991). Ion transport mechanisms along the large intestine differ not only gradually and quantitatively, but qualitatively, and are organized in a way to enable the

[1] Institut für Tierphysiologie, Justus-Liebig-Universität Giessen, Wartweg 95, W-6300 Giessen, Germany

Advances in Comparative and Environmental Physiology, Vol. 16

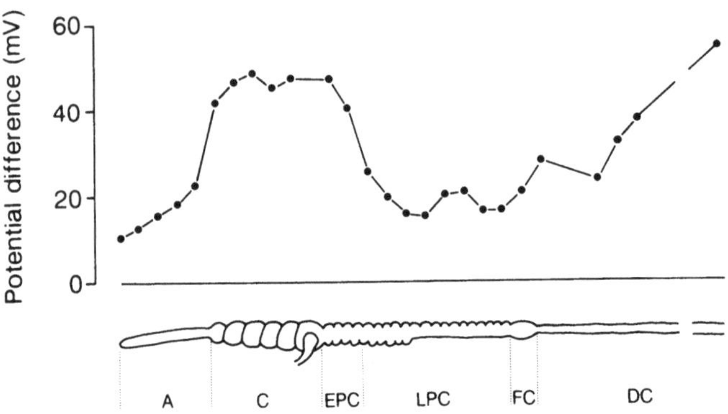

Fig. 1. Segmental organization of the rabbit large intestine, and gradient of the transepithelial electrical potential difference (serosal positive), measured in vivo during the hard faeces period (afternoon). *A* Appendix; *C* caecum; *EPC* early proximal colon; *LPC* late proximal colon; *FC* fusus coli; *DC* dictal colon

particular specificity of digestive functions in various animals. The rabbit large intestine is a special example of such a complex diversity, contributing to a unique digestive mechanism.

The present chapter does not attempt to give a detailed review on the cellular mechanisms of ion transport in rabbit colon. Readers interested in this topic are referred to excellent, recent reviews (Schultz 1984; Halm and Frizzell 1991; Turnheim 1991). Instead, this chapter presents an integrated view of the heterogeneity of transport parameters, emphasizes special aspects of regulation, and attempts to stimulate both cellular physiologists to work out more details of yet unknown transport properties and comparative physiologists to put these findings in perspective with the digestive physiology of the rabbit and other species. Rabbit large intestine is also an interesting organ for research on systemic physiology, and due to its unique features is only partly suited as a general model for mammalian or clinical viewpoints.

2 Anatomy and Histology

The anatomy of the mammalian large intestine is usually divided into the appendix, caecum, and proximal and distal colon. The terms ascending, transverse and descending colon are assigned to the human colon, and should not be used in context with the rabbit colon in order to avoid confusion. The rabbit large intestine has a clear-cut segmentation, and from the exterior segments like appendix, caecum, three-fold haustration, one-fold haustration, and distal colon are easy to distinguish. These distinctions are based on the visual appearance of the muscular wall, and do not consider the diversity of the mucosal organization. Internally, the rabbit colon is organized differently, and has at least six different mucosal areas, which obviously serve different ion transport functions (Fig. 1). Figure 1 illustrates the transmural, electrical potential difference lumen-blood, which roughly indicates electrogenic ion transport in vivo, measured at different sites along the rabbit large intestine. It can be readily seen that no gradual proximal-distal gradient occurs, but that each segment has its own characteristic electrical profile, which will be discussed later in more detail in the respective sections of this chapter. From the viewpoint of epithelial transport, and what is known to date, the rabbit large intestine is organized in the following segments: appendix, caecum, early proximal colon, late proximal colon, fusus coli, and distal colon. Readers who are interested in particular anatomical and morphological details are referred to Snipes (1978) and Snipes et al. (1982).

3 Segmental Differences in Ion Transport

3.1 Caecum

In the digestive tract of the rabbit, the caecum has a prominent morphological and physiological role. The caecum is a site of microbial fermentation of non-digestible carbohydrates, and plays a major role in water and electrolyte conservation. The caecal contents are rich in short chain fatty acids and electrolytes (Parker 1976; Leng 1978; Leng-Peschlow and Marty 1979), and these contents are readily exchanged across the caecal epithelium (Garcia et al. 1982, 1983; Marty and Vernay 1984; Marty and Abravanel 1985; Hatch 1987; Hatch and Freel 1988). Without this organ, the special digestive mechanism of *caecotrophy* (Björnhag 1972, 1987; Pickard and Stevens 1972; Hörnicke and Batsch 1977; Jilge 1976; Hörnicke et al. 1985; Jilge et al. 1987) would not be possible. A circadian rhythm of food intake, caecal motility, and ion transport generates in the large intestine the separation of two types of faeces (Eden 1940; Ruckebusch and Hörnicke 1977; Ehrlein et al. 1983; Camara and Prieur 1984; Pairet et al. 1986; Jilge et al. 1987). The so-called *caecotrophes* (or soft faeces) contain high amounts of short chain fatty acids of caecal origin (Henning and Hird 1972). In contrast to the rumen forestomach system, rabbits are hindgut fermentors, and therefore the caecotrophes have to be re-ingested for a second passage through the gastrointestinal tract in order to be utilized. Segmental differences in ion transport within the caecum may occur (Clauss 1978), but have not yet been investigated in detail. For instance, the appendix seems to be involved in the secretion of a bicarbonate-rich fluid (Williams et al. 1961; Blackwood et al. 1973). In the present chapter all further details on caecal ion transport refer, however, to the caecal corpus, which has a spiral folding. For details on the caecal anatomy of the rabbit, readers are referred to Snipes (1978).

The basic electrophysiological parameters of the caecal epithelium have been characterized by Clauss et al. (1985a, 1989). This tissue is in the range between a leaky and a moderately tight epithelium, and has a transepithelial resistance of about $150\,\Omega \cdot cm^2$. It has, however, a surprisingly high paracellular resistance of $800-1200\,\Omega \cdot cm^2$, and a low transcellular resistance of about $100\,\Omega \cdot cm^2$. Therefore the main ion current (and high rate of ion transport) occurs via the transcellular route. This considerable transport activity is underlined by the high short-circuit current I_{sc} of about 130–250 $\mu A/cm^2$, which is mainly generated by electrogenic sodium transport (Clauss et al. 1985a, 1989; Hatch and Freel 1988; Sellin et al. 1988), and is equivalent to a net Na^+ absorption of $6-12\,\mu Eq\,cm^{-2}\,h^{-1}$, depending on the time of day.

The mechanism of electrogenic Na^+ transport in rabbit caecum is of special importance, because it has several new and interesting features (Fig. 2). The I_{sc} is entirely dependent on the mucosal Na^+ concentration, and increases linearly up to 100 mM without any saturation kinetics (Sellin et al. 1988). This is an unusual phenomenon which has only recently been shown in *Xenopus* colon (Krattenmacher et al. 1990; see also Chap. 2, this Vol.). Sodium enters the cell via several apical transport systems. An electroneutral Na^+-H^+ exchange system, which is sensitive to high doses of amiloride (≥ 1 mM), accounts only for a minor fraction of the mucosa to serosa Na^+ flux (Hatch and Freel 1988; Clauss et al. 1989). The major part of the sodium ions enter the cell via a yet unknown system, presumably a special type of Na^+ channel (Sellin et al. 1988). This entry is insensitive to low doses of amiloride (Clauss et al. 1985a), but can be blocked by the amiloride analogue phenamil (Sellin et al. 1988), although at higher doses and different affinity needed by as these analogues to bind to the classical epithelial Na^+ channel (Garty and Benos 1988).

The caecal epithelium is also highly permeable to chloride, and the net chloride absorption of about $0.8\,\mu Eq\,cm^{-2}\,h^{-1}$ does not seem to be dependent on the presence of sodium. Also, an electrogenic chloride secretion is present, which can be elicited by cAMP or secretagogues as theophylline (Hatch and Freel 1988; Sellin et al. 1988; Tai et al. 1989), and also inhibited by loop diuretics as bumetanide (Hatch and Freel 1988). Potassium movements across this epithelium were usually assessed by Rb^+ fluxes, and showed a net secretion of $-1.5\,\mu Eq\,cm^{-2}\,h^{-1}$. Serosal ouabain reduces this K^+ secretion and also the Na^+ absorption to zero, which gives evidence for a basolateral Na^+/K^+-ATPase. The basolateral membrane also contains a considerable K^+ conductance (Clauss et al. 1989). By the magnitude and direction of the residual flux a bicarbonate secretion was postulated by several investigators (Hatch and Freel 1988; Sellin et al. 1988).

There is presently no information on whether all these ion transport mechanisms reside in one cell or are allocated to different cell types. There-

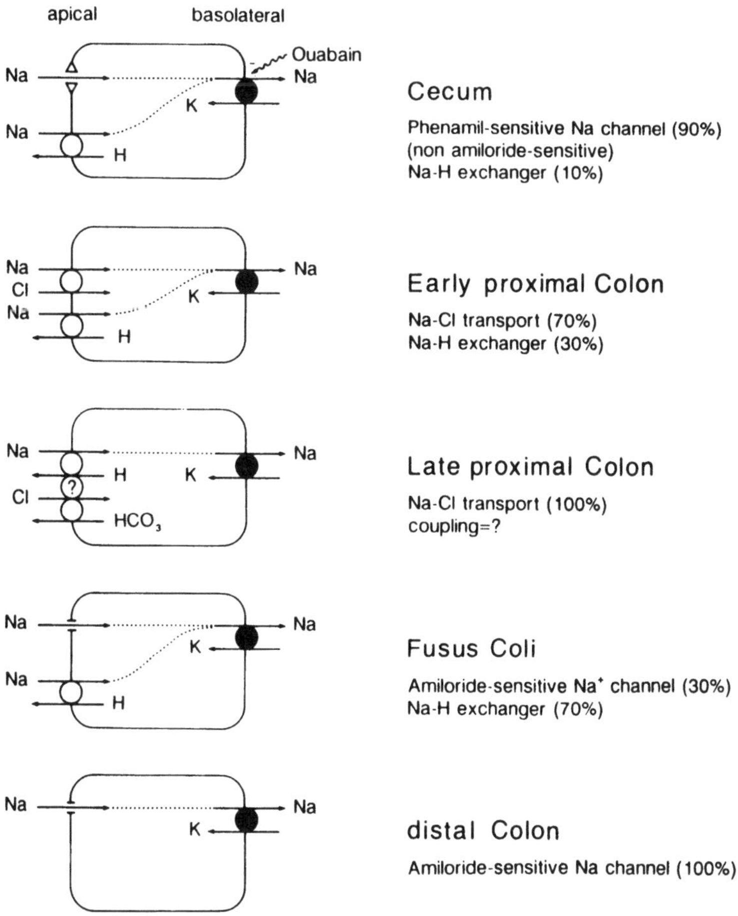

Fig. 2. Segmental organization of sodium transport mechanisms in the rabbit large intestine. The participation in net absorption is given in percentage

fore a conclusive model arising from all these investigatins has not yet been firmly established. Many more studies are needed to establish all ion transport routes and mechanisms in this prominent, large intestinal segment.

3.2 Early Proximal Colon (Three-Fold Haustration)

The proximal colon of the rabbit has a complex morphological and histological organization (Snipes et al. 1982). From the exterior several segments can be distinguished by their muscular organization. The first 10 cm of the proximal colon have three longitudinal taenia with three rows of haustra in between. This segment was named P1 (Snipes et al. 1982) or 3F-colon (Clauss et al. 1985a). Although such a distinction seems clear at first glance, the situation is complicated by the fact that internally the first 5 cm of the P1 segment has pronounced "wart-like" protrusions of the mucosa, whereas the second portion of the P1 segment (and also the following P2 segment) has a different mucosal appearance and organization.

So far only relatively few ion transport studies have been carried out in the rabbit proximal colon. These have found transport properties markedly distinct to the ileum or distal colon (Sellin and DeSoignie 1984). Unfortunately, most investigators have been unaware of the fact that in the proximal colon the external and the internal segmentation is different, and have therefore often chosen a longer portion (with two different mucosal organizations) for their studies. In this section we therefore consider only studies which have used portions no longer than 5 cm, beginning from the ampulla caecalis coli. It seems, furthermore, reasonable to abandon the externally based nomenclature, and to name this first functionally distinct segment of the proximal colon the "early proximal colon" (Clauss et al. 1987), based only on the internal segmentation of the mucosa. Functionally, this epithelium has many more similarities to the caecal epithelium than to the following "late proximal colon" (Clauss et al. 1987).

The basic electrophysiological properties and ion fluxes of the early proximal colon have been characterized by Clauss et al. (1987). This epithelium has a low transepithelial resistance of $133\,\Omega \cdot cm^2$, a large I_{sc} of $178\,\mu A/cm^2$, and a transepithelial potential difference of 21 mV, serosal side positive. The apical and basolateral membrane potentials were −21 and −42 mV, so that the epithelium has a trough-like potential profile (Clauss et al. 1987). The individual membrane resistances have not yet been determined.

The transport of Na^+, K^+, and Cl^- were all inhibited by serosal addition of ouabain, suggesting a dependence on the electrochemical gradient for Na^+ entry, and a basolateral Na^+/K^+-ATPase. Lower doses of amiloride had no effect on the I_{sc} and Na^+ fluxes (Clauss et al. 1985a), demonstrating the absence of apical Na^+ channels. Addition of 1 mM amiloride suppressed about one third of the Na^+ absorption, suggesting an apical Na^+-H^+ exchange system, mediating at least a portion of the Na^+ absorption (Fig. 2). Another clear finding was the mutual dependence of Na^+ and Cl^- absorption. The mechanism of electrogenic Cl^- secretion does not seem to be present in this epithelium, as secretagogues such as theophylline had no stimulatory effects on I_{sc} or Cl^- fluxes. The epithelium is capable of alkalinizing the mucosal or serosal compartment (Campos et al. 1991) by a Na^+-dependent, active transcellular HCO_3^- transport, which is activated by adrenergic stimuli (Sullivan and Smith 1986b). Thus, the epithelium can both absorb or secrete HCO_3^-. The mechanism of K^+ transport was investigated in detail by Sullivan and Smith (1986a) and Clauss et al. (1987). Like the distal colon, this epithelium secretes K^+ actively, but has no active K^+ uptake mechanism at the apical cell membrane. K^+ secretion results from an active, ouabain-sensitive K^+ uptake at the basolateral membrane, which elevates the intracellular K^+ activity to about 70 mM (Clauss et al. 1987). Potassium exit at the apical cell membrance occurs by a Cl^--independent mechanism, which could not be blocked by barium (Sullivan and Smith 1986a). In contrast, barium has an effect on the basolateral side and indicates the presence of K^+ channels (Clauss et al. 1987). In vivo, in the presence of large electrochemical gradients, the situation may be different, and K^+ absorption was observed (Clauss and Hörnicke 1984). Under in vivo conditions the epithelium also absorbs water (Clauss 1985; Campos et al. 1991), and this fluid absorption can be stimulated by aldosterone (Clauss 1985). This hormone also influences the absorption of short chain fatty acids in the proximal colon (Vernay 1987).

3.3 Late Proximal Colon (One-Fold Haustration)

The late proximal colon is externally characterized by a single, longitudinal taenia, and one row of haustra. It is about 20 cm long, and the mucosal structure is also organized in "wart-like" protrusions, although smaller than in the early proximal colon (Snipes et al. 1982). The electrophysiological properties characterize this epithelium as moderately leaky, with a transepithelial resistance of about $90-120\,\Omega \cdot cm^2$, an I_{sc} of about $1.3-1.5\,\mu Eq/cm^2$, and a low transepithelial potential of 4 mV, serosal side positive (Clauss et al. 1985c; Sellin and De Soignie 1985). The individual membrane potentials and conductances have not yet been characterized, as this epithelium was not accessible to microelectrode experiments.

Under basal conditions the absorption rates for Na^+ and Cl^- are near zero, although there are considerable unidirectional fluxes of these ions (Clauss et al. 1985c; Sellin and De Soignie 1985). Serosal ouabain inhibits Na^+ absorption, and this suggests a basolateral Na^+/K^+-ATPase (Clauss et al. 1985c). Epinephrine stimulates an electrically silent Na^+-Cl^- coupled absorption (Fig. 2), which is partially inhibited by high doses of amiloride (Sellin and De Soignie 1984, 1987a) and is under control of α_2 adrenergic receptors (Sellin and De Soignie 1987a). Loop diuretics such as furosemide or bumetanide had no effect of the stimulated Na^+-Cl^- transport, suggesting that no single Na^+-Cl^- or Na^+-$2Cl^-$-K^+ transport system is involved (Sellin and De Soignie 1987a). The studies by Ahn et al. (1985) favour a Na^+-H^+ exchange system which is inhibited by protein kinase C. A novel regulatory system for Na^+ transport is described by Sellin and De Soignie (1987b). They found a cation-specific stimulation of Na^+ absorption at lower external Na^+ concentrations with a maximum at 50 mM Na^+. Their experiments suggest that the serosal rather than the mucosal Na^+ concentration determines Na^+ absorption secondarily by a yet unknown intracellular mediator. The late proximal colon does not seem to have an electrogenic Cl^- secretory mechanism, as secretagogues like theophylline do not stimulate Cl^- secretion, but act in an anti-absorptive manner (Sellin and De Soignie 1984), and inhibit the epinephrine-enhanced Na^+ absorption (Sellin and De Soignie

1987b). The regulation of NaCl transport by gluco- and mineralocorticoids was investigated by Sellin and De Soignie (1985). Steroids did not alter the basal Na^+ and Cl^- transport rates in the proximal colon, but they stimulated the epinephrine-induced NaCl absorption.

The cellular mechanism of K^+ transport in the late proximal colon has not yet been investigated in detail, although some experimental manouvres demonstrate the presence of a transepithelial K^+ conductance, especially in the basolateral membrane (Clauss et al. 1985c). Clauss and Hörnicke (1984) compared the net K^+ transport under in vivo and in vitro conditions. Under both experimental conditions they found a net K^+ absorption of about 1.5 $\mu Eq\,cm^{-1}\,h^{-1}$. K^+ absorption was also found by Vernay et al. (1984) in isolated loops of the proximal colon. An increase in plasma aldosterone modulated K^+ absorption into K^+ secretion (Clauss 1985).

3.4 Fusus Coli

The fusus coli is a functionally distinct part at the end of the proximal colon (Auer 1925; Snipes et al. 1982). It is about 5 cm in length and has a spindle-like shape. The mucosal structure is very different from either the proximal or the distal colon. The thickness of the mucosa (about 630 μm) is four to five times than that of the proximal colon (about 120 μm). The crypts are rather large and deep with many goblet cells, and the surface cells have prominent microvilli (Snipes et al. 1982). The fusus coli plays a special role in the circadian formation of the two types of faeces, and has the function of a pacemaker area (Ruckebusch and Fioramonti 1976). Originally, it was thought that this short, spindle-shaped segment acts like a "pellet machine", squeezing out the faecal water to be absorbed, leaving the contents to be formed as the faecal pellets for the hard faeces period. Such a short, functionally distinct colonic segment is not known in any other species except the rabbit.

The basic electrophysiological characteristics and the mechanisms of Na^+, K^+ and Cl^- transport in the fusus coli have been investigated in vitro by Dürr (1988). The fusus coli has a low transepithelial potential of 3–8 mV, serosal side positive, an I_{sc} of 15–49 $\mu Eq/cm^2$, and a transepithelial resistance of about 150 $\Omega \cdot cm^2$. The potential and the I_{sc} as well as the Na^+ transport rate were strongly dependent on the time of day. They were low in the morning hours and high in the late afternoon and night (Dürr 1988). Sodium and chloride were absorbed, whereas potassium was secreted. Ion substitutions, isotope fluxes and amiloride effects revealed electrogenic Na^+ transport through apical Na^+ channels as well as electroneutral Na^+ absorption via an apical Na^+-H^+ exchange system. The transport of Na^+ and Cl^- was not directly coupled. Dürr (1988) suggested that cAMP modulated the diurnal differences in electrolyte transport. Obviously, the fusus coli has an intermediate organization of transport systems, as it has transport systems found in both the proximal colon (Na^+-H^+ exchanger) and in the distal colon (Na^+ channels) (see Fig. 2).

3.5 Distal Colon

The remaining portion of the large intestine, about 50 cm in length, is referred to as the descending or distal colon. The external shape is flat without any taenia or haustrations. The epithelium is characterized by surface cells, and cells in deeply invaginated crypts. The ultrastructure of this epithelium complicates investigations of ion transport, although it is this epithelium in which the most detailed electrophysiological studies of mammalian colonic ion transport have been performed.

The rabbit distal colon is a tight epithelium, in which the electrical parameters and ion transport are strongly dependent on the time of day, or on the sodium content of the diet (Turnheim et al. 1986), and vary in a diurnal rhythm (for details, see Sect. 4). Under in vitro conditions the transepithelial resistance is about 180–350 $\Omega \cdot cm^2$. The spontaneous transepithelial potential varies between 7 and 40 mV, serosal side positive. The I_{sc} varies between 40 and 130 $\mu A/cm^2$ (Clauss and Van Cleve 1987; Hoffmann et al. 1990), and is largely generated by the electrogenic absorption of Na^+ (see Fig. 2; Frizzell et al. 1976; Schultz et al. 1977), which is closely coupled to the oxidative metabolism of the cells (Durand et al. 1988). Electroneutral Na^+-Cl^- coupled absorption does not occur. The distal colon has instead a mechanism for electroneutral absorption of Cl^- (Frizzell et al. 1976), which is independent of the parallel Na^+ absorption, and operates presumably via an apical Cl^--HCO_3^- exchanger, which is inhibited by SITS (Hatch et al. 1984) and thiazide diuretics (Ferriola et al. 1986). The influences of intracellular pH and bicarbonate activities have been investigated by Duffey (1984) and Duffey and Devor (1990). The presence of a downhill electrochemical potential gradient for bicarbonate exit was established, which may drive and energize the tansepithelial Cl^- absorption. An increase in intracellular cAMP or Ca^{2+} inhibits Cl^- absorption and induces electrogenic Cl^- secretion (Frizzell et al. 1976; Heintze et al. 1983). The site of this secretion was located in the crypt region (Welsh eh al. 1982). The distal colon has a mechanism of active potassium secretion (Yorio and Bentley 1977; Mc Cabe et al. 1982; Wills and Biagi 1982) by a transcellular route (Wills and Biagi 1982; McCabe et al. 1984; Halm and Frizzell 1986), and an apical K^+ conductance in the surface cells (Wills et al. 1982; Wills 1985). Serosal ouabain abolishes K^+ secretion and unmasks a mechanism of K^+ absorption (Wills and Biagi 1982), presumably via a K^+-H^+ exchanger located in the apical membrane (Gustin and Goodman 1981; Kaunitz and Sachs 1986). A detailed survey of the mechanisms of potassium transport was given by Smith and McCabe (1984) and Wills et al. (1987).

The cellular electrophysiological properties of the distal colon have been investigated in greater detail than any other mammalian colonic segment

(Schultz et al. 1977; Wills et al. 1979; Thompson et al. 1982a,b; Turnheim et al. 1983). Therefore rabbit distal colon is often referred to as a classical model for the analysis of ion transport mechanisms in the large intestine (Schultz 1984; Turnheim 1991). The epithelium has a trough-like potential profile (Schultz et al. 1977; Wills et al. 1979). The individual membrane resistances have been determined by electrical equivalent analysis, using conventional or time domain-based methods (Wills and Clausen 1987; Wills et al. 1987; Hoffmann and Clanss 1989; Hoffmann et al. 1990). Readers who are interested in theory of the electrophysiological equivalent analysis of epithelial tissues are referred to a review by Helman and Thompson (1982). By using biochemical techniques and membrane vesicles it has been possible recently to isolate viable epithelial cells (Kaunitz 1988), and even to separate crypt and surface cells for the study of the transport mechanisms across their individual cell membranes (Wiener et al. 1989a,b, 1990). Patch-clamp and reconstitution techniques have been applied to a calcium-activated K^+ channel in the basolateral membrane (Loo and Kaunitz 1989; Turnheim et al. 1989). The development of the ion transport mechanisms has been studied by Potter and Burlingame (1986) and Potter et al. (1987), and the ontogenic and morphological development has been related to the overall digestive functions of the rabbit by Buddington and Diamond (1989, 1990).

4 Regulation of Ion Transport

4.2 Diurnal Differences

As briefly mentioned in Section 3.1, rabbit digestive physiology is characterized by the special mechanism of *caecotrophy*, in which the large intestine plays a central role. The production of two different types of faeces with remarkable differences in their water and electrolyte contents involves, besides alterations in motility, a time-dependent regulation of absorptive and secretory processes in all large intestinal segments. To date, the precise nature of this regulation is mostly unknown, although there are some indications which will be discussed in this section. It has been demonstrated very clearly that the regulation is of a circadian nature, governed by an internal clock, which in turn is synchronized to the normal day and night shift by an external "synchronizer"such as the light and food regime (Jilge 1976; Hörnicke and Batsch 1977; Jilge et al. 1987). Under a normal, light-dark rhythm (L:D = 12:12, with light on at 06.00 h), the rabbit produces soft faeces in the morning hours, and hard faeces in the afternoon and night. Soft faeces (caecotrophes) are rich in water, electrolytes, and short chain fatty acids, and are covered by a mucous envelope. They are immediately re-ingested and taken directly from the anus, so that they are normally not recognized in the cage. In contrast, the hard faeces contain less water,

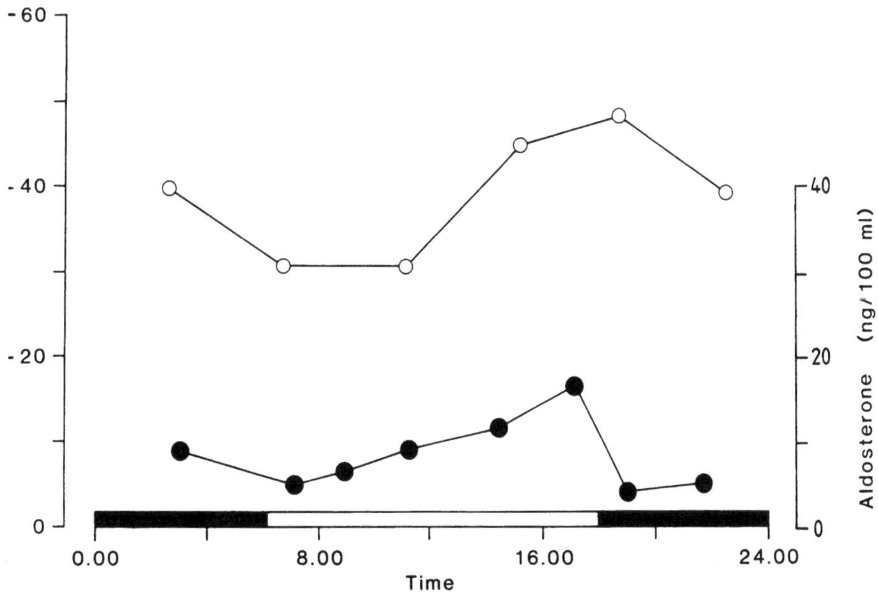

Fig. 3. Diurnal rhythm of transmural electical potential difference (○) of the rabbit distal colon (*PD*, serosal positive), and plasma aldosterone concentration (●) of one rabbit under a normal daily light-dark cycle (L:D = 12:12, light on at 06.00 h). (Clauss et al. 1988)

electrolytes, and short chain fatty acids, and are discarded. These differences in ionic contents are readily reflected in the sodium/potassium ratio of the faeces (Clauss 1978; Clauss and Hörnicke 1979; Hörnicke et al. 1985).

Experiments in the unrestrained rabbit under a controlled light-dark period demonstrated a rhythmicity of the transmural electrical potential difference (PD) in the rectum (Clauss 1978, 1984). In this segment the PD is predominantly generated by electrogenic Na^+ absorption. This rhythm was paralleled by a profound rhythm of the plasma aldosterone concentration (Clauss 1978; Clauss and Hörnicke 1979; Vernay et al. 1984; Clauss et al. 1988). Figure 3 shows an example of such a diurnal rhythm. It is obvious that low aldosterone concentrations and low PD occur during the morning hours, exactly during the time when caecotrophes with a high Na^+ content are produced. This led to the hypothesis that a diurnal variation in plasma aldosterone modulates Na^+ absorption in the distal colon in a way, which leads to low absorption in the morning hours and high absorption in the late afternoon. Obviously, the PD is an excellent indicator of the absorptive processes. In vivo and in vitro experiments later confirmed this hypothesis, and showed that Na^+ absorption is doubled during the hard faeces period (Vernay et al. 1984; Clauss et al. 1988). The involvement of the light-dark rhythm and the mediation by aldosterone were confirmed by Clauss et al.

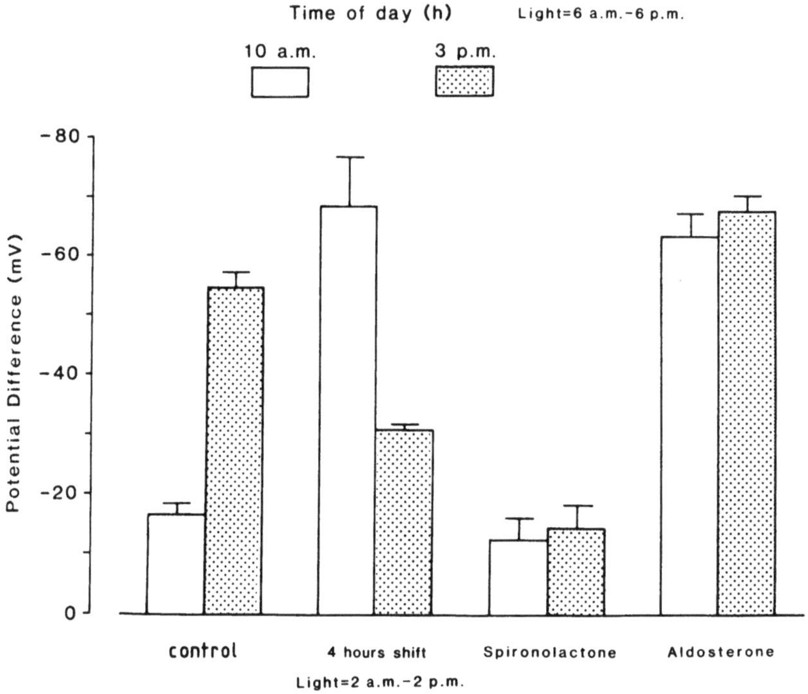

Fig. 4. Transmural electrical potential difference (colonic lumen-blood, blood side positive) of the rabbit distal colon under various experimental conditions. *Columns* represent mean values ±SEM of 5–9 rabbits in each group. (Clauss et al. 1988)

(1985b, 1988). They showed that with a time shift of the light-dark rhythm, or by using the aldosterone antagonist spironolactone, the maximum or minimum PD rhythm can also be time-shifted or even reversed (Fig. 4).

4.2 Cellular Mechanisms

The circadian regulation of the cellular processes of the distal colon, which are involved in this absorptive rhythm, is not yet fully clear. Target sites for the regulation of transepithelial Na^+ transport are the transcellular and paracellular pathways. The transcellular pathway involves Na^+ entry through amiloride-sensitive Na^+ channels in the apical membrane (Zeiske et al. 1982), and Na^+ extrusion by the Na^+/K^+-ATPase across the basolateral cell membrane. Whereas the maximal transport capacity of the Na^+/K^+-ATPase does not differ between the transport states, the apical Na^+ entry is regulated by the number of conducting Na^+ channels (Clauss et al. 1988). The single channel current was found to be similar in both transport states at about 1 pA, but the number of Na^+ channels varied

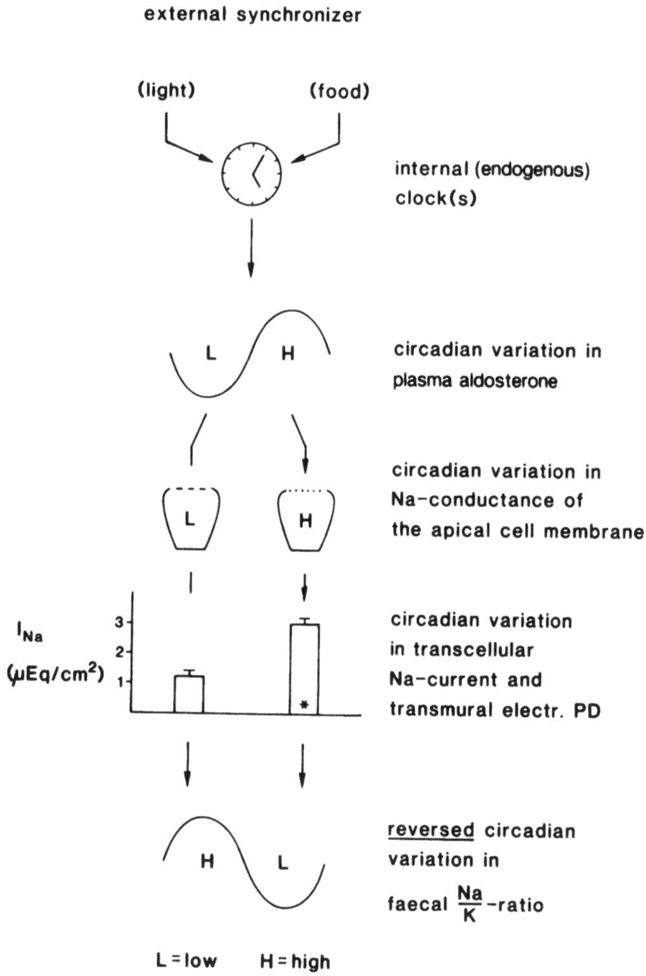

Fig. 5. Current hypothesis of the regulation of colonic Na^+ transport involved in the circadian digestive mechanism of caecotrophy in rabbits

between 25 and 60 million/cm^2, and was significantly higher during the hard faeces period, when more Na^+ is absorbed.

Investigations of the paracellular pathway showed that the electrical resistance of this pathway is also different and dependent on the time of day (Hoffmann and Clauss 1989; Hoffmann et al. 1990). It was shown that aldosterone significantly increased the paracellular resistance, presumably by regulation of the tight junctions. This in turn leads to a diminished back-diffusion of Na^+ down its steep gradient into the colonic lumen (Clauss and Van Cleve 1987), and therefore optimizes the transepithelial Na^+ absorption. From all these in vivo and in vitro experiments, a regulatory cascade can be shown from the internal clock to the Na^+ channel in the apical membrane of the colonic enterocyte (Fig. 5).

5 Conclusions

From all the work cited above it is clear that the rabbit large intestine is not a representative model for the mammalian colon, but instead a complex species-dependent system, which involves the interaction of many morphologically distinct segments with time- and hormone-dependent changes in motility and absorptive processes. Many epitheliologists in the field of gastrointestinal research are interested in "their" particular piece of tissue of a certain gut segment, because the elaboration of detailed transport mechanisms is a lengthy and time-consuming task. Not all information concerning these widely unknown segments of the rabbit large intestine is significant, however, the real challenge for us will be to show that our particular subject is of importance for the overall organ function. With regard to the rabbit large intestine, we know a great deal about the cellular ion transport mechanisms in the distal colon, but very little about ion transport in the proximal colon and caecum, and almost nothing about the interaction of these special absorptive capacities in the digestive function of this organ. Therefore, future work on the cellular aspects of ion transport in any portion of the rabbit large intestine should be aware of the time dependence of these processes in order to integrate new pieces to the puzzle.

Acknowledgements. Work from the author's lab was supported by the Deutsche Forschungsgemeinschaft, the European Community, and the H. Wilhelm Schaumann Stiftung.

References

Ahn J, Chang EB, Field M (1985) Phorbol ester inhibition of Na-H exchange in rabbit proximal colon. Am J Physiol 249: C527–C530

Auer J (1925) Further notes on the fusus coli of the rabbit. Proc Soc Exp Biol Med 22: 301–303

Björnhag G (1972) Separation and delay of contents in the rabbit colon. Swed J Agric Res 2:125–136

Björnhag G (1987) Comparative aspects of digestion in the hindgut of mammals. The colonic separation mechanism (CSM). (A review). Dtsch Tierärztl Wochenschr 94:33–36

Blackwood WD, Bolinger RA, Lifson N (1973) Some characteristics of the rabbit vermiform appendix as a secreting organ. J Clin Invest 52:143–151

Buddington RK, Diamond JM (1989) Ontogenic development of intestinal nutrient transporters. Annu Rev Physiol 51: 601–619

Buddington RK, Diamond JM (1990) Ontogenic development of monosaccharide and anino acid transporters in rabbit intestine. Am J Physiol 259: G544–G555
Camara VM, Prieur DJ (1984) Secretion of colonic isoenzyme of lysozyme in association with coecotrophy of rabbits. Am J Physiol 274: G19–G23
Campos MS, Galindo MC, Garcia JA, Lisbona F, Lopez-Aliaga I (1991) A comparative study of electrolyte and water transport in the rabbit ileum and colon in vitro and in vivo: influence of D-glucose. Biomed Biochim Acta 50: 279–283
Clauss W (1978) Resorption und Sekretion von Wasser und Elektrolyten im Colon des Kaninchens im Zusammenhang mit der Bildung von Weichkot und Hartkot. Dissertation, Universität Hohenheim
Clauss W (1984) Circadian rhythms in Na transport. In: Skadhauge E, Heintze K (eds) Intestinal absorption and secretion. MTP Press, Lancaster, pp 273–283
Clauss W (1985) Segmental action of aldosterone on water and electrolyte transport across rabbit colon in vivo. Comp Biochem Physiol 81A: 873–877
Clauss W, Hörnicke H (1979) Circadian rhythms of electrical potential difference, plasma aldosterone and faecal Na/K-ratio in rabbits. Chronobiologica 2:112
Clauss W, Hörnicke H (1984) Segmental differences in K-transport across rabbit proximal and distal colon in vivo and in vitro. Comp Biochem Physiol 79A: 267–269
Clauss W, Van Cleve I (1987) Electrophysiological analysis of cellular regulation of sodium transport in rabbit colon. Dtsch Tierärztl Wochenschr 94: 4–6
Clauss W, Schäfer H, Horch I, Hörnicke H (1985a) Segmental differences in electrical properties and Na^+ transport of rabbit caecum, proximal and distal colon in vitro. Pflügers Arch Eur J Physiol 403: 278–282
Clauss W, Dürr JE, Hörnicke H (1985b) Effects of aldosterone and dexamethasone on apical membrane properties and Na-transport of rabbit distal colon in vitro. Pflügers Arch Eur J Physiol 403:186–192
Clauss W, Horch I, Hörnicke H (1985c) Electrolyte transport across rabbit late proximal colon. Comp Biochem Physiol 82A: 71–75
Clauss W, Biehler KH, Schäfer H, Wills NK (1987) Ion transport and electrophysiology of the early proximal colon of rabbit. Pflügers Arch 408: 592–599
Clauss W, Dürr JE, Krattenmacher R, Hörnicke H, Van Driessche W (1988) Circadian rhythm of apical Na-channels and Na-transport in rabbit distal colon. Experientia 44: 608–610
Clauss W, Hoffmann B, Schäfer H, Hörnicke H (1989) Ion transport and electrophysiology in rabbit cecum. Am J Physiol 256: G1090–G1099
Duffey ME (1984) Intracellular pH and bicarbonate activities in rabbit colon. Am J Physiol 246:C558–C561
Duffey ME, Devor DC (1990) Intracellular pH and membrane potassium conductance in rabbit distal colon. Am J Physiol 258:C336–C343
Durand J, Durand-Arczynska, Wankmiller D (1988) Coupling of active sodium transport to oxidative metabolism in the rabbit distal colon. J Physiol 396: 55–64
Dürr J (1988) Der transepitheliale Elektrolyttransport am Fusus Coli des Kaninchens. Dissertation, Universität Hohenheim
Eden A (1940) Coprophagy in the rabbit: origin of night faeces. Nature 145: 628
Ehrlein HJ, Reich H, Schwinger M (1983) Colonic motility and transit of digesta during hard and soft faeces formation in rabbits. J Physiol 338:75–86
Ferriola PC, Acara MA, Duffey ME (1986) Thiazide diuretics inhibit chloride absorption by rabbit distal colon. J Pharmacol Exp Ther 238: 912–915
Frizzell RA, Schultz SG (1978) Effect of aldosterone on ion transport by rabbit colon in vitro. J Membr Biol 39:1–26
Frizzell RA, Turnheim K (1978) Ion transport by rabbit colon: II. Unidirectional sodium influx and the effects of amphotericin B and amiloride. J Membr Biol 40: 193–211
Frizzell RA, Koch MJ, Schultz SG (1976) Ion transport by rabbit colon. I. Active and passive components. J Membr Biol 27: 297–316
Garcia JA, Campos MS, Lopez MA (1982) A comparative study of electrolyte absorption from the caecum and colon of *Oryctolagus cuniculus*. I. Chloride and bicarbonate. Comp Biochem Physiol 73A: 459–462

Garcia JA, Campos MS, Lopez MA (1983) A comparative study of electrolyte absorption from the caecum and colon of *Oryctolagus cuniculus*. II. Sodium and potassiums. Comp Biochem Physiol 75A: 87–89
Garty H, Benos DJ (1988) Characteristics and regulatory mechanisms of the amiloride-blockable Na^+ channel. Physiol Rev 68: 309–373
Gustin MC, Goodman DB (1981) Isolation of brush-border membrane from the rabbit descending colon epithelium. Partial characterization of a unique K^+-activated ATPase. J Biol Chem 256: 10651–10656
Halm DR, Frizzell RA (1986) Active K transport across rabbit distal colon: relation to Na absorption and Cl secretion. Am J Physiol 251: C252–C267
Halm DR, Frizzell RA (1991) Ion transport across the large intestine. In: Field M, Frizzell RA (eds) Intestinal absorption and secretion, Handbook of physiology, section 6, vol IV. Am Physiol Soc, Oxford University Press, New York, pp 257–273
Hatch M (1987) Short-chain fatty acid transport and its effects on ion transport by rabbit cecum. Am J Physiol 253: G171–G178
Hatch M, Freel RW (1988) Electrolyte transport across the rabbit caecum in vitro. Pflügers Arch Eur J physiol 411: 333–338
Hatch M, Freel RW, Goldner AM, Earnest DL (1984) Oxalate and chloride absorption by the rabbit colon: sensitivity to metabolic and anion transport inhibitors. Gut 25: 232–237
Heintze K, Stewart CP, Frizzell RA (1983) Sodium-dependent chloride secretion across rabbit descending colon. Am J Physiol 244: G357–G365
Helman SI, Thompson SM (1982) Interpretation and use of electrical equivalent circuits in studies of epithelial tissues. Am J Physiol 243: 519–534
Henning SJ, Hird FJ (1972) Diurnal variations in the concentrations of volatile fatty acids in the alimentary tract of wild rabbits. Br J Nutr 27: 57–64
Hoffmann B, Clauss W (1989) Time-dependent effects of aldosterone on sodium transport and cell membrance resistances in rabbit distal colon. Pflügers Arch Eur J Physiol 415: 156–164
Hoffmann B, Nagel I, Clauss W (1990) Aldosterone regulates paracellular pathways in rabbit distal colon. J Comp Physiol B 160: 381–388
Hörnicke H, Batsch F (1977) Caecotrophy in rabbits – a circadial function. J Mammal 58: 240–242
Hörnicke H, Clauss W, Römer B (1985) Circadian variations of the sodium/potassium ratio in rabbit faeces as related to aldosterone effects on large intestinal electrolyte transport. J Interdiscip Cycle Res 16: 135–136
Jilge B (1976) The entrainment of circadial soft faeces excretion in the rabbit. J Interdiscip Cycle Res 7: 229–235
Jilge B, Hörnicke H, Staehle H (1987) Circadian rhythm of rabbits during restrictive feeding. Am J Physiol 253: R46–R54
Kaunitz JD (1988) Preparation and characterization of viable epithelial cells from rabbit distal colon. Am J Physiol 254: G502–G512
Kaunitz JD, Sachs G (1986) Identification of a vanadate-sensitive potassium-dependent proton pump from rabbit colon. J Biol Chem 261: 14005–14010
Krattenmacher R, Voigt R, Clauss W (1990) Ca-sensitive sodium absorption in the colon of *Xenopus laevis*. J Comp Physiol B 160: 161–165
Leng E (1978) Absorption of inorganic ions and volatile fatty acids in the rabbit caecum. Br J Nutr 40: 509–519
Leng-Peschlow E, Marty J (1979) Absorption of water, electrolytes and volatile fatty acids in the rabbit caecal pouch. J Comp Physiol 133: 205–210
Loo DDF, Kaunitz JD (1989) Ca^{2+} and cAMP activate K^+ channels in the basolateral membrane of crypt cells isolated from rabbit distal colon. J Membr Biol 110: 19–28
Marty JF, Abravanel GM (1985) Acetate absorption and metabolism in the rabbit hindgut. Gut 26: 562–569
Marty JF, Vernay MY (1984) Absorption and metabolism of the volatile fatty acids in the hindgut of the rabbit. Br J Nutr 51: 265–277

McCabe RD, Cooke H, Sullivan L (1982) Potassium transport by rabbit descending colon. Am J Physiol 242: C81–C86
McCabe RD, Smith PL, Sullivan LP (1984) Ion transport by rabbit distal colon: mechanisms of transepithelial potassium transport. Am J Physiol 246: G594–G602
Pairet M, Boyssou T, Ruckebusch Y (1986) Colonic formation of soft faeces in rabbits: a role for endogenous prostaglandins. Am J Physiol 250: G302–G308
Parker DS (1976) The measurement of production rates of volatile fatty acids in the caecum of the conscious rabbit. Br J Nutr 36: 61–70
Pickard DW, Stevens CE (1972) Digesta flow through the rabbit large intestine. Am J Physiol 222: 1161–1166
Potter GD, Burlingame SM (1986) Ion transport by neonatal rabbit distal colon. Am J Physiol 250: G754–G759
Potter GD, Lester R, Burlingame SM, Mitchell PA, Schmidt KL (1987) Taurodeoxycholate and the developing rabbit distal colon: absence of secretory effect. Am J Physiol 253: G483–G488
Ruckebusch Y, Fioramonti Y (1976) The fusus coli of the rabbit as a pacemaker area. Experientia 32: 1023–1024
Ruckebusch Y, Hörnicke H (1977) Motility of the rabbits colon and caecotrophy. Physiol Behav 18: 871–878
Schultz SG (1981) Homocellular regulatory mechanisms in sodium-transporting epithelia: avoidance of extinction by "flush-through". Am J Physiol 241: F579–F590
Schultz SG (1984) A cellular model for active sodium absorption by mammalian colon. Annu Rev Physiol 46: 435–451
Schultz SG, Frizzell RA, Nellans HN (1977) Active sodium transport and the electrophysiology of rabbit colon. J Membr Biol 33: 351–384
Sellin JH, De Soignie RC (1984) Rabbit proximal colon: a distinct transport epithelium. Am J Physiol 246: G603–G610
Sellin JH, De Soignie RC (1985) Steroids alter ion transport and absorptive capacity in proximal and distal colon. Am J Physiol 249: G113–G119
Sellin JH, De Soignie RC (1987a) Regulation of Na-Cl absorption in rabbit proximal colon in vitro. Am J Physiol 252: G45–G51
Sellin JH, De Soignie RC (1987b) Ionic regulation of Na absorption in proximal colon: cation inhibition of electroneutral Na absorption. Am J Physiol 252: G100–G108
Sellin JH, Oyarzabal H, Cragoe EJ (1988) Electrogenic sodium absorption in rabbit cecum in vitro. J Clin Invest 81: 1275–1283
Smith PL, McCabe RD (1984) Mechanism and regulation of transcellular potassium transport by the colon. Editorial review. Am J Physiol 247: G445–G456
Snipes RL (1978) Anatomy of the rabbit cecum. Anat Embryol 155: 57–80
Snipes RL, Clauss W, Weber A, Hörnicke H (1982) Structural and functional differences in various divisions of the rabbit colon. Cell Tissue Res 225: 331–346
Sullivan SK, Smith PL (1986a) Active potassium secretion by rabbit proximal colon. Am J Physiol 250: G475–G483
Sullivan SK, Smith PL (1986b) Bicarbonate secretion by rabbit proximal colon. Am J Physiol 251: G436–G445
Tai Y-H, Gage TP, McQueen C, Formal SB, Boedeker EC (1989) Electrolyte transport in rabbit cecum. I. Effect of RDEC-1 infection. Am J Physiol 256: G721–G726
Thompson SM, Suzuki Y, Schultz SG (1982a) The electrophysiology of rabbit descending colon. I. Instantaneous transepithelial current-voltage relations and the current-voltage relations of the Na-entry mechanism. J Membr Biol 66: 41–54
Thompson SM, Suzuki Y, Schultz SG (1982b) THe electrophysiology of rabbit descending colon. II. Current-voltage relations of the apical membrane, the basolateral membrane, and the parallel pathways. J Membr Biol 66: 55–61
Turnheim K (1991) Intrinsic regulation of apical sodium entry in epithelia. Physiol Rev 71: 429–445
Turnheim K, Frizzell RA, Schultz SG (1977) Effect of anions on amiloride-sensitive, active sodium transport across rabbit colon in vitro: evidence for transinhibition of the Na entry mechanism. J Membr Biol 37: 63–85

Turnheim K, Frizzell RA, Schultz SG (1978) Interaction between cell sodium and amiloride-sensitive sodium entry step in rabbit colon. J Membr Biol 39: 233–256

Turnbeim K, Thompson SM, Schultz SG (1983) Relation between intracellular sodium and active sodium transport in rabbit colon: current-voltage relations of the apical sodium entry mechanism in the presence of varying luminal sodium. J Membr Biol 76: 299–309

Turnheim K, Plass H, Grasl M, Krivanek P, Wiener H (1986) Sodium absorption and potassium secretion in rabbit colon during sodium deficiency. Am J Physiol 250: F235–F246

Turnheim K, Hudson RL, Schultz SG (1987) Cell Na activities and transcellular Na absorption by descending colon from normal and Na-deprived rabbits. Pflügers Arch Eur J Physiol 410: 279–283

Turnheim K, Constantin J, Chan S, Schultz SG (1989) Reconstitution of a calcium-activated potassium channel in basolateral membranes of rabbit colonocytes into planar lipid bilayers. J Membr Biol 112: 247–254

Ussing HH, Zerahn K (1951) Active transport of sodium as the source of electric current across the short-circuited isolated frog skin. Acta Physiol Scand 23: 110–127

Vernay MY (1987) Effects of plasma aldosterone on butyrate absorption and metabolism in the rabbit proximal colon. Comp Biochem Physiol 86A: 657–662

Vernay MY, Marty JF, Moatti J-P (1984) Absorption of electrolytes and volatile fatty acids in the hindgut of the rabbit. Circadian rhythm of hindgut electrolytes and plasma aldosterone. Br J Nutr 52: 419–428

Welsh MJ, Smith PL, Fromm M, Frizzell RA (1982) Crypts are the site of intestinal fluid and electrolyte secretion. Science 218: 1219–1221

Wiener H, Turnheim K, Van Os CH (1989a) Rabbit distal colon epithelium. 1. Isolation and characterization of basolateral plasma membrane vesicles from surface and crypt cells. J Membr Biol 110: 147–162

Wiener H, Turnheim K, Van Os CH (1989b) Rabbit distal colon epithelium. 2. Characterization of (Na^+, K^+, Cl^-)-cotransport and (H-3)-bumetanide binding. J Membr Biol 110: 163–174

Wiener H, Turnheim K, Van Os CH (1990) Rabbit distal colon epithelium. 3. Ca^{2+}-activated K^+ channels in basolateral membrane vesicles of surface and crypt cells. J Membr Biol 117: 275–283

Williams JA, Griffen WC, Sharma A, Wangensteen OH (1961) Composition and source of secretion from lymphatic aggregations in the rabbit gut. Br J Exp Pathol 42: 153–157

Wills NK (1985) Apical membrane potassium and chloride permeabilities in surface cells of rabbit descending colon epithelium. J Physiol 358: 433–445

Wills NK, Biagi B (1982) Active potassium transport by the rabbit descending colon epithelium. J Membr Biol 64: 195–203

Wills NK, Clausen C (1987) Transport-dependent alterations of membrane properties of mammalian colon measured using impedance analysis. J Membr Biol 95: 21–35

Wills NK, Lewis SA, Eaton DC (1979) Active and passive properties of rabbit descending colon: a microelectrode and nystatin study. J Membr Biol 45: 81–108

Wills NK, Zeiske W, Van Driessche W (1982) Noise analysis reveals K channel conductance fluctuations in the apical membrane of rabbit colon. J Membr Biol 69: 187–197

Wills NK, Clausen C, Clauss W (1987) Electrophysiology of active potassium transport across the mammalian colon. Curr Top Membr Transp 28: 441–456

Yorio T, Bentley PS (1977) The permeability of the rabbit colon in vitro. Am J Physiol 232: F5–F9

Zeiske W, Wills HK, Van Driessche W (1982) Na channels and amiloride induced noise in the mammalian colon epithelium. Biochim Biophys Acta 688: 201–210

Chapter 6

Ion Transport in Rat Colon

Vazhaikkurichi M. Rajendran and *Henry J. Binder*[1]

Contents

1 Introduction

The rat large intestine has been the object of multiple studies of fluid and electrolyte absorption for more than 30 years. Initial experiments performed under in vivo conditions provided an overall description of electrolyte movement, while more recent in vitro studies with both intact tissue and membrane vesicles have provided increasing insight into both the cellular processes responsible for net fluid movement and the mechanisms regulating these transport events. Luminal perfusion experiments that studied the entire colon revealed that sodium (Na) and chloride (Cl) are absorbed, while potassium (K) and bicarbonate (HCO_3) are secreted (Curran and Schwartz 1960; Edmonds 1967b; Powell et al. 1968). Water absorption was considered secondary to the net electrolyte movement. Evaluation of the mechanism(s) responsible for ion movement (based on concentration and potential differences) suggested that Na absorption and HCO_3 secretion are active and energy-dependent, while K secretion is the result of existing electrochemical concentration gradients.

[1]Department of Internal Medicine, Yale University School of Medicine, 89 LMP-333 Cedar Street, New Haven, CT 06510, USA

Advances in Comparative and Environmental Physiology, Vol. 16

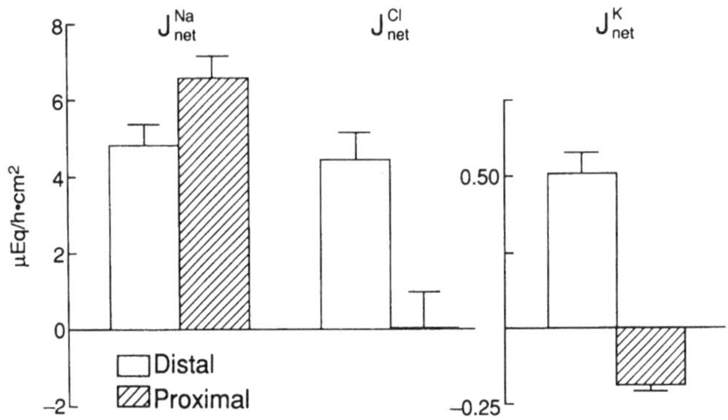

Fig. 1. Net Na, K and Cl transport in proximal and distal colon under voltage-clamp conditions. In distal colon, net Na,K and Cl absorption are present. In proximal colon, net Na absorption and net K secretion are observed, but net Cl transport is absent. (After data in Foster et al. 1983, 1986; Turnamian and Binder 1990)

In vitro studies performed under voltage-clamped conditions demonstrate significant qualitative differences in Na, K and Cl transport processes between proximal and distal segments of the rat colon (Foster et al. 1983, 1986; Turnamian and Binder 1990; Fig.1). Thus, although net Na absorption is Cl-dependent and amiloride-insensitive in both proximal and distal colon, net Cl absorption, which is equivalent to Na absorption in the distal colon, is *not* present in proximal colon. There are also significant segmental differences in active K movement in that active K secretion and active K absorption have been identified in proximal and distal colon, respectively.

Aldosterone is a major regulator of colonic electrolyte transport but has qualitatively *different* effects on electrolyte transport in proximal and distal segments (Fig. 2). Aldosterone induces amiloride-sensitive, Cl-independent, electrogenic Na absorption in distal colon but stimulates Cl-dependent electroneutral Na absorption in the proximal colon (Foster et al. 1983, 1986; Sect. 3). In addition, aldosterone converts net K absorption to net K secretion in the distal colon (Sweiry and Binder 1989), but enhances net K secretion in the proximal colon (Turnamian and Binder 1990; Fig. 3). The qualitative nature of these segmental differences is such that it is necessary to study separately individual portions of the colon and that the results of those studies which have treated the colon as a single organ require reassessment.

The description of Na absorption in the rabbit distal colon has often been considered the prototype model of Na absorption in the mammalian colon. However, the mechanism of Na absorption in rabbit distal colon deffers substantially from that of rat distal colon (and rabbit proximal colon) (Frizzell et al. 1976; Sellin and DeSoignie 1984; see Clauss, this volume). Na absorption in rabbit colon is amiloride-sensitive, Cl-independent, equivalent to Isc, and is believed to occur via an electrogenic Na channel. In contrast,

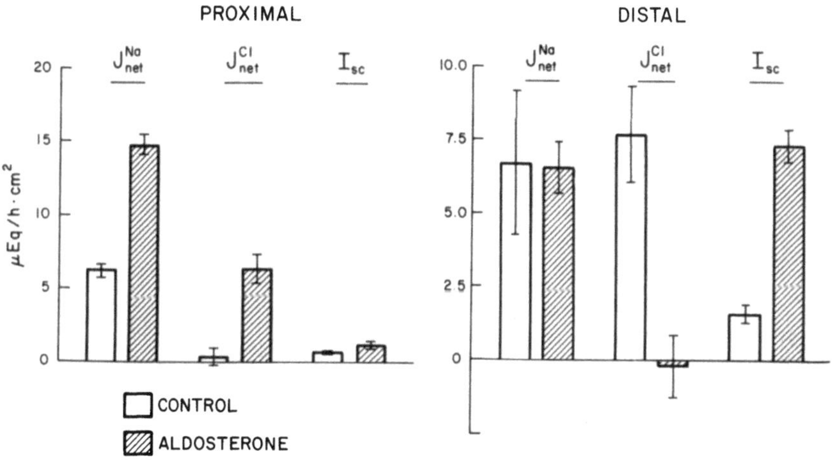

Fig. 2. Effect of aldosterone in net Na and Cl transport in proximal and distal colon under voltage-clamp conditions. In proximal colon, aldosterone stimulates electroneutral Na and Cl absorption. In distal colon, aldosterone induces electrogenic Na absorption and inhibits electroneutral Na and Cl absorption. (After data in Foster et al. 1983, 1986)

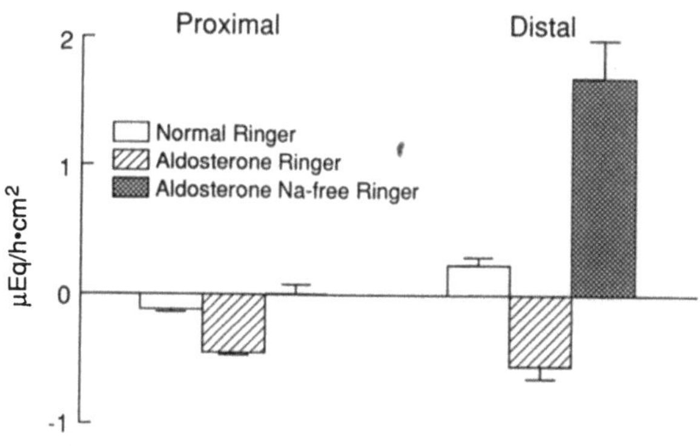

Fig. 3. Effect of aldosterone on net K transport in proximal and distal colon under voltage-clamp conditions. In proximal colon, aldosterone stimulates net K secretion and in distal colon induces net K secretion. Net K secretion is abolished by the removal of Na from the bath solution in both proximal and distal colon but in the distal colon net K absorption is unmasked. (After data in Foster et al. 1984)

Na absorption in the rat distal colon is amiloride-insensitive, Cl-dependent, substantially greater than Isc, and occurs via an electroneutral Na-H exchange. Moreover, aldosterone inhibits electroneutral Na absorption and induces electrogenic Na absorption in rat distal colon. As a result, the model of Na absorption in the aldosterone-treated rat distal colon is almost indistinguishable from that present in the normal rabbit distal colon. As aldosterone regulation of ion transport in the rat colon is unique, this present review will emphasize both the mechanisms of colonic electrolyte transport in the normal rat and aldosterone regulation of these transport processes; many of these studies have been performed in the authors' laboratory during the past 20 years.

2 Distal Colon

2.1 Na-Cl Transport

In vivo perfusion studies of entire colon with an isotonic solution established a relationship between water and the Na absorption (Curran and Schwartz 1960; Powell et al. 1968). Net Na absorption occurs against its electrochemical potential (5–15 mV lumen negative) gradient and decreases as Na concentration in the perfusate decreases. Net water absorption, which is zero in the absence of Na, is directly proportional to the rate of Na absorption. As a result, water absorption is not considered an active process but secondary to active Na absorption. This study also concluded that Cl absorption occurred as a result of passive diffusion. It should be noted that this and other earlier in vivo studies treated the colon as a single organ and did not recognize regional differences.

2.1.1 Na Transport

In rat distal colon net Na absorption is greater than Isc, a suggestion that Na absorption is electroneutral process (Foster et al. 1983). Electroneutral Na and Cl absorption may represent either (1) Na-K-2Cl cotransport; (2) Na-Cl cotransport; (3) Na-H exchange; or (4) coupled dual ion [i.e., Na-H and Cl-HCO_3 (OH)] exchanges (Fig. 4B). Recent observations suggest that coupled dual ion exchanges are the mechanism for electroneutral NaCl absorption in rat distal colon (Binder et al. 1987; Goldfarb et al. 1988; Fig. 4A,B). Net Na absorption is equivalent to net Cl absorption. Both Na and Cl absorption is: (1) HCO_3-dependent and inhibited by a carbonic anhydrase inhibitor, acetazolamide; (2) inhibited by 1 mM amiloride (a concentration that inhibits electroneutral Na-H exchange in other epithelia); and (3) not affected by either removal of mucosal K or addition of furosemide, a loop diuretic that inhibits coupled cotransport processes.

Although coupled dual ion exchange processes have been identified as a mechanism for Na and Cl absorption in other epithelia, direct demonstration, which is possible only in isolated membrane vesicle studies, is lacking for the rat distal colon. Studies with membrane vesicles are a potent tool to study electrolyte transport in other epithelia (Murer and Kinne 1980), however, isolation of apical membranes from colon have long been hampered

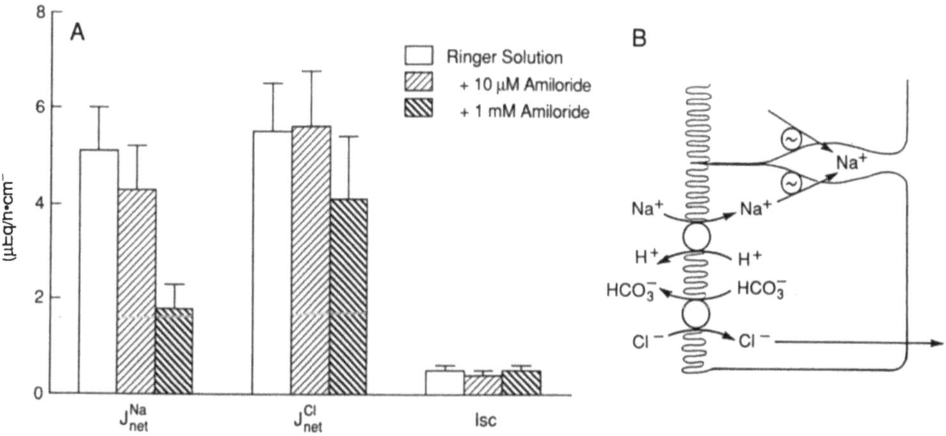

Fig. 4A,B. Effect of amiloride on net Na and Cl absorption under voltage-clamp conditions in distal colon. **A** 1 mM amiloride (which inhibits Na-H exchange), but not 10 µM amiloride (which blocks electrogenic Na transport), substantially inhibits net Na and Cl absorption (after data in Binder et al. 1987) **B** Model of Cl-dependent, electroneutral net Na absorption: dual ion (Na-H and Cl-HCO_3) exchange processes

due to the absence of specific marker enzyme(s). Recently, Stieger et al. developed a novel morphological isolation method, independent of marker enzymes, to isolate apical membrane vesicles (AMV) from colon (Stieger et al. 1986). The origin of these vesicles from the apical but not from the basolateral membrane has been confirmed by immunofluorescent studies and by Western blot analysis using a monoclonal antibody that was developed against isolated apical membranes. This methodological approach has been adapted successfully for colonic AMV to characterize Na uptake mechanisms in both proximal and distal segments of rat large intestine (Rajendran et al. 1989; Rajendran and Binder 1990).

In AMV isolated from distal colon Na uptake was not stimulated by inward-directed gradients of Na and/or K – observations that indicate the absence of both coupled Na-Cl and Na-K-2Cl cotransport processes. In contrast, Na uptake in AMV was significantly stimulated by an outward-directed proton gradient (Rajendran et al. 1989; Fig. 5). This proton gradient-stimulated Na uptake represents electroneutral Na-H exchange, as it was not inhibited by voltage clamping but was inhibited by 1 mM amiloride and EIPA, an amiloride analog which is a specific inhibitor of Na-H exchange.

Electroneutral Na-H exchange identified in several other epithelial and nonepithelial cells has several important intracellular functions, such as regulation of cell volume, cell pH (pHi), and cell proliferation (Grinstein and Dixon 1989). Regulation of intracellular pH_i may be associated with Na-H exchange activation by intracellular (intravesicular) proton (i.e., activation at an internal proton modifier site) (Aronson et al. 1982). In

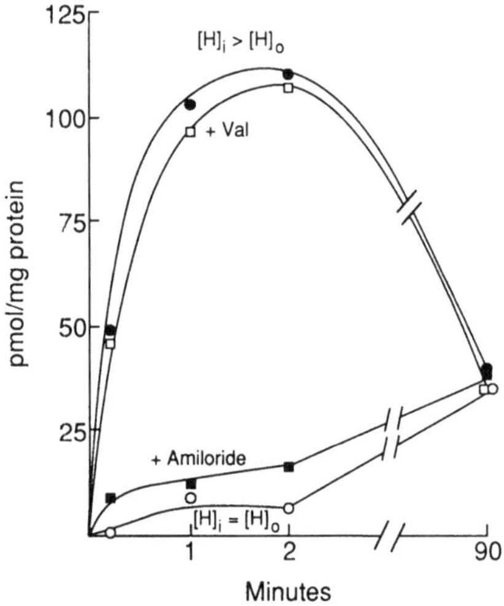

Fig. 5. Effect of proton gradient on Na uptake in apical membrane vesicles of distal colon. An outward-directed proton gradient significantly stimulates Na uptake more than that in the absence of a proton gradient. The proton gradient-stimulated Na uptake is not affected by voltage clamping but is inhibited almost completely by 1 mM amiloride, indicating the presence of electroneutral Na-H exchange

contrast to exponential activation of Na-H exchange by intravesicular proton in apical membrane vesicles prepared from rabbit proximal tubules, Na uptake in rat distal colonic AMV saturates with increasing intravesicular proton concentrations (Rajendran and Binder 1990; Fig. 6). Additional observations of identical rates of Na uptake by both a tenfold outward Na gradient (i.e., Na-Na exchange) and an outward proton gradient at different pHs confirm that Na-H exchange of AMV rat colon is not activated by pH (Fig. 7A,B). Thus, no evidence of an intravesicular proton modifier (activation) site was identified. These observations indicate that the properties of Na-H exchange of AMV from rat distal colon differ substantially from Na-H exchanges that have been studied in other epithelial and nonepithelial cells. Thus, it is likely that the primary role of Na-H exchange of AMV rat distal colon is not pH_i regulation, but rather vectorial Na movement.

If the function of colonic Na-H exchange is mainly associated with vectorial Na movement and not pH_i regulation, it is likely that another proton or hydroxyl-linked transport system is located on either the apical or basolateral membrane of distal colonocytes which is sensitive to pH. Thus, not unexpectedly, proton gradient-stimulated Na uptake is present in basolateral membrane vesicles (BLMV) of rat distal colon (Rajendran et al. 1991b; Fig. 8A). However, despite its inhibition by amiloride, 1 mM DIDS (4,4′-isothiocyanatostilbene-2-2′-disulfonicacid), an inhibitor of Na-HCO_3 cotransport (as well as anion exchange processes), also significantly inhibited proton gradient-stimulated Na uptake – an observation that suggests that

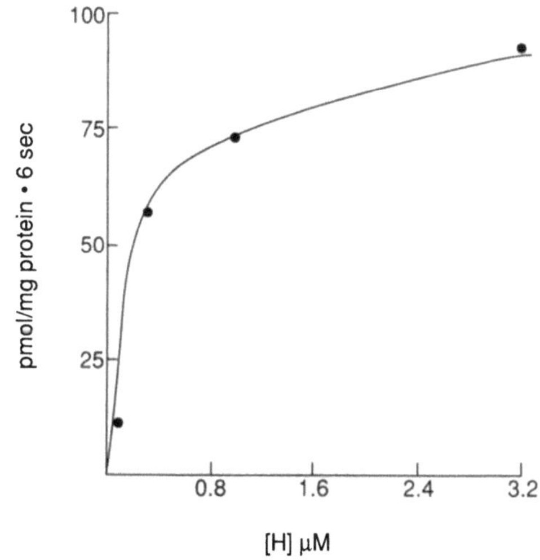

Fig. 6. Effect of intravesicular proton oncentration on Na uptake in apical membrane vesicles of distal colon. Increasing intravsicular proton concentration resulted in saturation of Na uptake

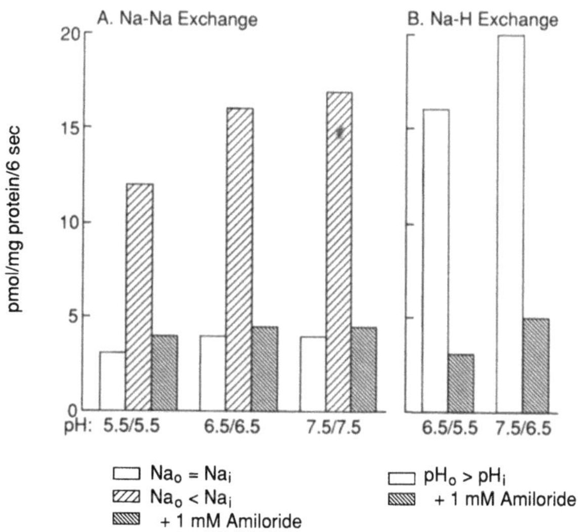

Fig. 7A,B. Effect of pH on Na-Na (**A**) and Na-H exchanes (**B**) in apical membrane vesicles of proximal colon. Neither outward-directed Na gradient-stimulated Na uptake (i.e., Na-Na exchange) nor outward-direced proton gradient-stimulated Na uptake (i.e., Na-H exchange) are activated by higher proton oncentrations (i.e., at lower pH). (Rajenran and Binder 1990)

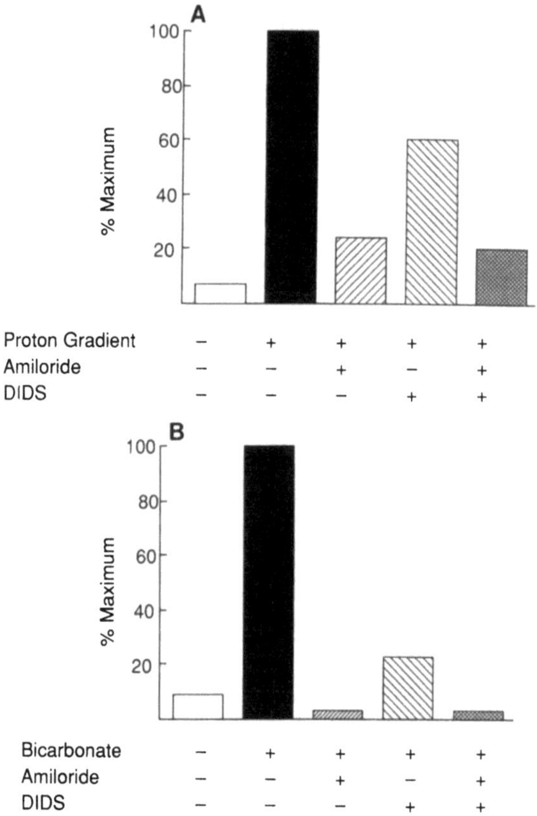

Fig. 8A,B. Proton gradient and HCO_3 gradient stimulate Na uptake in basolateral membrane vesicles of distal colon. **A** An outward-directed proton gradient-stimulated Na uptake is inhibited 80 and 40% by 100 μM amiloride and 1 μM DIDS, respectively. **B** An inward-directed HCO_3 gradient substantially stimulates Na uptake which is inhibited by 98 and 76% by amiloride and DIDS, respectively. (Rajendran et al. 1991a)

this DIDS-sensitive Na uptake occurred as a result of Na-OH cotransport, perhaps analogous to Na-HCO_3 cotransport (Fig. 8A). Results shown in Fig. 8B demonstrate the presence of an electrogenic DIDS-sensitive Na-HCO_3 cotransport in BLMV of rat distal colon. Based on these studies we have also proposed that there is an electrogenic Na-anion cotransport process in the BLMV with varying affinity for both hydroxyl and bicarbonate. In addition, BLMV Na-HCO_3 cotransport, but not the BLMV Na-H exchange, is activated by pH (V.M. Rajendran and H.J. Binder, unpubl. observ.), suggesting that this transport process may be a primary regulator of pH_i in colonocytes from the rat distal colon.

2.1.2 Cl Transport

Cl absorption in rat colon has been studied in experiments that have employed both in vivo luminal perfusion and in vitro transepithelial flux techniques. Initial in vivo studies performed in rat colon described Cl absorption as a passive diffusion process (Curran and Schwartz 1960; Powell et al. 1968). In vitro studies indicate that active net Cl movement is absent in proximal colon but present in distal colon, and that net Cl absorption in the distal colon is electroneutral, Na-dependent, HCO_3-dependent and is inhibited by acetazolamidc (Binder et al. 1987; Fig. 4). These observations are consistent with the possibility that Cl absorption is the result of Cl-HCO_3 exchange.

In studies with AMV isolated from distal colon, an inward Na and/or K gradient failed to stimulate Cl uptake which is an indication that Na-Cl, KCL, or Na-K-2Cl cotransport processes are absent. In contrast, both an outward HCO_3 gradient (absence of pH gradient) and pH gradient ($pH_i > pH_o$) in the absence of HCO_3 significantly stimulated Cl uptake (Rajendran et al. 1991a). Both HCO_3 and pH gradient-stimulated Cl uptake was inhibited by DIDS, an anion exchange inhibitor. These results suggest that HCO_3 and pH gradient-stimulated Cl uptake occurs via anion (Cl-HCO_3 and/or Cl-OH) exchange processes. pH gradient-stimulated Cl uptake is three-to-fourfold higher than that of HCO_3 gradient-stimulated Cl uptake, and HCO_3 gradient-stimulated Cl uptake was enhanced significantly by a pH gradient. However, pH gradient-stimulated Cl uptake was not altered by HCO_3. These observations indicate that hydroxyl has a higher affinity than HCO_3 for the colonic apical Cl-anion exchange mechanism.

Recent studies have also provided additional characterization of Cl-anion exchange in rat distal colon. pH gradient-stimulated Cl uptake saturated as a function of extravesicular Cl concentration with an apparent Km of 15 mM for Cl. In contrast, Cl uptake as a function of increasing intravesicular hydroxyl concentration demonstrated a sigmoidal curve, an indication that an intravesicular pH (OH) activation site was present (unpubl. observ.). We predict that in rat colonocytes pH-activated Cl-OH exchange present in AMV together with the Na-HCO_3 cotransport that is located in BLMV play predominant roles in pH_i regulation, while Na-H exchange of both AMV and BLMV are primarily involved in vectorial Na movement. Na-HCO_3 cotransport may also be important in HCO_3 secretion.

2.1.3 K Transport

Initial studies of K transport that had been performed in vivo revealed net K secretion (Edmonds 1967a,b). As the potential difference (PD) is usually between 10 to 25 mV (lumen negative), it was uncertain whether this existing lumen-negative PD was the sole mechanism for the observed net K secretion or whether there was an additional active transport process present. In vivo

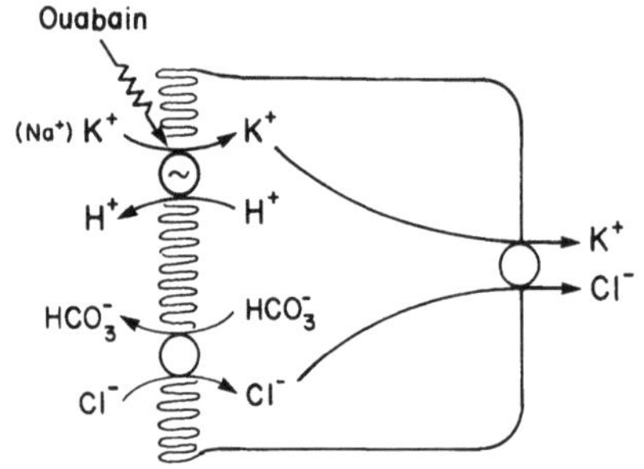

Fig. 9. Cellular model for active K absorption of distal colon. See text for further discussion

and in vitro studies in the rat colon have established that K transport is not uniform throughout the colon but rather that an active K secretory process is present in the proximal colon and an active K absorptive process in the distal colon (Kliger et al. 1981; Foster et al. 1984). Since dietary K loading and aldosterone administration (see below) converts active K absorption to active secretion in the distal colon (as well as stimulating active K secretion in the proximal segment) (Foster et al. 1983; Turnamian and Binder 1990), the distal colon has the capacity to both absorb and secrete K in response to various regulatory signals. It is also likely that an active K secretory mechanism is present in the basal state in the distal colon but that it is less than the basal active K absorptive process.

The active K secretary process is dependent on both serosal Na and Cl, as K secretion is inhibited in Na and/or Cl-free Ringer solution (Foster et al. 1984). Active K secretion is inhibited by mucosal tetraethylammonium (TEA) and barium, K channel blockers, and by serosal ouabain and bumetanide, inhibitors of the Na pump and Na-K-2Cl cotransport, respectively (Sweiry and Binder 1989). These observations indicated that a mucosal TEA and barium-sensitive K channel, and a basolateral Na pump and Na-K-2Cl cotransport process are essential for K secretion.

In the distal colon of normal rat active K absorption is present, as already noted. It is likely that in the basal state there is a small active K secretory component present as experimental maneuvers that affect K secretion (e.g., serosal ouabain, serosal bumetanide, mucosal TEA, or mucosal barium) reduce serosal-to-mucosal K fluxes without affecting mucosal-to-serosal K movement. This K absorptive process is electroneutral, Na-independent and amiloride-insensitive (Foster et al. 1984), and is consistent with a K-H exchange. These findings are very similar to studies of K-dependent mucosal acidification in the guinea pig distal colon which Suzuki and Kaneko (1984) also concluded represented an apical K-H exchange. Additional observations of active K absorption indicate that K absorption is inhibited by orthovanadate and by mucosal ouabain and is partially Cl-dependent and led to the hypothesis that active K absorption was energized by an apical K-activated ATPase and coupled to an apical Cl-HCO_3

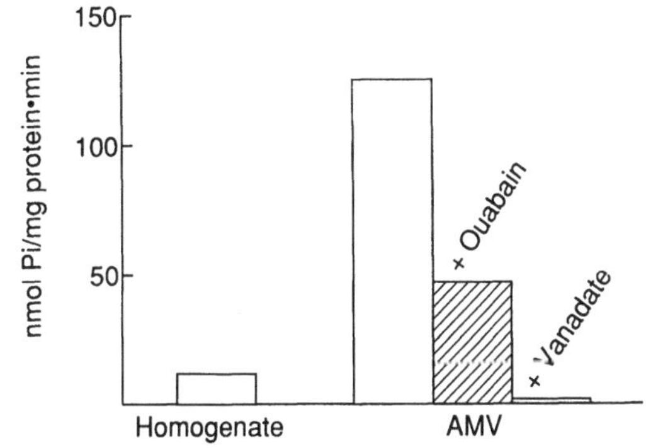

Fig. 10. K-activated ATPase in homogenate and apical membranes of distal colon. K-activated ATPase activity in isolated apical membranes is enriched approximately tenfold over that in the homogenate. K-activated ATPase is inhibited 60 and 94% by ouabain (1 mM) and vanadate (100 μM), respectively. (After data in del Castillo et al. 1991)

exchange and/or a basolateral K-Cl cotransport mechanism (Sweiry and Binder 1990; Fig. 9).

This conclusion is further confirmed by our recent demonstration of the presence of K-activated ATP-hydrolase (K-ATPase) in apical membranes of distal colon (del Castillo et al. 1991). K-ATPase activity is partially ouabain-sensitive and is not present in the basolateral membrane of distal colon (Fig. 10). Its activity in apical membranes of proximal colon is approximately 20% of that in distal apical membrane (see Fig. 20B). It should be remembered that active K absorption is absent in proximal colon.

2.1.4 Short Chain Fatty Acid Transport

Short chain fatty acids (SCFA) are derived from nonabsorbed carbohydrate by enteric bacteria and are the major stool anions in mammalian colon. The primary SCFA are butyrate, propionate, and acetate. Initially, SCFA were thought to induce Na secretion and were responsible for diarrhea in patients with carbohydrate digestive disorder (i.e., lactose intolerance). However, recent studies indicate that SCFA are well absorbed by mammalian colon and probably represent an adaptive mechanism for conservation of carbohydrate, fluid, and electrolytes in patients with impaired small intestinal carbohydrate absorption.

The mechanism of SCFA absorption in rat colon has not been established. Nonionic diffusion and an anion exchange (i.e., SCFA-HCO_3 exchange) process have been suggested as mechanisms to explain SCFA absorption (Powell 1986). Nonsaturable SCFA transport and increasing SCFA absorption with reduction in luminal pH have been described in in

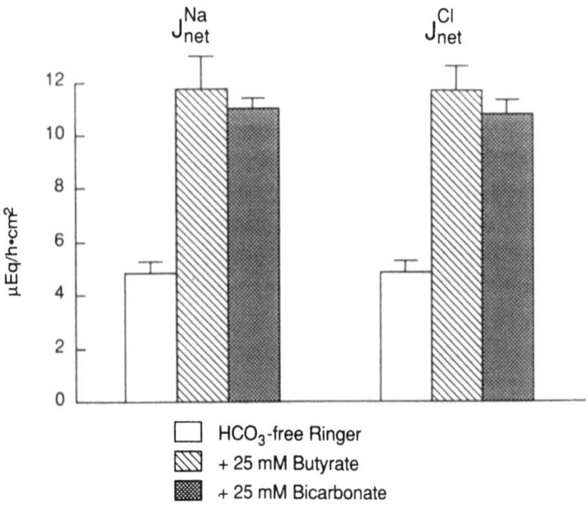

Fig. 11. Role of HCO_3 and butyrate on electroneutral Na and Cl absorption in distal colon. Both 25 mM mucosal butyrate and bicarbonate stimulate net Na and net Cl absorption. (After data in Binder and Mehta.1989)

vivo perfusion studies and are compatible with nonionic diffusion as an important mechanism of SCFA transport (Ruppin et al. 1980).

Although SCFA enhance fluid and electrolyte absorption in vivo, the mechanism of this stimulation has also not been identified. Under voltage-clamp conditions butyrate stimulates electroneutral Na-Cl absorption in rat distal colon (Binder and Mehta 1989; Fig. 11). The characteristics of this process suggested involvement of nonionic diffusion of butyrate, which at a relatively higher intracellular pH, dissociates to H and butyrate anion which may, in part, back diffuse across the apical membrane via two electroneutral (i.e., Na-H and Cl-butyrate) exchange processes resulting in stimulation of both Na and Cl absorption.

This model is attractive to explain the enhancement of Na-Cl absorption by SCFA and is similar to studies in gallbladder and renal proximal tubule (Heintze et al. 1981; Karniski and Aronson 1985). Unfortunately, studies with AMV have not confirmed the presence of a Cl-butyrate exchange (Mascolo et al. 1991). Such studies of butyrate uptake in isolated AMV have demonstrated that neither inward Na, outward pH, nor outward Cl gradients stimulate butyrate uptake, indicating the absence of Na-butyrate cotransport and butyrate-Cl(OH) exchange processes. However, an outward HCO_3 gradient significantly stimulated butyrate uptake which was further enhanced three- to fourfold by the additional presence of a pH gradient ($pH_i > pH_o$) (Fig. 12). HCO_3 gradient-stimulated butyrate uptake is a saturable process suggesting the presence of an anion (butyrate-HCO_3) exchange

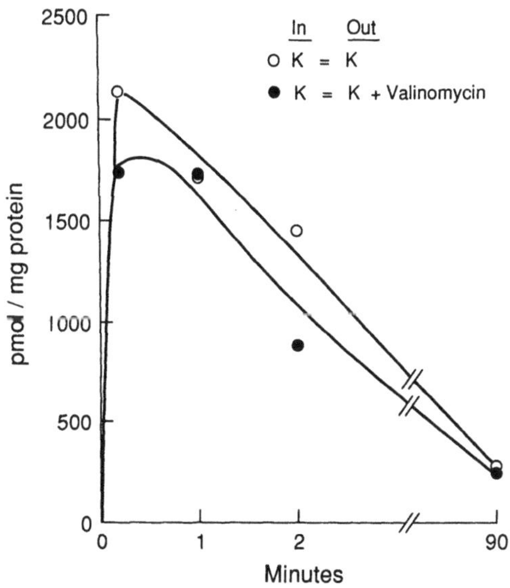

Fig. 12. Butyrate uptake in apical membrane vesicles of distal colon. An outward-directed HCO_3 gradient in the presence of pH gradient (pH_i/pH_o = 7.5/6.0) significantly stimulates butyrate uptake. This HCO_3 gradient-stimulated butyrate uptake is not altered by voltage clamping. (After data in Mascolo et al. 1991)

process. Butyrate-HCO_3 exchange appears distinct from the DIDS-sensitive Cl-OH exchange which is also present in apical membranes as butyrate-HCO_3 exchange is insensitive to DIDS. Thus, there appears to be at least two distinct anion exchange mechanisms in AMV of rat distal colon.

In contrast to the few studies that describe SCFA transport across the apical membrane of rat colon, studies of its exit mechanism across the basolateral membrane have been lacking. Therefore, we examined the transport process for butyrate across BLMV isolated from rat colon. Our studies identified an active butyrate-HCO_3 exchange process in BLMV (Reynolds et al. 1991). HCO_3 gradient-stimulated butyrate uptake of BLMV differs from the AMV butyrate-HCO_3 exchange process in the following aspects: (1) the BLMV exchange process is sensitive to DIDS and (2) it saturates with a higher affinity (Km of 6.5 mM) for butyrate.

Based on the observations obtained with AMV and BLMV studies, we have proposed a cellular model in which the transcellular transport of butyrate occurs as a result of two separate active anion exchange processes that are localized on both apical and basolateral membranes (Fig. 13). Since nonionic diffusion has been observed in both in vitro studies with intact tissue and in vivo studies and since pH enhancement of butyrate uptake is limited in AMV, we propose that SCFA movement occurs as a result of nonionic diffusion via a tight junction and not via a transcellular route.

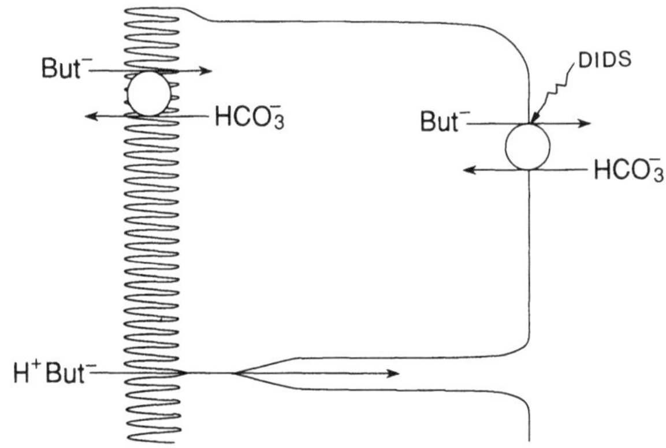

Fig. 13. Transcellular model of butyrate transport in distal colon. Transcellular butyrate transport occurs as a result of apical and basolateral localized DIDS-insensitive and DIDS-sensitive carrier-mediated butyrate HCO_3 exchange processes, respectively. In contrast, diffusion of nonionic protonated butyrate occurs via a paracellular (tight junction) pathway

3 Role of Steroid Hormones

The colon is a corticosteroid-responsive epithelium, although initially a role for glucocorticoids in the regulation of colonic ion transport was not accepted. Edmonds demonstrated in in vivo studies of the entire colon that dietary Na depletion (i.e., secondary hyperaldosteronism) resulted in increased Na absorption and K secretion (Edmonds 1967a,b). Studies during the past decade have provided increased understanding of the effect of both mineralocorticoids and glucocorticoids on ion transport in the rat colon. Further, these recent studies emphasize that aldosterone has distinctly different effects on ion transport in proximal and distal segments of the colon.

The in vivo studies of Edmonds (1981) and in vitro experiments of Will et al. (1980) established that aldosterone (and other corticosteroids) induced amiloride-sensitive Na absorption but did not provide a detailed description of the effect of steroids on basal electroneutral Na-Cl absorption. Aldosterone has several effects on Na and Cl transport in the rat distal colon as examined by both in vitro voltage-clamp and AMV uptake studies (Foster et al. 1983; Rajendran et al. 1989). Not only does aldosterone induce electrogenic Na absorption, but it also inhibits electroneutral Na-Cl absorption (Fig. 14). Thus, in normal animals Na absorption is Cl-dependent, amiloride-insensitive, and is inhibited by cAMP; in contrast, Na absorption in aldosterone animals is Cl-dependent, amiloride-sensitive, and is not inhibited by cAMP (cAMP inhibits electroneutral Na absorption but not electrogenic Na absorption). These phenomena have also been confirmed by our recent studies with AMV isolated from aldosterone animals

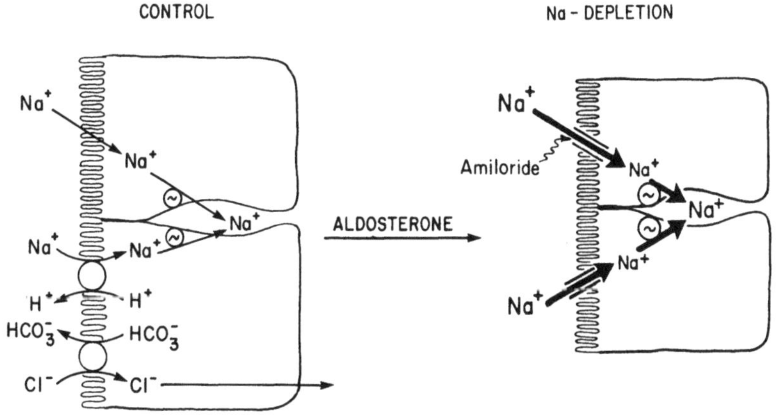

Fig. 14. Aldosterone alteration of Na and Cl transport in distal colon. Aldosterone induces Na transport that is electrogenic, amiloride-sensitive, and Cl-independent, and inhibits electroneutral Na and Cl absorption

(Rajendran et al. 1989). In AMV isolated from aldosterone animals both an outward proton gradient and an intravesicular membrane negative potential stimulated Na uptake which is inhibited by 10 μM amiloride but is insensitive to EIPA, an amiloride analog that is a specific inhibitor of Na-H exchange. These observations suggest that Na uptake stimulated by a proton gradient occurs as a result of an intravesicular membrane negative potential that was created by a proton diffusion (Fig. 15). Ten μM amiloride-sensitive and EIPA-insensitive Na uptake is absent in AMV of normal rat distal colon. These results indicate that aldosterone induces an electrogenic Na absorption and inhibits electroneutral Na-H exchange.

The mechanism by which aldosterone inhibits electroneutral Na-H exchange is not known. However, there are distinctly different time courses for aldosterone alteration of these two Na transport processes: amiloride-sensitive Na uptake is observed within 4 of aldosterone administration and is maximal by 24–36 h (Halevy et al. 1986). In contrast, inhibition of electroneutral Na-Cl absorption is not evident for approximately 48 h and requires more than 72 h for maximal effect. Since crypt to surface cell migration in the colon requires 48 to 72 h, such observations suggest that aldosterone affects crypt but not surface cells, and that its regulatory message is not manifest until the crypt cells migrate and become surface epithelial cells.

Although aldosterone has specific effects on Na transport in the colon, there is now abundant evidence that glucocorticoids also alter Na transport and that glucocorticoid-mediated changes are distinct from those of aldosterone. High doses of dexamethasone and methylprednisolone, corticosteroids that have been generally considered glucocorticoid-specific agonists, produce changes that mimic, in part, the effect of aldosterone (Charney et al. 1975; Foster et al. 1983; Halevy et al. 1986). In contrast,

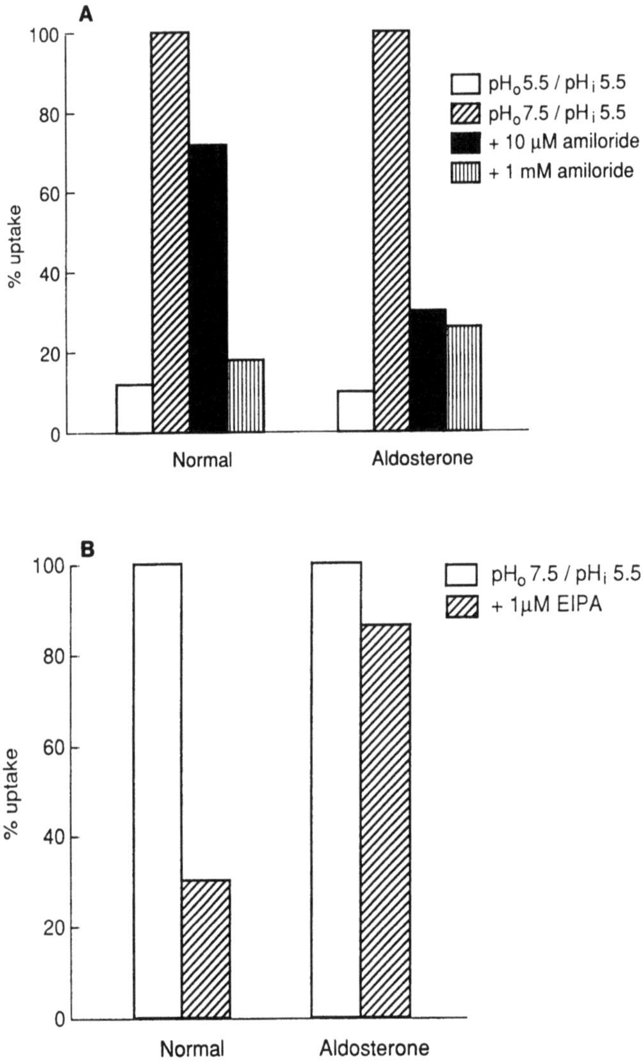

Fig. 15A,B. Effect of amiloride (**A**) and EIPA (**B**) on proton gradient-stimulated Na uptake in apical membrane vesicles (AMV) of normal and aldosterone-treated distal colon. An outward-directed proton gradient stimulates Na uptake in both normal and aldsterone AMV. Proton gadient-stimulated Na uptake of normal AMV is insensitive both to 10 µM amiloride and to EIPA. (Rajendran et al. 1989)

Bastl (1980, 1984; Bastl et al. 1987) proposed that basal Na absorption is exclusively regulated by glucocorticoids, not mineralocorticoids. These observations were further supported by studies with RU-28362, a glucocorticoid receptor-specific agonist, which does not inhibit ^{3}H-aldosterone binding to the aldosterone receptor, and with a low dose of dexamethasone,

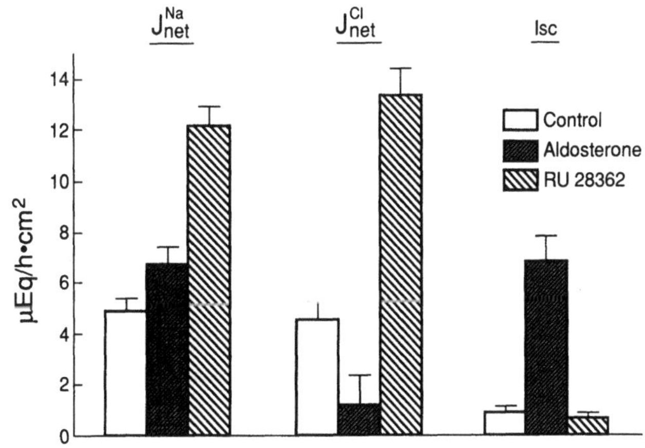

Fig. 16. Role of aldosterone and RU 28362 (a glucocorticoid receptor-specific agonist) on net Na and Cl absorption of distal colon under voltage-clamp conditions. Aldosterone induces Na absorption that is equivalent to the increase in Isc and inhibits Cl absorption. However, RU 28362 stimulates electroneutral Na absorption that is equivalent to Cl absorption. (After data in Turnamian and Binder 1989)

which corrected the effect of adrenalectomy. Other in vitro studies have established that RU-28362 enhanced electroneutral Na-Cl absorption, but did not stimulate electrogenic Na absorption or inhibit electroneutral Na-Cl absorption (Turnamian and Binder 1989; Fig. 16). As a consequence, we believe that pure glucocorticoids activate the glucocorticoid receptor which stimulates electroneutral Na-Cl absorption (not electrogenic Na absorption). Further, dexamethasone at higher concentrations has significant effects on electrogenic Na absorption via the mineralocorticoid receptor. Thus, the two corticosteroid receptors mediate two distinctly different transport systems: the mineralocorticoid receptor regulates electrogenic Na absorption, while the glucocorticoid receptor enhances electroneutral Na-Cl absorption (Fig. 17).

Aldosterone stimulates net K secretion in in vivo studies (Edmonds 1967a,b). As this enhancement is associated with an increase in lumen-negative PD, it was initially uncertain whether this change represented stimulation of active K secretion or only potential-dependent K secretion. As noted above, active K absorption is present in the distal colon. Studies under voltage-clamp conditions have established that aldosterone induces active K secretion (i.e., converts net K absorption to net K secretion) (Foster et al. 1983; Sweiry and Binder 1989; Fig. 18A). The failure of RU-28362 to induce K secretion provides additional support for separate mineralocorticoid and glucocorticoid receptor-mediated mechanisms (Turnamian and Binder 1989).

A model of aldosterone-induced K secretion has evolved from recent in vitro studies (Fig. 19). Uptake of K across the basolateral membrane is

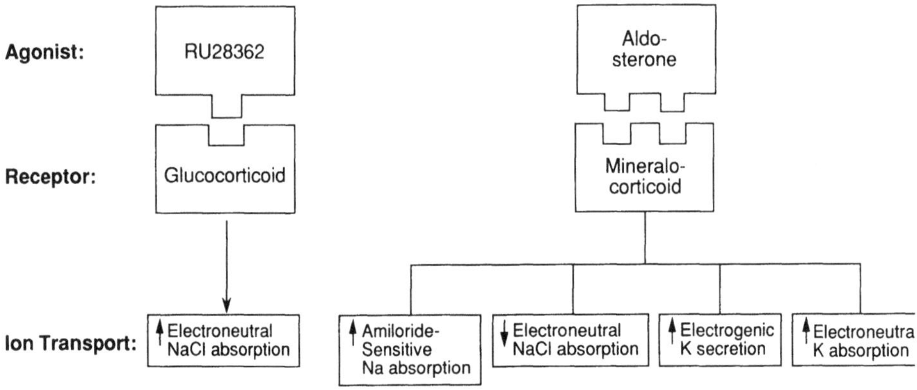

Fig. 17. Model of aldosterone and RU 28362 action on electrolyte transport in distal colon. Aldosterone and RU 28362 mediate electrolyte transport by binding to specific mineralocorticoid and glucocorticoid receptors, respectively. Mineralocorticoid-specific receptor activation induces electrogenic Na absorption, electrogenic K secretion, and electroneutral K asorption, as well as inhibiting electroneutral Na-Cl absorption. Glucocorticoid-specific receptor activation stimulates electroneutral NaCl absorption. Model is based on data in Turnamian and Binder (1989)

primarily the result of Na-K-2Cl cotransport rather than the Na,K-pump as luminal amiloride alters K secretion only when basolateral Na-K-2Cl cotransport process is inhibited. K exit across the apical membrane is a result of its movement via TEA-sensitive K channels. It is likely that aldosterone induces distinct apical K channels as TEA completely abolishes K secretion in both normal and aldosterone animals, while barium abolishes K secretion in normal but only modestly in aldosterone rats. More important is the observation that inhibition of the K secretory process not only abolishes active K secretion but also unmasks a rate of K absorption that is considerably greater than that present in normal amimals (Fig. 18A,B). As a result, *aldosterone*, when K secretion is inhibited, *also* markedly *stimulates active K absorption*. Interestingly, RU-28362 does not stimulate active K absorption. Thus, in the distal colon the glucocorticoid receptor has no effect on either K absorption or K secretion.

Studies of active K absorption in normal and aldosterone-treated animals have yielded a model in which active K absorption occurs via an apical K-H exchange that is energized by a K-activated ATPase (Sweiry and Binder 1990; del Castillo et al. 1991). Both active K absorption and K-activated ATPase are inhibited by vanadate and ouabain but not by omeperozole or SCH 28080, inhibitors of gastric parietal cell H,K-ATPase. Aldosterone stimulates both active K absorption (when K secretion is inhibited) and K-activated ATPase activity (Sweiry and Binder 1990; del Castillo et al. 1991). Thus, it is likely that this apical K-activated ATPase resembles, but is not identical to, both Na,K-ATPase and gastric parietal cell H,K-ATPase.

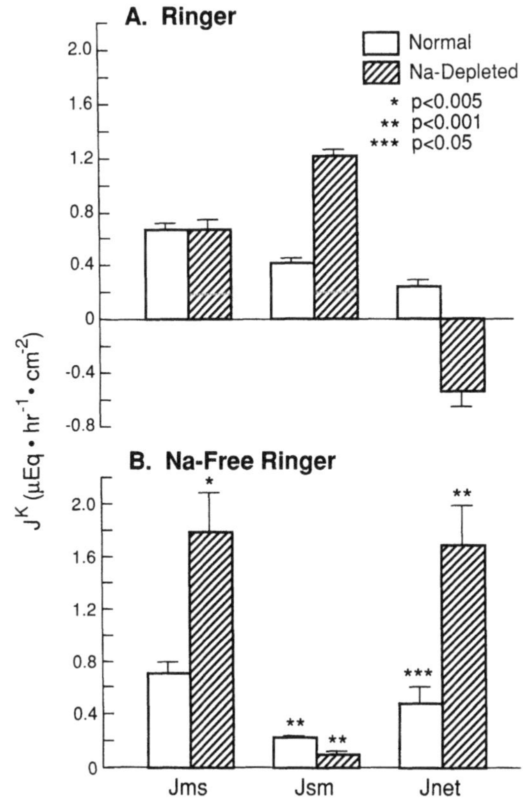

Fig. 18A,B. Aldosterone induction of net K secretion (**A**) and net K absorption (**B**) in distal colon. Aldosterone induces net K secretion in Ringer solution by stimulating serosal to mucosal K movement. In Na-free Ringer solution net K absorption is induced by stimulating mucosal to serosal K movement. (Sweiry and Binder 1990)

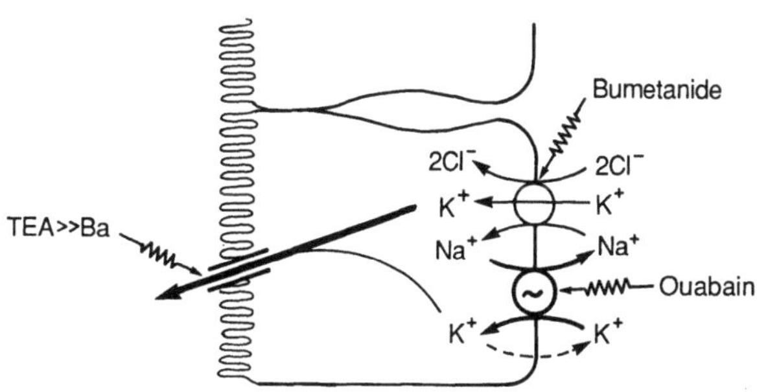

Fig. 19. Cellular model for active K secretion of distal colon. See text for further disussion

4 Proximal Colon

As previously outlined, electrolyte transport in the proximal colon differs significantly from that in the distal colon. Compared to the distal colon, there are three major differences in electrolyte transport in the proximal colon: (1) net Cl absorption is absent; (2) net K secretion, not K absorption is present; and (3) aldosterone enhances electroneutral Na-Cl absorption but does not induce electrogenic Na absorption (Foster et al. 1986; Turnamian and Binder 1990). As a consequence, it is necessary to study and describe the electrolyte transport processes of proximal and distal colon separately.

4.1 Na-Cl Absorption

In the basal state Na absorption in the proximal colon is indistinguishable from that of the distal colon: net Na absorption is substantially greater than Isc, is Cl-dependent, and is not inhibited by 10 μM amiloride but is inhibited by 1 mM amiloride and by cAMP. The presence of electroneutral Na-H exchange has been confirmed by studies with AMV from proximal colon.

The mechanism of Cl movement in the proximal also differs from the distal colon. Net Cl absorption is zero and in in vitro studies this absence of net Cl movement does not appear to be the result of equal rates of opposing absorptive and secretory processes (Foster et al. 1986). Despite the absence of net Cl absorption, Na absorption is completely Cl-dependent. Although the exact nature of Cl movement in the proximal colon is uncertain, it appears to be most consistent with exchange diffusion. Despite the absence of net Cl absorption, cAMP induces active Cl secretion in the proximal colon (Foster et al. 1986). It is not known whether this absence of active Cl absorption is due to an absence of the anion exchange protein in the apical membrane of proximal colon or to changes in its regulatory control.

4.2 K Transport

Overall K movement in the basal state of the in vivo proximal colon is secretory in direction (Hirsch et al. 1985). Although the resting PD in the proximal colon is considerably less than that in the distal colon (5 mV vs. 15 to 25 mV), when K transport is studied under voltage-clamp conditions, active K secretion is always observed (Foster et al. 1984; Fig. 20A). Similar to distal colon, aldosterone increases active K secretion in the proximal colon (Foster et al. 1984; Turnamian and Binder 1990). However, when K secretion is inhibited (e.g., by removal of bath Na), net K secretion is abolished but active K absorption is *not* observed. Thus, it appears that the K absorptive process is not present in the proximal colon either under basal or aldosterone-stimulated conditions. Not surprisingly, K-activated ATPase

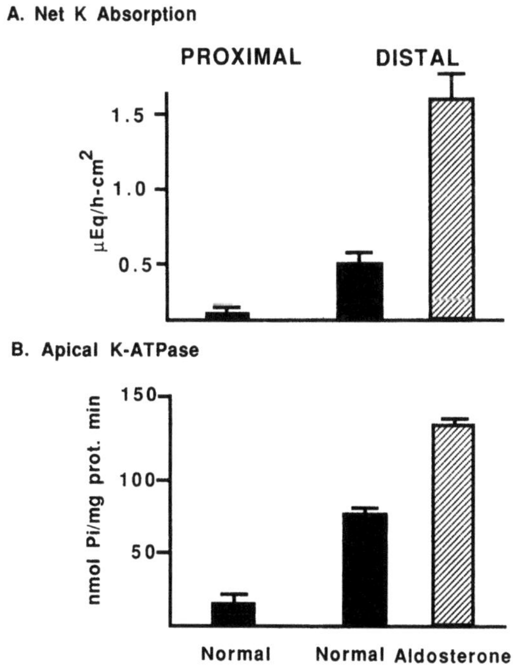

Fig. 20A,B. Relationship between active K absorpton (**A**) and apical K-ATPase (**B**) in proximal and distal colon. K-ATPase activity is markedly reduced in proximal colon compared to net K absorption, while that in distal colon is absent. Aldosterone stimulates both active K absorption (in the absence of Na) and K-ATPase activity in distal colon. (After data in Foster et al. 1984; Sweiry and Binder 1990 and unpubl. observ.)

in apical membrane of rat proximal colon is approximately 20% of that in distal colon (del Castillo et al. 1991; Fig. 20B).

4.3 Role of Steroid Hormones

The response of the proximal colonic mucosa to corticosteroids represents another major difference in the transport processes in these two segments of the rat colon.

Aldosterone enhances Na absorption in the proximal colon, like that in the distal colon. However, analysis of this effect reveals that the response of the proximal colon to aldosterone is unique and is not typical of the changes in Na transport usually ascribed to aldosterone. Aldosterone enhances *electroneutral* Na-Cl absorption without any evidence of induction of amiloride-sensitive Na channels (Foster et al. 1986; Turnamian and Binder 1990; Fig. 2). Not only is net Na absorption stimulated, but net Cl absorption is also induced, and limited studies indicate that aldosterone-induced electroneutral Na-Cl absorption represents the coupling of Na-H and Cl-HCO_3 exchanges.

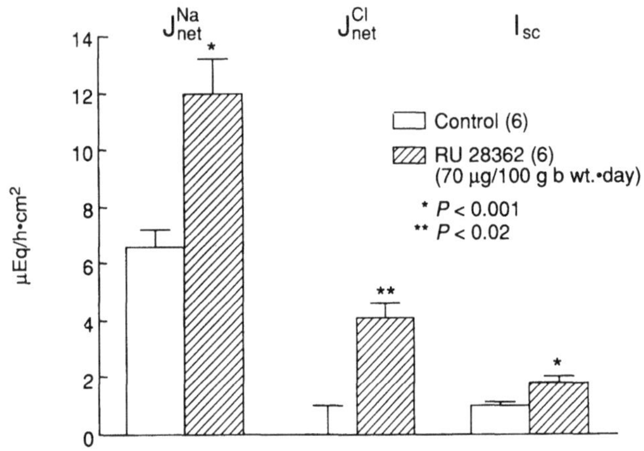

Fig. 21. Role of RU 28362 on net Na and Cl absorption in proximal colon. RU 28362, a glucorticoid receptor-specific agonist, stimulates Na and Cl absorption without altering Isc findings that are consistent with stimulation of electroneutral Na-Cl absorption. (After data in Turnamian and Binder 1990)

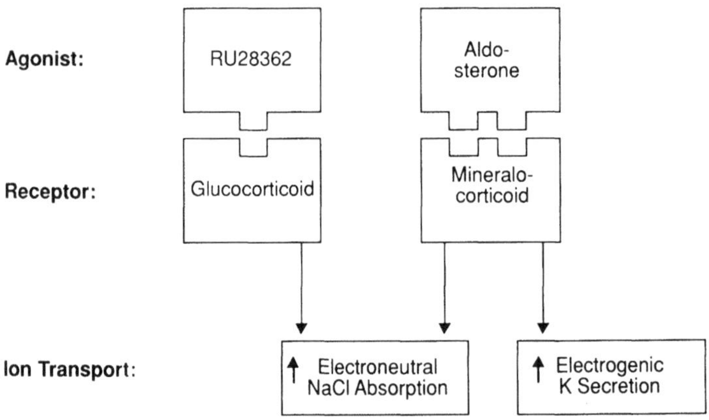

Fig. 22. Model of aldosterone and RU 28362-mediated electrolyte transport in proximal colon. Similar to distal colon, both aldosterone and RU 28362 mediate electrolyte transport by binding to mineralocorticoid and glucocorticoid receptors, respectively. Mineralocorticoid-specific receptor activation stimulates electroneutral NaCl absorption and electrogenic K secretion. Glucocorticoid-specific receptor activation stimulates electroneutral NaCl absorption by a mineralocorticoid receptor-independent mechanism. Model is based on data in Turnamian and Binder. (1990)

Glucocorticoids also alter Na and Cl transport in proximal colon in that RU-28362, the glucocorticoid receptor-specific agonist, also enhances electroneutral Na-Cl absorption in a manner very similar to that of aldosterone (Turnamian and Binder 1990; Fig. 21). However, aldosterone-induced

Na-Cl absorption is prevented by spironolactone, an inhibitor of the mineralocorticoid receptor. Since spironolactone does not alter the increase in Na-Cl absorption produced by RU-28362, it is possible that a single postreceptor mechanism is responsible for the regulation of electroneutral Na-Cl absorption but that this regulatory process is mediated by two distinct receptor-mediated processes (Fig. 22). Additional support for this concept is provided by the observations that aldosterone, but not RU-28362, stimulates active K secretion in the proximal colon.

Acknowledgments. The experimental studies performed in the authors' laboratory described in this chapter were supported by USPHS research grants (DK 14669 and DK 18777) from the National Institute of Diabetes, Digestive and Kidney Diseases. Ms. Irene Pollard provided expert secretarial assistance in the preparation of this chapter.

References

Aronson PS, Nee J, Suhm MA (1982) Modifier role of internal H^+ in activating the Na^+-H^+ exchanger in renal microvillus membrane vesicles. Nature 299: 161–163

Bastl C (1987) Regulation of cation transport by low doses of glucocorticoids in in vivo adrenalectomized rat colon. J Clin Invest 80: 348–356

Bastl CP, Binder HJ, Hayslett JP (1980) Role of glucocorticoids and aldosterone in maintenance of colonic cation transport. Am J Physiol 238: F181–F186

Bastl CP, Barnett CA, Schmidt TJ, Litwack G (1984) Glucocorticoid stimulation of sodium absorption in colon epithelia is mediated by corticosteroid IB receptor. J Biol Chem 259: 1186–1195

Binder HJ, Mehta P (1989) Short-chain fatty acids stimulate active Na and Cl absorption in vitro in the rat distal colon. Gastroenterology 96: 989–996

Binder HJ, Foster E, Budinger ME, Hayslett JP (1987) Mechanism of electroneutral sodium-chloride absorption in distal colon of the rat. Gastroenterology 93: 449–455

Charney AN, Kinsey MD, Myers L, Giannella RA, Gots RE (1975) Na^+-K^+-activated adenosine triphosphatase and intestinal electrolyte transport. Effect of adrenal steroids. J Clin Invest 56: 653–660

Curran PF, Schwartz GF (1960) Na, Cl and water transport by rat colon. J Gen Physiol 43: 555–571

del Castillo JR, Rajendran VM, Binder HJ (1991) Apical membrane localization of ouabain-sensitive K^+-activated ATPase activities in rat distal colon. Am J Physiol 261: G1005–G1011

Edmonds CJ (1967a) Transport of potassium by the colon of normal and sodium-depleted rats. J Physiol (Lond) 193: 603–617

Edmonds CJ (1967b) Transport of sodium and secretion of potassium and bicarbonate by the colon of normal and sodium depleted rats. J Physiol (Lond) 193: 589–602

Edmonds CJ (1981) Amiloride sensitivity of the transepithelial electrical potential and of sodium and potassium transport in rat distal colon in vivo. J Physiol (Lond) 313: 547–549

Foster ES, Zimmerman TW, Hayslett JP, Binder HJ (1983) Corticosteroid alteration of active electrolyte transport in rat distal colon. Am J Physiol 245: G668–G675

Foster ES, Hayslett JP, Binder HJ (1984) Mechanism of active potassium absorption and secretion in the rat colon. Am J Physiol 246: G611–G617

Foster ES, Budinger ME, Hayslett JP, Binder HJ (1986) Ion transport in proximal colon of the rat: sodium depletion stimulates neutral sodium chloride absorption. J Clin Invest 77: 228–235

Frizzell RA, Koch MJ, Schultz SG (1976) Ion transport by rabbit colon. I. Active and passive components. J Membr Biol 27: 297–316
Goldfarb DS, Egnor RW, Charney AN (1988) Effects of acid-base variables on ion transport in rat colon. J Clin Invest 81: 1903–1910
Grinstein S, Dixon SJ (1989) Ion transport, membrane potential, and cytoplasmic pH in lymphocytes: changes during activation. Physiol Rev 69: 417–471
Halevy J, Budinger ME, Hayslett JP, Binder HJ (1986) The role of aldosterone in the regulation of sodium and chloride transport in the distal colon of sodium-depleted animals. Gastroenterology 91: 1227–1233
Heintze K, Petersen KU, Wood JR (1981) Effects of bicarbonate on fluid and electrolyte transport by guinea pig and rabbit gallbladder: stimulation of absorption. J Membr Biol 62: 175–181
Hirsch D, Pace P, Binder HJ, Hayslett JP (1985) Evidence that aldosterone influences transport-target tissues by dissimilar mechanisms. Am J Physiol 248: F507–F512
Karniski LP, Aronson PS (1985) Chloride/formate exchange with formic acid recycling: a mechanism of active chloride transport across epithelial membranes. Proc Natl Acad Sci USA 82: 6362–6365
Kliger A, Binder HJ, Bastl C, Hayslett JP (1981) Demonstration of active potassium transport in the mammalian colon. J Clin Invest 67: 1189–1196
Mascolo N, Rajendran VM, Binder HJ (1991) Mechanism of short chain fatty acid uptake by apical membrane vesicles of rat distal colon. Gastroenterology 101: 331–338
Murer H, Kinne R (1980) The use of isolated membrane vesicles to study epithelial transport processes. J Membr Biol 55: 81–95
Powell DW (1986) Ion and water transport in the intestine. In: Andreoli TE (ed) Physiology of membrane disorders 2nd edn. Plenum Medical, New York, pp 559-596
Powell DW, Malawer SJ, Plotkin GR (1968) Secretion of electrolytes and water by the guinea pig small intestine in vivo. Am J Physiol 215: 1226–1233
Rajendran VM, Binder HJ (1990) Characterization of Na-H exchange in apical membrane vesicles of rat colon. J Biol Chem 265: 8408–8414
Rajendran VM, Kashgarian M, Binder HJ (1989) Aldosterone induction of electrogenic sodium transport in the apical membrane vesicles of rat distal colon. J Biol Chem 264: 18638–18644
Rajendran VM, Oesterlin M, Binder HJ (1991a) pH-regulated Cl-uptake in basolateral membrane vesicles of rat distal colon. Gastroenterology 100: A700 (Abstr).
Rajendran VM, Oesterlin M, Binder HJ (1991b) Sodium uptake across basolateral membrane of rat distal colon: evidence for Na-H exchange and Na-anion cotransport. J Clin Invest 88: 1379–1385
Reynolds D, Rajendran VM, Binder HJ (1991) Butyrate transport in rat distal colon basolateral membrane vesicles: evidence for bicarbonate/butyrate exchange. Gastroenterology 100: A838
Ruppin H, Bar-Meir S, Soergel K, Wood CM, Schmitt MG Jr (1980) Absorption of short-chain fatty acids by the colon. Gastroenterology 78: 1500–1507
Sellin J, DeSoignie R (1984) Rabbit proximal colon: a distinct transport epithelium. Am J Physiol 246: G603–G610
Stieger B, Marxer A, Hauri HP (1986) Isolation of brush-border membranes from rat and rabbit colonocytes: is alkaline phosphatase a marker enzyme? J Membr Biol 91: 19–31
Suzuki Y, Kaneko K (1989) Acid secretion in isolated guinea pig colon. Am J Physiol 256: G979–G988
Sweiry J, Binder HJ (1989) Characterization of aldosterone-induced potassium secretion in rat distal colon. J Clin Invest 83: 844–851
Sweiry J, Binder HJ (1990) Active potassium absorption in rat distal colon. J Physiol (Lond) 423: 155–170
Turnamian SG, Binder HJ (1989) Regulation of active sodium and potassium transport in the distal colon of the rat: role of the aldosterone and glucocorticoid receptors. J Clin Invest 84: 1924–1929

Turnamian SG, Binder HJ (1990) Aldosterone and glucocorticoid receptor-specific agonists regulate ion transport in rat proximal colon. Am J Physiol 258: G492–G408
Will PC, Lebovitz FI, Hopfer U (1980) Induction of amiloride-sensitive sodium transport in the rat colon by mineralocorticoids. Am J Physiol 238: F261–F268

Chapter 7

Absorption and Secretion of Electrolytes and Short-Chain Fatty Acids in the Guinea Pig Large Intestine

Gerhard Rechkemmer and *Wolfgang von Engelhardt*[1]

Contents

1 Introduction

The guinea pig (*Cavia porcellus*) is a herbivorous, monogastric mammal. Taxonomically, it is classified as a New World hystricomorph rodent. However, recent phylogenetic analysis suggests that the Caviomorpha should be elevated in taxonomic rank and regarded as a separate mammalian order distinct from the Rodentia (Graur et al. 1991).

[1]Physiologisches Institut, Tierärztliche Hochschule, Bischofsholer Damm 15, W-3000 Hannover, Germany

Advances in Comparative and Environmental Physiology, Vol. 16

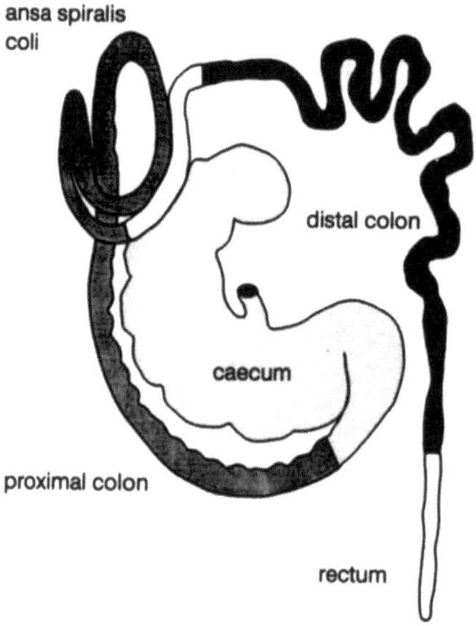

Fig. 1. Anatomy and terminology of the guinea pig large intestine

1.1 Anatomy

Anatomically, three segments of the guinea pig large intestine can be distinguished: the caecum, the colon, and the rectum (Fig. 1).

The caecum is the largest dilatation of the alimentary canal. It is a large, thin-walled, 15–20 cm long semi-circular sac with three taenia and many haustra (Cooper and Schiller 1975; Snipes 1982).

The colon is about 80–90 cm long and may be arbitrarily subdivided into two segments, the proximal and distal colon. The proximal colon is approximately 20 cm long and starts at the caeco-colonic junction and includes a tightly wound coil (ansa spiralis coli). Up to the coil two rows of taenia are present on the mesenterial side and small haustra are formed between taeniae. The contents have a paste-like appearance, after passage through the ansa spiralis coli contents are formed into ellipsoid-shaped fecal pellets. The section of the colon containing fecal pellets may be designated as the distal colon.

1.2 Histology

Mucosal folds and protrusions into the lumen provide some surface enlargement in the guinea pig large intestinal epithelium (Snipes 1982). The number and appearance of crypts vary between segments. In the caecum crypts are short and few in number, in the proximal colon crypts have a funnel-like transition region between the surface and the crypt epithelium, designated as foveola, and crypts often show basal bifurcations. In the distal colon, crypts are longer than in the more proximal segments and the number of crypts per macroscopic surface area is higher (Luciano et al. 1984). Epithelial cells on the surface and in the crypts are joined by tight junctions (zonulae occludentes). The morphological characteristics of tight junctions between epithelial cells have been extensively studied with freeze-fracture methods in the guinea pig caecal epithelium (Mora-Galindo 1986) and in the proximal and distal colonic epithelium (Luciano et al. 1984). Generally, the height, the number of strands and the amount of interconnections between strands are higher between surface epithelial cells than between crypt cells, probably indicating a higher paracellular permeability in the crypt as compared to the surface epithelium (Claude and Goodenough 1973; Marcial et al. 1984). Another interesting observation is the difference in the number of strands between surface epithelial cells in the different segments. In the caecum and distal colon the number of strands is similar (6.9 ± 1.6, Mora-Galindo 1986 and 7.4 ± 0.2, Luciano et al. 1984, respectively), whereas in the proximal colon much less strands (4.4 ± 0.5, Luciano et al. 1984) have been observed. The lipid composition, i.e., lipid fluidity, and particularly the cholesterol content of epithelial cell membranes have been implicated in the modulation of transport properties of cells. Upon treatment of biomembranes with the polyene antibiotic filipin the formation of large complexes between cholesterol and filipin is induced. These aggregates can be observed with freeze-fracture scanning electron microscopy. A higher number of filipin-cholesterol complexes was shown in the apical membrane of surface epithelial cells in the proximal colon as compared to the distal colon, indicating a greater lipid fluidity in membranes of the distal colon (Luciano et al. 1989). Histochemical studies have demonstrated regional differences in lectin binding in the proximal and distal colon and between surface and crypt epithelium goblet cells (Gorelick et al. 1982).

2 Inorganic Electrolyte Transport

Transport of inorganic ions across epithelia is accomplished by two different pathways, transcellular and paracellular. The primary focus of this chapter will be on the transcellular processes involved in transepithelial transport of ions. The vectorial nature of ion transport, the hallmark of epithelia, is caused by the presence of specific transport systems in the apical and basolateral membranes of differentiated epithelial cells. In contrast to other laboratory animal species (e.g., rat and rabbit), only few data on the electrical characteristics of the different segments of the guinea pig large intestine have been published (Table 1). Particularly data for the caecum and the proximal colon are sparse. Electrical parameters measured in the guinea pig distal colon showed a high variability. Under control conditions most tissues (80–90%) had a negative I_{sc}, indicative of either electrogenic cation

Table 1. Electrophysiological characteristics of different segments of the guinea pig large intestine

Tissue	Ψ_t (mV)	I_{sc} (μA/cm²)	G_t (mS/cm²)	References
Caecum	10.2 ± 1.9	58 ± 22	nd	Ussing and Andersen (1956)
	5.3 ± 0.9	113 ± 35	nd	Watkins and Pintauro (1979)
	nd	43 ± 6	6.9 ± 1.2	Rechkemmer (1988)
	11.6 ± 0.4	89 ± 5	7.6 ± 0.3	Engelhardt and Rechkemmer (1992)
Proximal colon	3.4 ± 0.1	39 ± 4	12.7 ± 0.8	Kuwahara and Radowicz-Cooke (1988)
	nd	26 ± 1	9.4 ± 0.2	Rechkemmer et al. (1988b)
	2.3 ± 0.1	28 ± 1	12.9 ± 0.4	Engelhardt and Rechkemmer (1992)
Distal colon	6.3 ± 1.3	40 ± 8	8.6 ± 0.9	Clauss et al. (1985)
	−4.1 ± 2.1	−42 ± 5	12.1 ± 1.9	
	nd	−39 ± 7	12.8 ± 1.3	Ishida and Suzuki (1987)
	nd	−33 ± 17	12 ± 1	Kuwahara et al. (1987b)
	nd	−6 ± 3	7.4 ± 2.2	Yajima et al. (1988)
	nd	17 ± 10	5.6 ± 0.6	Rechkemmer and Halm (1989)
	−4.9 ± 0.5	−69 ± 6	14.3 ± 0.8	Russell and Castro (1989)
	7.6 ± 2.7	95 ± 29	13.5 ± 0.7	
	nd	−32 ± 4	8.9 ± 0.7	Kuwahara and Cooke (1990)
	nd	−15 to −42	4 to 15	Wang et al. (1990)
	10.0 ± 0.7	85 ± 6	9.3 ± 0.4	Engelhardt and Rechkemmer (1992)

Ψ_t with reference to the mucosal side, I_{sc} positive for cation flow from mucosal to serosal side, nd: not determined.

secretion or anion absorption (Clauss et al. 1985). After treatment of tissues with the cyclooxygenase blocker indomethacin, during preparation and subsequent measurements, electrical values were more consistent and I_{sc} was generally positive (Rechkemmer 1988), indicating that a variable degree of endogenous prostaglandin formation, inducing electrogenic transport processes in this segment, was responsible for the inconsistency.

2.1 Sodium Transport

2.1.1 Caecum

Not much information is available so far about the ion transport mechanisms of the guinea pig caecum, although in the first study applying the Ussing-chamber technique to mammalian tissues, the guinea pig caecum had been used (Ussing and Andersen 1956). In this early study ^{22}Na and ^{24}Na unidirectional tracer fluxes were simultaneously measured across the guinea pig caecum and in parallel the I_{sc} was estimated. Net Na transport approximately equalled I_{sc} and thus it was concluded that Na transport was electrogenic and accounted fully for the I_{sc} (Ussing and Andersen 1956). In the 1970s the diuretic drug amiloride was developed and since then has been widely used as a pharmacological tool in the study of electrogenic Na transport. Amiloride very specifically blocks the Na channel in the apical membrane of epithelial cells (Benos 1982). Permeation of Na through the Na channel is the rate-limiting step in transepithelial Na transport and is responsible for the generation of I_{sc} in different tight epithelia. Amiloride in concentrations ($<0.1\,mmol\,l^{-1}$) used to block electrogenic Na transport had no effect on Na transport or I_{sc} in the guinea pig caecum (Rechkemmer 1988), although net Na transport approximately equals I_{sc} (Ussing and Andersen 1956; Rechkemmer 1988; Rechkemmer and Engelhardt, unpubl. data) and I_{sc} varies with luminal Na concentrations (Watkins and Pintauro 1979). At an amiloride concentration of $1\,mmol\,l^{-1}$, however, J_{ms}^{Na} was reduced by 40%, J_{sm}^{Na} was slightly decreased and J_{net}^{Na} was inhibited by approximately 55% in parallel with an inhibition of I_{sc} by a similar magnitude, in addition G_t was also decreased (Rechkemmer and Engelhardt, unpubl. data). Thus, as in the rabbit caecum, it appears that an apical membrane Na channel with a rather low sensitivity to inhibition by amiloride exists in guinea pig caecal epithelial cells.

2.1.2 Proximal Colon

In the guinea pig proximal colon J_{ms}^{Na} by far exceeds J_{sm}^{Na}, resulting in a significant net Na absorption in this segment, and J_{net}^{Na} is much higher than the corresponding I_{sc} (Kuwahara and Radowicz-Cooke 1988; Rechkemmer et al. 1988b). Unidirectional Na transport and I_{sc} are insensitive to amiloride at concentrations up to $0.1\,mmol\,l^{-1}$, indicating that Na transport in this segment is not related to passage of Na through typical amiloride-sensitive Na channels. However, net Na transport can be completely abolished in the proximal colon by higher concentrations of amiloride ($1\,mmol\,l^{-1}$) in the mucosal solution without a significant change in I_{sc} (Rechkemmer 1988). Na and Cl transport is of similar magnitude (Kuwahara and Radowicz-Cooke 1988; Rechkemmer et al. 1988b) and seems to be closely linked. Inhibition of Na transport with amiloride also reduces Cl transport by a similar magnitude (Rechkemmer 1988). These observations indicate that in the guinea pig proximal colon Na transport is most likely accomplished by an electroneutral Na,H-exchange mechanism present in the apical membrane of Na-absorbing cells.

Electrical field stimulation, aminophylline or the addition of prostaglandin E_2 (PGE_2) or forskolin to the serosal side of the guinea pig proximal colon resulted in a marked parallel decrease in J_{ms}^{Na} and J_{ms}^{Cl}, and no significant change in J_{sm}^{Na} or J_{sm}^{Cl}, respectively; moreover, G_t and I_{sc} were increased (Kuwahara and Radowicz-Cooke 1988; Rechkemmer et al. 1988b). Serosal application of bumetanide largely (70–80%) reversed the changes described above, indicating that a basolateral Na,K,2Cl-cotransport system may be involved in the transport of Na across the proximal colonic epithelium. Tetrodotoxin (TTX) almost completely and atropine partially inhibited the changes in G_t and I_{sc} evoked by electrical field stimulation (Kuwahara and Radowicz-Cooke 1988). Carbachol-induced changes in the electrical parameters of the guinea pig proximal colon were very similar to the effects of electrical field stimulation (Kuwahara and Radowicz-Cooke 1988). All these findings signify the importance of submucosal neurons in the regulation of electrolyte transport in the guinea pig proximal colon. The specific neurotransmitter(s) responsible for the atropine-resistant portion of the response to electrical field stimulation is presently unknown (Kuwahara and Radowicz-Cooke 1988). A large number of chemically specified neurons have been demonstrated in recent years in the guinea pig colon (cf. Messenger and Furness 1990).

High plasma aldosterone levels induced by feeding a low-Na diet significantly stimulated Na absorption in the guinea pig proximal colon by about 25% along with a much smaller change in I_{sc} and no difference in G_t (Rechkemmer 1988). Low doses of amiloride under these conditions again had no effect, suggesting that high aldosterone concentrations did not induce electrogenic Na transport in the proximal colon but may have rather stimulated Na transport through increased basolateral Na,K-ATPase activity (Hayslett and Binder 1982).

2.1.3 Distal Colon

Na absorption in the guinea pig distal colon is completely abolished by low concentrations of amiloride concomitant with a reduction in I_{sc} numerically equivalent to the change in Na transport (Clauss et al. 1985; Rechkemmer et al. 1988b; Rechkemmer and Halm 1989). Thus, Na transport across the guinea pig distal colon is electrogenic and occurs through typical amiloride-sensitive epithelial Na channels in the apical membrane of transporting colonocytes. This conclusion is supported by studying the concentration-dependent inhibition of I_{sc} with amiloride (Fig. 2). Half-maximal inhibition of I_{sc} is already achieved at submicromolar amiloride concentrations, which is again a typical characteristic of the epithelial Na channel. Using distal colonic epithelia in which the basolateral membrane potential was deplorized by a high K concentration and then recording an apical membrane current-voltage relationship in the absence or presence of amiloride allowed the calculation of the electromotive force (EMF: $61 \pm 6\,mV$), the intracellular Na concentration ($[Na]_c$: $12 \pm 3\,mmol/l$) and the apical membrane Na permeability (P_{Na}: $0.020 \pm 0.002\,cm/h$) (Clauss et al. 1985). Na transport is highly variable in the guinea pig distal colon and seems to be rigorously regulated by aldosterone (Clauss et al. 1985; Rechkemmer and Halm 1989). Secondary hyperaldosteronism induced by feeding a low-Na diet leads to a marked increase in Na absorption in vivo (Rechkemmer 1988; Sander 1990).

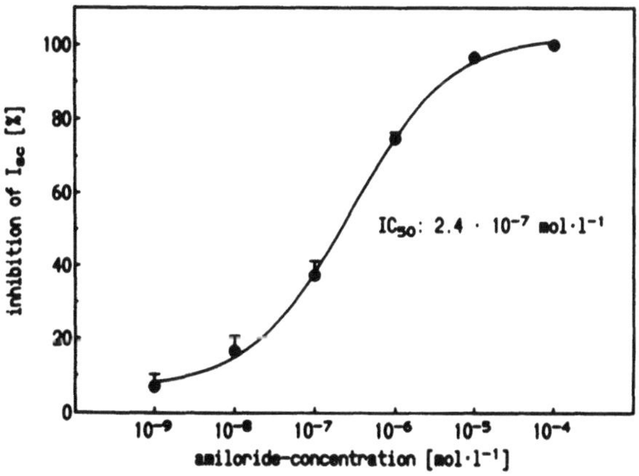

Fig. 2. Amiloride dose-response curve for the inhibition of I_{sc}, indicating electrogenic Na transport in the guinea pig distal colon. Data shown are means ± SEM from 8 tissues. (Rechkemmer unpubl.)

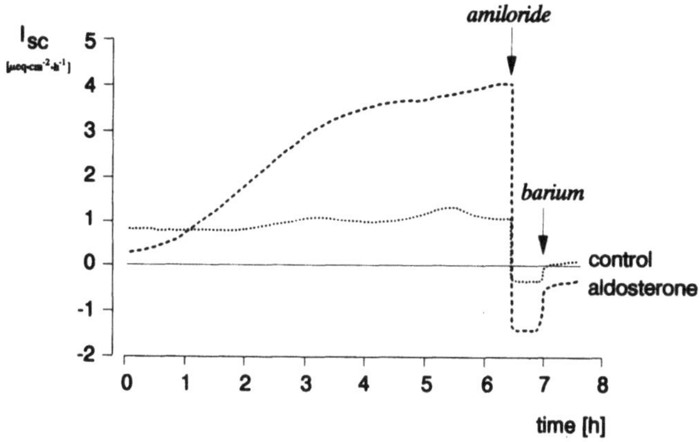

Fig. 3. Representative recordings of I_{sc} from a pair of tissues. Untreated guinea pig distal colon epithelium (control: *dotted line*) and aldosterone-treated epithelium (aldosterone: *dashed line*; aldosterone added at time 0 to both sides of the preparation at a concentration of 10^{-6} mol l^{-1}). Amiloride (10^{-4} mol l^{-1}) and barium (5×10^{-3} mol l^{-1}) where added to the mucosal solution. (After Rechkemmer and Halm 1989)

K-canrenoate, a specific aldosterone antagonist, obliterates the effects on Na transport despite an even larger increase in plasma aldosterone under these conditions (Rechkemmer 1988). Aldosterone applied in vitro to isolated distal colonic epithelia of the guinea pig induced a sustained increase in I_{sc} reaching a new steady state after about 5 h (Fig. 3). The increase in I_{sc} is due to the stimulation of electrogenic Na transport as is demonstrated by the effect of amiloride (Rechkemmer and Halm 1989).

Thus, with respect to Na transport, the guinea pig distal colon appears to be similar to the rabbit distal colon, and different to that of the rat.

2.2 Potassium Transport

2.2.1 Caecum

Transport of potassium across the guinea pig caecum has not been studied in detail. However, from experiments using diets differing in the Na and K content, a linear inverse relationship between Na and K concentrations in caecal contents was estimated, e.g. at a high Na concentration, the K concentration is low and vice versa (Rechkemmer et al. 1987a; Sander 1990; Rechkemmer 1991a). A small net K secretion across the caecal epithelium was observed (Rechkemmer 1988).

2.2.2 Proximal Colon

A small but significant K secretion has been observed under short-circuit current conditions (Rechkemmer 1988; Rechkemmer et al. 1988b). K secretion is stimulated about two- to three-fold by serosal application of epinephrine ($5\,\mu mol\,l^{-1}$) or almost four-fold by serosal PGE_2 ($2\,\mu mol\,l^{-1}$). While PGE_2 also leads to a small but significant increase in I_{sc}, caused by simultaneous stimulation of Cl secretion, epinephrine caused a decrease in I_{sc} consistent with the stimulation of electrogenic K secretion. The change in I_{sc} is numerically equal to the change in J^{K}_{net}. K secretion and the associated changes in electrical parameters elicited by either PGE_2 or epinephrine are abolished by serosal bumetanide, indicating the involvement of a basolateral Na,K,2Cl-cotransport system for transepithelial K movement across the proximal colonic epithelium (Rechkemmer 1988).

Secondary hyperaldosteronism or in vitro aldosterone treatment stimulated K secretion about three-fold; under this condition epinephrine did not further stimulate K secretion (Rechkemmer 1988). At present, the link between the action of aldosterone and epinephrine on electrogenic K secretion is not apparent.

2.2.3 Distal Colon

Potassium transport and its regulation have been intensively studied in the distal colon. Under basal conditions the guinea pig distal colon shows a remarkably high K absorption in contrast to rats (Foster et al. 1984, 1986) and rabbit (Halm and Frizzell 1986). While potassium movement across the rabbit distal colon has until recently been considered as a purely passive paracellular transport, dependent on the establishment of favourable electrochemical gradients by electrogenic Na transport (Fromm and Schultz 1981), it is now firmly established that K transport for the most part occurs by transcellular mechanisms and involves absorptive and secretory cellular pathways (Wills and Biagi 1982; Smith and McCabe 1984; Halm and Frizzell 1986), as has already been suggested for the rat distal colon long ago (Edmonds and Smith 1979).

K Absorption

The measurement of unidirectional fluxes using ^{42}K or ^{86}Rb as radioactive tracers revealed a much higher J^{K}_{ms} than J^{K}_{sm}, resulting in a J^{K}_{net} of approximately $3\,\mu Eq\,cm^{-2}\,h^{-1}$ (Rechkemmer et al. 1986b, 1987b; Suzuki and Kaneko 1989; Watanabe et al. 1990). This rate of K absorption is several times higher than estimated in the rat and rabbit distal colon, respectively (Foster et al. 1984; Smith and McCabe 1984; Halm and Frizzell 1986; Sweiry and Binder 1989). K absorption is electroneutral and independent of the presence of Na in the mucosal or serosal solution. Inhibition of the basolateral Na,K-ATPase by serosal application of ouabain does not interfere with K absorption (Rechkemmer et al. 1987b; Watanabe et al. 1990). K absorption in the distal colon of guinea pig and rat, however, is blocked by the gastric K,H-ATPase inhibitors omeprazole and SCH28080 (Rechkemmer 1988; Sweiry and Binder 1990). The K,H-pump of the rabbit and guinea pig distal colon is inhibited by vanadate, indicating the involvement of a phosphorylated intermediate in the reaction cycle (Kaunitz and Sachs 1986; Watanabe et al. 1990). Inhibition of cellular metabolism by dinitrophenol (DNP) greatly inhibits K absorption pointing to the requirement of ATP (Suzuki and Kaneko 1989). H secretion rates were found to be similar to K absorption rates (Suzuki and Kaneko 1989), indicating similar characteristics to the gastric K,H-ATPase (Suzuki 1983, 1984; Suzuki and Kaneko 1987, 1989). The measurement of ATPase activity demonstrated the presence of an ouabain-sensitive K,H-ATPase in the apical membrane of guinea pig distal colonic epithelium (Watanabe et al. 1990). It was also demonstrated that the K,H-ATPase was linked to the intracellular supply of H ions because inhibition of carbonic anhydrase with acetazolamide reduced ouabain-sensitive Rb influx as well as the H-secretion rate significantly (Suzuki and Kaneko 1989). Although the colonic K,H-ATPase is sensitive to the gastric proton pump inhibitors omeprazole and SCH28080 the inhibition is only about 30–50% and much higher concentrations are required to inhibit the colonic pump compared to the gastric pump (Rechkemmer 1988; Watanabe et al. 1990). Recently, evidence for distinct species differences with regard to K,H-ATPase was provided by the use of monoclonal antibodies raised against the gastric K,H-ATPase (Takeguchi et al. 1990). From four monoclonal antibodies raised against hog gastric K,H-ATPase, three inhibited the ouabain-insensitive gastric K,H-ATPase activity. Using indirect immunofluorescence all four monoclonal antibodies stained K,H-ATPase in gastric mucosae of various animal species (Takeguchi et al. 1990). Two monoclonal antibodies also cross-reacted with K,H-ATPase located in crypts of the transverse, descending colon and rectum of the rabbit. However, none of the antibodies inhibited ouabain-sensitive K,H-ATPase activity in the guinea pig distal colon, and the antibodies did not stain the enzyme in this tissue (Takeguchi et al. 1990). Earlier biochemical studies have already indicated that the rabbit distal colonic K-ATPase is insensitive to ouabain (Gustin and Goodman 1981, 1982). In the rat the Na,K-ATPase is remarkably resistant to inhibition by ouabain, however, K absorption in the distal colon is inhibited by mucosal ouabain (Perrone and McBride 1988; Sweiry and Binder 1990). In conclusion, an electroneutral K,H-ATPase located in the apical membrane of distal colonic enterocytes is responsible for active K absorption and H secretion. The properties of this transport system are different between species, particularly with reference to the ouabain sensitivity. K,H-ATPase in the guinea pig distal colon possesses the highest ouabain sensitivity of all species studied so far (IC_{50} about 1–4 μM) (Rechkemmer 1988; Watanabe et al. 1990).

Since net K transport in the distal colon of various species is regulated by aldosterone (Perrone and McBride 1988; Rechkemmer and Halm 1989; Sweiry and Binder 1989, 1990) it is important to consider the involvement of the K,H-ATPase in hormonal adaptation. In contrast to the rat (Wolffram et al. 1985; Foster et al. 1986; Sweiry and Binder 1989), changes in dietary K content and subsequent modulation of plasma aldosterone concentration do not change the rate of active K absorption in the guinea pig (Rechkemmer, unpubl. data). In the rat distal colon secondary hyperaldosteronism stimulated luminal acidification (McLaughlin et al. 1990), probably due to increased K,H-ATPase activity. In secondary hyperaldosteronism the rate of active K secretion is increased, J^K_{sm} is increased and J^K_{ms} is decreased and thus J^K_{net} appears to be greatly decreased. However, inhibition of the enhanced K secretion with bumetanide reveals that active K absorption in the guinea pig distal colon is not altered under these conditions (Rechkemmer 1988; Rechkemmer and Halm 1989). Enhanced acidification as measured in the rat distal colon in secondary hyperaldosteronism has not been evaluated in the guinea pig distal colon yet. However, since the rate of active K absorption does not change under high aldosterone conditions, it appears unlikely that the rate of acidification would change. At least in the guinea pig distal colon it remains questionable whether K,H-ATPase activity is regulated by mineralocorticoids.

K Secretion

K secretion across the distal colonic epithelium of the guinea pig is electrogenic and occurs along with K absorption. Under physiological conditions K absorption by far exceeds K secretion, such that the net movement of K is in the absorptive direction. The small electrogenic K secretion present in nonstimulated tissue can be blocked by Na-free solution or serosal ouabain application (Rechkemmer 1988). K secretion is stimulated in vitro by aldosterone (Rechkemmer and Halm 1989) or high plasma aldosterone concentrations induced by feeding of low-Na diets (Rechkemmer 1988). K

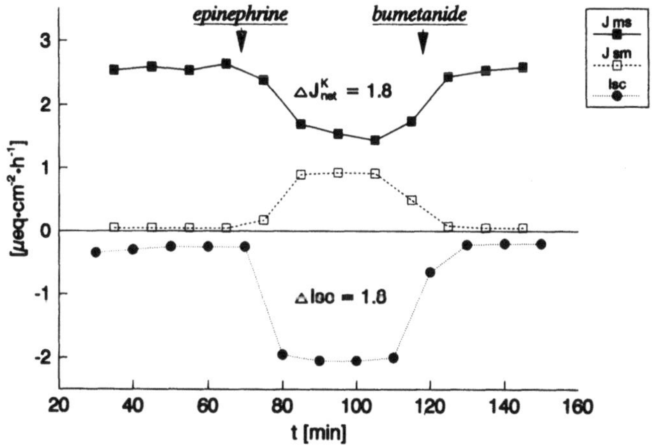

Fig. 4. Representative data of J_{ms}^{K}, J_{sm}^{K} and I_{sc} from a tissue pair of guinea pig distal colon. At the time points indicated epinephrine ($5\,\mu mol\,l^{-1}$) and bumetanide ($0.1\,mmol\,l^{-1}$) where added to the serosal bathing solution, respectively. (Rechkemmer 1988)

secretion also seems to be controlled neurohormonally, since electrogenic K secretion can be evoked by β-adrenergic agonists (Ishida and Suzuki 1987; Rechkemmer 1989a,b; Rechkemmer and Halm 1989). In contrast to the rabbit distal colon, where induction of K secretion is mediated through β_1-receptors (Smith and McCabe 1986), in the guinea pig distal colon K secretion is under the control of β_2-receptors (Rechkemmer 1989a,b). K secretion stimulated by β-adrenergic agonists is electrogenic, the change in I_{sc} is numerically equivalent to the change in J_{net}^{K} (Fig. 4). K secretion is completely abolished by serosal bumetanide, demonstrating the importance of the basolateral Na,K,2Cl-cotransport system for K secretion (Ishida und Suzuki 1987; Rechkemmer 1988). PGE_2 also induces K secretion (Rechkemmer and Halm 1989) but in contrast to β-adrenergic stimulation also electrogenic Cl secretion, thus the resulting change in I_{sc} is much less than with β-adrenergic agonists. Stimulation of β-adrenergic receptors and PGE_2 are both thought to stimulate adenylate cyclase activity and increase intracellular cAMP, the second messenger of their action. K secretion is partially inhibited by the K-channel blockers TEA and Ba and thus seems to involve K channels in the apical membrane of colonocytes (Clauss et al. 1985; Ishida and Suzuki 1987; Rechkemmer and Halm 1989). In the dietary Na-depleted rat amiloride has profound effects on unidirectional K fluxes, resulting in a large increase in net K absorption (Sweiry and Binder 1989). This indicates that in the rat electrogenic K secretion is coupled to electrogenic Na absorption and these two processes may be localized in the same cell type. Inhibition of electrogenic Na absorption with amiloride could hyperpolarize the apical membrane potential, reduce the driving force for K exit and thus decrease K secretion. In the guinea pig distal colon no such interaction between Na and K movement was observed, amiloride had no effect on unidirectional K fluxes even in aldosterone-treated tissues (Rechkemmer and Halm 1989). In the guinea pig distal colon Na and K movement appear to be localized in different cells or cell types. The striking mirror-image decrease in J^{K}_{ms} and increase in J^{K}_{sm} may indicate that active K absorption by the K,H-ATPase and active K secretion through an apical membrane K channel reside in the same cell, and thus opening of the K channel would induce an increase in apical membrane K recycling and a less efficient K absorption.

2.3 Chloride Transport

Absorptive and secretory pathways are involved in transepithelial Cl transport. Net transport thus depends on the magnitudes of these two oppositely directed mechanisms. In unstimulated tissue from the caecum and proximal colon usually net Cl absorption is present. Cl absorption is electroneutral and most likely involves Cl,HCO_3 exchange like in other species.

2.3.1 Caecum

Cl transport in the caecum is not well understood. In the guinea pig caecum in vitro net Cl absorption is about three to four times higher than the corresponding Na absorption (Rechkemmer 1988). Cl absorption is electroneutral and most likely in exchange for Cl bicarbonate is secreted into the caecal lumen. In addition to the bicarbonate-rich fluid flowing into the caecum from the ileum, the high rate of bicarbonate secretion from Cl,HCO_3 exchange may be important for buffering the caecal contents to a pH favourable for microbial metabolism and subsequent degradation of dietary fiber. A change from Cl absorption to Cl secretion can be induced by PGE_2. PGE_2 causes a decrease in J^{Cl}_{ms} and an increase in J^{Cl}_{sm}, in addition, the I_{sc} is significantly elevated. Thus, it appears that PGE_2 inhibits electroneutral Cl absorption and stimulates electrogenic Cl secretion (Rechkemmer 1988).

2.3.2 Proximal Colon

In the proximal colon Cl is absorbed. Cl transport is closely linked to Na transport as already described. In contrast to the caecum, PGE_2 only reduced J^{Cl}_{ms} in the proximal colon with no change in J^{Cl}_{sm}. However, the I_{sc} was also elevated by about 1 $\mu eq\,cm^{-2}\,h^{-1}$ (Rechkemmer 1988). These results can only be explained by a parallel decrease in paracellular Cl permeability and an increase in cellular electrogenic Cl secretion, balancing each other such that J^{Cl}_{sm} does not change. Certainly this phenomenon has to be studied in more detail to be understood. It is obvious, however, that the major effect of PGE_2, and probably of other cAMP-mediated secretagogues, in the proximal large intestine is the inhibition of electroneutral NaCl absorption.

2.3.3 Distal Colon

In contrast to other species, no net Cl absorption has been observed under short-circuit current conditions in the guinea pig distal colon (Kuwahara et al. 1987a). In the guinea pig distal colon the relation between the activity of submucosal neurons, the effects of different secretagogues and the involvement of immunological responses to the induction of electrogenic Cl secretion and resulting secretory diarrhea have been studied in more detail than in other species. Substances which either directly interact with receptors on epithelial cells or indirectly affect ion transport through the modulation of submucosal neuronal activity or the release of modulating substances from submucosal immune-competent cells and mast cells have been compiled (Table 2).

Submucosal Neurons and Chloride Secretion
Basal I_{sc} is not affected by blockers of neuronal activity [tetrodotoxin (TTX), hexamethonium], indicating that basal electrolyte transport is not under neuronal control (Kuwahara et al. 1987a). Stimulation of submucosal neuronal activity with electrical field stimulation (EFS), however, leads to a tremendous increase in I_{sc} of 300–400 $\mu A\,cm^{-2}$, corresponding to the induction a secretory Cl flux in the order of 12–15 $\mu Eq\,cm^{-2}\,h^{-1}$ (Kuwahara et al. 1987a; McCulloch et al. 1987). This secretory capacity in the guinea pig distal colon is much higher than that reported in other species. The electrogenic Cl secretion evoked by EFS is partially due to stimulation of acetylcholine release from submucosal neurons as demonstrated by the inhibition of the response with muscarinic antagonists with a rank order of potency: atropine = 4-diphenyl acetoxy-N-methylpiperidine > pirenzepine (Kuwahara et al. 1987b). Carbachol and bethanechol mimicked the action of EFS and the changes in I_{sc} were only partially inhibited by TTX, indicating the direct action of muscarinic agonists on epithelial cells (Kuwahara et al. 1987b). In addition to responses mediated by cholinergic transmission and muscarinic receptors on epithelial cells, histamine secreted from mucosal mast cells also affects electrogenic Cl secretion (Wang and Cooke 1990; Wang et al. 1990). Histamine evoked a transient, concentration-dependent increase in I_{sc}. The response to histamine was reduced by indomethacin and piroxicam, blockers of prostaglandin formation, but not by nordihydroguaiaretic acid, an inhibitor of lipoxygenase products, indicating that the action of histamine is mediated by the release of prostaglandins. Furthermore, the I_{sc} response to histamine was reduced by the H_1-receptor blocker pyrilamine but not by the H_2-receptor antagonist cimetidine, suggesting that histamine acts at H_1 receptors on cells within the gut wall. The initial

Table 2. Factors implicated in the induction of electrogenic Cl secretion across the guinea pig distal colon

Stimulus (agonists)	Antagonists	Mode of action	References
Carbachol	TTX = 4-DAMP > hexamethonium > pirenzepine, atropine	Muscarinic, cholinergic receptors on epithelial cells and submucosal neurons	Kuwahara et al. (1987b) Wang et al. (1990)
Substance P = physalaelemin > eledoisin = neurokinin A	TTX, atropine, mecamylamine, pyrilamine	Activation of cholinergic and non-cholinergic submucosal neurons, release of histamine	Kuwahara and Cooke (1990)
Histamine	TTX, hexamethonium, pyrilamine, cimetidine, NDGA, piroxicam, indomethacin	Release of PGE_2, action on H_1 receptors on epithelial cells (initial transient increase in I_{sc}) and on H_2 receptors on submucosal neurons (cyclical I_{sc} response)	Wang et al. (1990)
5-Hydroxytryptamine (5-HT), 2-methyl-5-HT	TTX, ICS 205–930, atropine, mecamylamine, bumetanide	Activation of 5-HT receptors causing release of ACh from cholinergic interneurons	Cooke et al. (1991)
VIP, PGE_2, Br-cAMP, bethanechol, A23187	Bumetanide	Intracellular Ca and cAMP, synergistic action on intracellular Ca and cAMP	Yajima et al. (1988)
VIP, bethanechol	Furosemide, neuropeptide Y	Intracellular Ca and cAMP	McCulloch et al. (1987)
Electrical field stimulation (EFS)	TTX, atropine	Non-specific stimulation of submucosal neurons, release of ACh	Kuwahara et al. (1987a) McCulloch et al. (1987)
β-Lactoglobulin	H_1 blockers, indomethacin	Immunoglobulin G, release of PGE_2 and histamine	Baird et al. (1987) Baron et al. (1988)
Antigen against *Trichinella spiralis*	Pyrilamine + indomethacin, furosemide, TTX	Release of PGE_2 and histamine	Russel and Castro (1989)

increase in I_{sc}, mediated by stimulation of H_1 receptors, is followed by recurrent cyclic increases in I_{sc} (Wang and Cooke 1990). This action of histamine was attributed to the activation of H_2 receptors on submucosal neurons involved in the regulation of Cl transport (Wang and Cooke 1990). Tachykinins (substance P, neurokinin A, eledoisin and physalaemin) also induce a concentration-dependent increase in I_{sc} (Kuwahara and Cooke 1990). Anion secretion evoked by the tachykinins was attributed to the activation of cholinergic and non-cholinergic neurons and the release of histamine (Kuwahara and Cooke 1990).

Immune System and Chloride Secretion

Immunological mechanisms play a role in the regulation of Cl secretion. Challenging of the distal colonic epithelium from *Trichinella spiralis* infected guinea pigs with parasite-derived antigen elevated I_{sc}, whereas the antigen had no effect in uninfected animals (Russell and Castro 1989). The antihistamine pyrilamine and indomethacin reduced the I_{sc} response by 40% when used in combination but had insignificant effects when used singly (Russell and Castro 1989). Antigenic stimulation of sensitized colon released histamine and PGE_2, but the PGE_2 released was only one-tenth of that stimulated by bradykinin. Thus, these autacoids contribute to the antigen-induced I_{sc} but are not the major factors in the distal colon (Russell and Castro 1989). In guinea pigs sensitized to cow's milk β-lactoglobulin evokes a concentration-dependent increase in I_{sc}, a response which could be mimicked in non-sensitized tissue by the application of purified immunoglobulin G (IgG). Colonic ultrastructure was significantly changed under these conditions (Baird et al. 1987; Baron et al. 1988).

Second Messengers and Chloride Secretion

Electrogenic Cl secretion across the guinea pig distal colon can be evoked by either Ca- or cAMP-mediated agonists (Yajima et al. 1988). If Cl secretion is induced by carbachol or bethanechol (Ca-mediated), or VIP, PGE_2 or forskolin (cAMP-mediated), the I_{sc} responses are usually small and transient (Fig. 5). However, the combination of Ca- and cAMP-mediated agonists leads to a prolonged increase in I_{sc} comparable in magnitude to EFS, demonstrating marked synergistic effects (Yajima et al. 1988). Thus, Cl secretion may be evoked through a variety of different neuronal, hormonal and immunological pathways, as described above, but all routes seem to converge at the intracellular second-messenger level. The ion transport mechanisms involved in electrogenic Cl secretion are (1) the basolateral Na,K-ATPase responsible for the generation of a transmembrane Na gradient; (2) the basolateral Na,K,2Cl-cotransport system required for transporting Cl into the cell utilizing the Na gradient; (3) the apical membrane Cl channel which is regulated by cAMP, and maybe by intracellular Ca; and (4) the basolateral K channel for maintaining the electrical driving force for Cl exit at the apical membrane. These transport mechanisms have

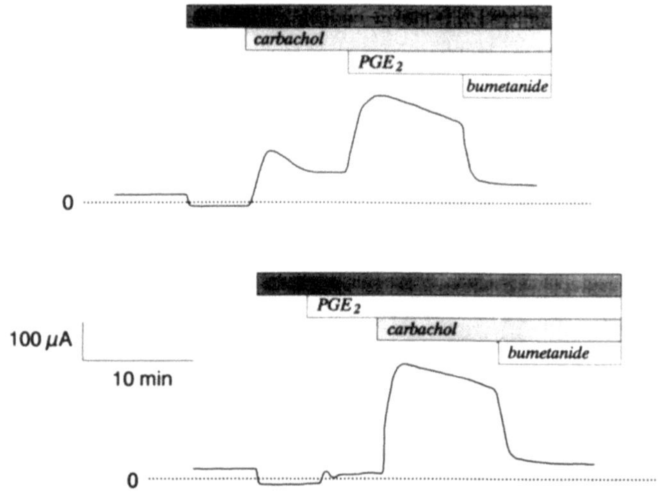

Fig. 5. Synergistic action of cAMP- and Ca-mediated secretagogues on I_{sc}, reflecting electrogenic Cl secretion, in the guinea pig distal colon. Representative recordings of I_{sc} from a pair of tissues. Tissues were incubated in a standard Ringer solution containing indomethacin (10^{-6} mol l^{-1}). Drugs were sequentially added in the order shown in the graph. Amiloride (10^{-4} mol l^{-1}) was added to the mucosal solution, carbachol (10^{-4} mol l^{-1}), PGE_2 (2×10^{-6} mol l^{-1}) and bumetanide (10^{-4} mol l^{-1}) were added to the serosal solution. (Rechkemmer unpubl. data)

to operate in a coordinated fashion to sustain Cl secretion. The biophysical properties of colonic apical membrane Cl channels will be discussed in detail in other contributions to this volume.

Cl secretion is blocked by any maneuvers inhibiting one of the components of the above described system. Loop diuretics, furosemide and more potently bumetanide, which block the Na,K,2Cl-cotransport system, thus largely decrease electrogenic Cl secretion independent of the way of activation (Fig. 5, Table 2).

Chloride Secretion and Fluid Transport

Cellular Cl secretion under open-circuit (in vivo) conditions is followed by paracellular Na movement along a favourable electrochemical gradient and accompanied by water flow, resulting in the secretion of an isotonic NaCl solution. The fluid secretion evoked by PGE_2 in the guinea pig distal colonic epithelium, at least partially if not completely, originates in the colonic crypts (Fig. 6), supporting the hypothesis that crypts are the site of electrolyte and fluid secretion in the large intestine (Welsh et al. 1982). This notion was further supported by a recent study assessing the cellular localization of K and Cl secretion in the guinea pig distal colon with electron microprobe analysis (Halm and Rick 1992). Stimulation of either K and Cl secretion with PGE_2 or K secretion alone with epinephrine increased intracellular Na in crypt cells but not in surface cells or cells in the crypt neck, supporting the

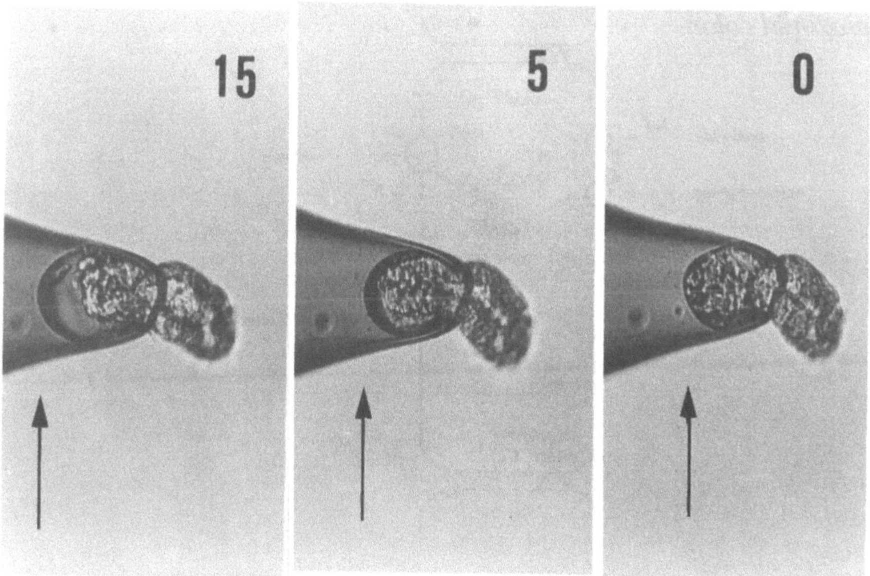

Fig. 6. Isolated crypt of the guinea pig distal colon attached to a glass pipette filled with paraffin oil. At time 0 PGE_2 ($2\,\mu mol\,l^{-1}$) was added to the bathing solution. Five and 15 minutes after PGE_2 the development of a fluid droplet at the mouth of the crypt is clearly evident. (Rechkemmer 1988) *Arrows* indicate the border between paraffin oil and the emerging fluid droplet

location of ion secretory cells in the lower parts of the crypts (Halm and Rick 1992). However, the concept that crypts are responsible for fluid and electrolyte secretion has been questioned recently in a study measuring interstitial tissue osmolarity with video imaging of a cell-impermeant fluorescent probe where a high interstitial Na concentration at the bottom of the crypts and a decreasing Na concentration along the crypt lumen were demonstrated, indicating that crypt cells considerably absorb Na and do not secrete (Naftalin and Pedley 1990). To resolve this contradictory issue, further measurements are urgently needed in the intact epithelium as well as in isolated crypts to discern the importance of the colonic crypts for fluid and electrolyte secretion.

2.4 Model of Ion Transport Mechanisms in the Guinea Pig Large Intestine

In Fig. 7 the electrolyte transport systems demonstrated in the proximal and distal colon of the guinea pig are summarized. Qualitatively, transport systems in the caecum appear to be similar to the proximal colon, with the exception of an amiloride-insensitive Na channel, which may occur in the caecum. In the proximal segments (caecum and proximal colon) electroneutral, absorptive mechanisms dominate (Na, H and Cl,HCO_3 exchange),

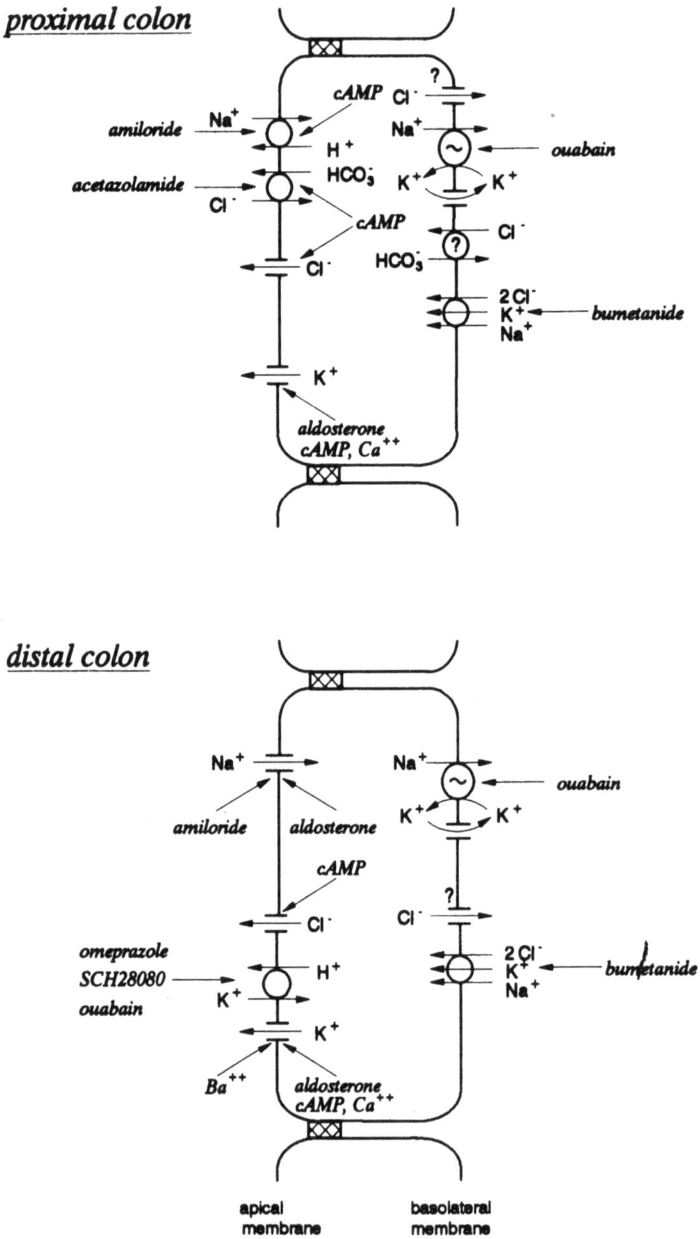

Fig. 7. Model for electrolyte transport in the proximal and distal colon of guinea pig. Transport mechanisms in the caecum are similar to the proximal colon. Receptors for hormones or neurotransmitters are not included in the model. The transport systems indicated in the model are probably not situated in the same cell, however, present knowledge does not allow a conclusive statement about the cellular localizations of the various transport systems

while in the distal colon only electrogenic Na absorption is present. Secretory mechanisms, on the other hand, appear to be qualitatively similar between the segments, probably indicating the functional similarity of undifferentiated crypt cells. Surface cells of the proximal and distal segments, however, appear to develop diverse transport mechanisms in the course of differentiation.

3 Transport of Short-Chain Fatty Acids

3.1 Production and Concentrations of Short-Chain Fatty Acids in the Large Intestine

The large intestine, particularly the caecum, serves as an anaerobic microbal fermentation compartment in which undigested plant material, mainly cellulose, is metabolized into short-chain fatty acids (SCFA). This aspect is particularly important in herbivores. SCFA are absorbed in the large intestine and subsequently utilized by the animal as substrates of energy metabolism (Rechkemmer et al. 1988a). The caecum and proximal colon from a functional compartment with respect to microbial fermentation (Sakaguchi et al. 1985). Acetate/acetic acid (Ac), propionate/propionic acid (Pr) and butyrate/butyric acid (Bu) are the major SCFA produced by anaerobic microbial fermentation in the large intestine of mammals. The pK_a of SCFA is around 4.8, and thus at the physiological pH in the large intestine more than 99% of SCFA are present in the intestinal fluid in the ionized form. SCFA constitute the major anions in hindgut contents. Total concentrations in guinea pig average 95 to 130 $mmol\,l^{-1}$, the relative ratios are Ac 70%, Pr 16%, and Bu 13%, respectively (Engelhardt and Rechkemmer 1983b; Sakaguchi et al. 1985). The energy contribution of SCFA metabolism to the basal metabolic rate of guinea pig was calculated to be 31% (Sakaguchi et al. 1986). Slightly lower SCFA concentrations are present in the caecum compared to the colon, this may be due to a high inflow of ileal fluid into the caecum (Rechkemmer et al. 1988a); from caecal dilution an inflow from the ileum of 206 $ml\,day^{-1}\,kg^{-1}$ body weight was estimated in guinea pig (Sakaguchi et al. 1986).

3.2 Absorption of Short-Chain Fatty Acids from the Caecum, the Proximal Colon and the Distal Colon in in Situ Experiments

Ninety-five to 99% of SCFA produced in the large intestine are rapidly absorbed. Net absorption is higher than net transport of Na or Cl at equimolar concentrations (Rechkemmer and Engelhardt 1982; Engelhardt and Rechkemmer 1983a). In the proximal colon a positve, linear relationship between Na and SCFA absorption could be seen; no such effect was

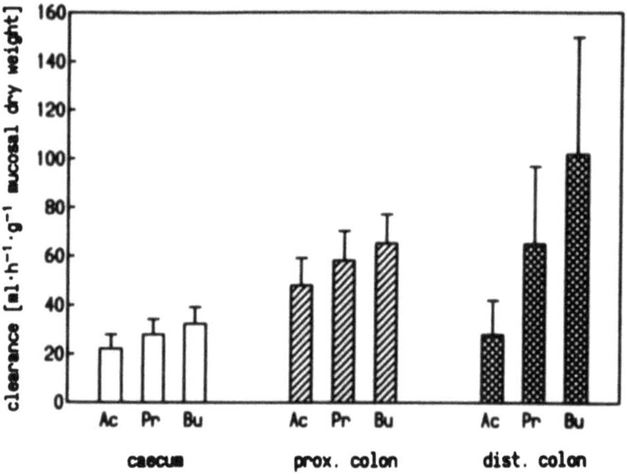

Fig. 8. Clearance of acetate (*Ac*), propionate (*Pr*) and butyrate (*Bu*) in the caecum, the proximal and the distal colon of guinea pig. Data for the proximal and the distal colon are taken from Engelhardt and Rechkemmer (1983a) and Luciano et al. (1984), for the caecum from Oltmer (unpubl.). Means ±SD are shown

apparent in the distal colon (Engelhardt and Rechkemmer 1983a). Replacement of Cl by SCFA resulted in a significantly higher bicarbonate secretion in the proximal colon and in a change from bicarbonate absorption to secretion in the distal colon in in situ perfusion studies (Rechkemmer and Engelhardt 1982).

The absorption of acetate was linearly dependent on acetate concentration in the proximal and distal colon of guinea pig, indicating that diffusion along favourable gradients is the primary pathway for SCFA transport (Rechkemmer and Engelhardt 1988).

Relative epithelial permeabilities for the different SCFA were obtained by calculating clearance rates (Engelhardt and Rechkemmer 1983a; Luciano et al. 1984; Oltmer, unpubl.). In the caecum and in the proximal colon only a small increase in clearance with SCFA chain length was observed (Fig. 8), indicating that at physiological pH lipid solubility is of minor importance. In the distal colon propionate clearance increased about two-fold with respect to acetate, and butyrate clearance about 1.5-fold with respect to propionate, denoting the importance of lipid solubility for SCFA permeability in this segment of the large intestine. No interference between the different SCFAs could be demonstrated (unpubl. observ.).

According to the pH-partition hypothesis only the undissociated, lipid-soluble fraction is able to permeate biological membranes easily. The general characteristics of weak electrolyte transport have been reviewed in detail recently (Rechkemmer 1991b). Thus, one would expect an increase in SCFA absorption with a decrease in luminal pH. However, clearance rates of SCFA did not change when the pH in the luminal solution was decreased

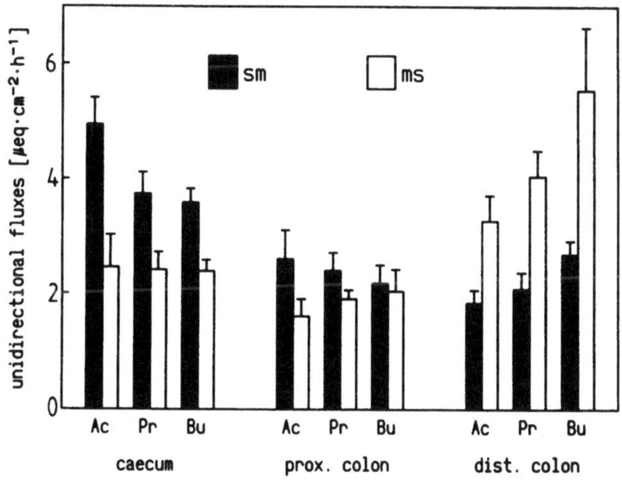

Fig. 9. Unidirectional fluxes (J_{sm}, J_{ms}) for acetate (*Ac*), propionate (*Pr*) and butyrate (*Bu*) across the mucosa of the caecum, proximal and distal colon of guinea pig. SCFA concentrations on both sides of the epithelium were 25 mmol l^{-1}. Means ±SD are shown. (Engelhardt and Rechkemmer 1992)

from pH 8 to pH 6 (Rechkemmer and Engelhardt 1988). Only at pH 5.3 was the clearance elevated in the distal colon, but the increase was much smaller than predicted by the pH-partition hypothesis. The unexpected pH independence of SCFA absorption can be attributed to a constant pH microclimate at the luminal surface of the proximal and the distal colon of guinea pig (Rechkemmer et al. 1986a). The pH in the microclimate region is close to neutral and largely independent from changes in bulk luminal pH. The pH in the microclimate dropped only in the distal colon at a luminal pH below 5.5, and this may explain the higher clearance rates at pH 5.3 in the distal colon (Rechkemmer and Engelhardt 1988).

3.3 Unidirectional Fluxes of Short-Chain Fatty Acids in in Vitro Studies Across Epithelia of Caecum, Proximal and Distal Colon

In the caecum with equal concentrations of SCFA on both sides of the mucosa, J_{sm} was nearly twice J_{ms} for all three SCFA (Fig. 9). In the distal colon, on the other hand, J_{ms} was always higher. Thus, in the caecum net secretion and in the distal colon net absorption were estimated under short-circuit current conditions (Engelhardt and Rechkemmer 1992). In the caecum J_{sm} decreased with chain length, J_{ms} was similar for the three SCFA. In contrast to the distal colon, J_{ms} increased with chain length similar to the increase in clearance rates of SCFA (Figs. 8 and 9). Unidirectional fluxes across the mucosa of the proximal colon were intermediate to those in the caecum and in the distal colon. The segmental differences in unidirectional fluxes of SCFA may be due partially to regional differences in membrane fluidity. Luciano et al. (1989) measured twice as many filipin-cholesterol complexes in the brush border membrane of the proximal colon than in the distal colon, indicating a greater lipid fluidity of the colonocyte membranes in the distal colon. Furthermore, the microvilli of the enterocytes in the proximal colon of guinea pigs are shorter than in the distal colon, and thus the surface area may be different between these two segments (Luciano et al. 1989).

For non-ionic diffusion of SCFA the SCFA anions have to be protonated. H ions may be gained at the apical membrane by three mechanisms: (1) amiloride-sensitive Na,H exchange, occurring in the caecum and the proximal colon of guinea pigs (Rechkemmer 1988); (2) K,H-ATPase present in the apical membrane of the distal colon of guinea pig (Suzuki and Kaneko 1987, 1989; Watanabe et al. 1990); and (3) CO_2 conversion to carbonic acid and subsequent dissociation into H ions and bicarbonate.

Inhibition of the amiloride-sensitive Na,H exchange in the apical membrane of the caecum and the proximal colon resulted in a decrease in J_{ms}^{SCFA}; J_{sm}^{SCFA} was not affected (Fig. 10). Inhibition of the K,H-ATPase in the apical membrane of the distal colon reduced J_{ms}^{SCFA} by 50%, J_{sm}^{SCFA} was slightly increased. We thus conclude that the H-ion antiport systems in the apical membrane do not markedly affect J_{sm} of SCFA. Flux studies with propionate being present on one side of the large intestine epithelium only indicated that the apical entry of SCFA into the enterocytes of the caecum and the proximal colon is primarily mediated by Na,H exchange, and in the distal colon by the K,H-ATPase (Engelhardt et al. 1993). The higher J_{ms} compared to J_{sm} in the distal colon is independent of Na and is related to the activity of the apical K,H-ATPase. When the K,H-ATPase was inhibited by mucosal ouabain J_{sm} was similar to J_{ms} (Fig. 10; Engelhardt et al. 1993). In the caecum and in the proximal colon the higher J_{sm} compared to J_{ms} seems to be coupled to the transmembrane Na gradient generated by the basolateral Na,K-ATPase. Inhibition of the basolateral Na,K-ATPase with serosal ouabain or incubation in nominally Na-free solution abolished net SCFA secretion in the proximal colon and largely also in the caecum (Engelhardt et al. 1993). Under physiological conditions SCFA are present in high concentrations (around $100\,mmol\,l^{-1}$) in the large intestinal lumen, while SCFA concentrations in peripheral blood are only about $1\,mmol\,l^{-1}$. Because of this huge gradient J_{ms}, and not J_{sm}, will be prominent in vivo.

It was supposed that the transport of SCFA anions may occur via the paracellular shunt in the more leaky proximal colon of guinea pig (Luciano et al. 1984; Rönnau et al. 1989). However, clamping the transepithelial electrical potential difference to $-30\,mV$, $0\,mV$ or $+30\,mV$ had no effect on unidirectional SCFA fluxes (Engelhardt and Rechkemmer 1992), thus, ruling out a major contribution of paracellular permeation of SCFA anions in the large intestine of guinea pigs.

The intracellular hydrogen ion gain from dissociation of carbonic acid seems to be a major source of hydrogen ions for the antiport systems;

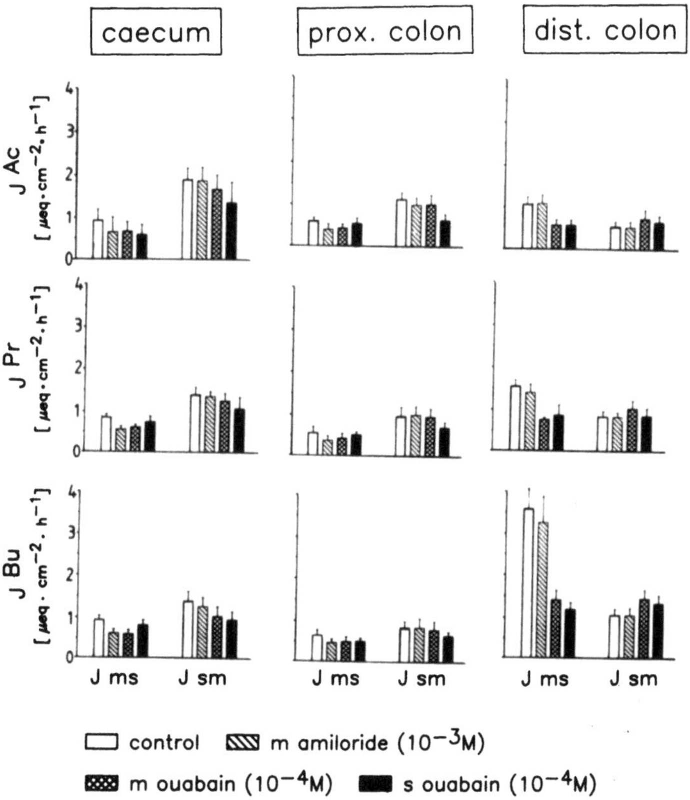

Fig. 10. Effects of amiloride and ouabain added to the solutions bathing the mucosal (*m*) or the serosal (*s*) surface of guinea pig large intestinal epithelia on unidirectional fluxes (*J*) of acetic acid (*Ac*), propionic acid (*Pr*), and butyric acid (*Bu*). Mean values from 5 to 9 epithelial tissue pairs were calculated. Means ±SD are shown. (Engelhardt et al. 1993)

inhibition of carbonic anhydrase with acetazolamide or ethoxzolamide reduced J_{ms}^{SCFA} fluxes significantly (Engelhardt and Rechkemmer unpubl.). Carbonic anhydrase activity is high in the proximal colon; in the distal colon, carbonic anhydrase activity is much less and seems to be caused by a different isoenzyme (Carter and Parsons 1970, 1972; Lönnerholm 1977).

In the rat the characteristics of an apical membrane Na,H exchange have been studied in detail using isolated brush-border membrane vesicles (Binder et al. 1986; Rajendran and Binder 1990), and it was demonstrated that SCFA stimulate active Na and Cl absorption (Binder and Mehta 1989). Such studies have not been performed in the guinea pig colon, because of the lack of a suitable apical membrane marker (Biber et al. 1983).

Recent studies using brush-border membrane vesicles of the rat and human colon (Harig et al. 1990; Mascolo et al. 1991) provided evidence for SCFA-bicarbonate exchange in the apical membrane. The importance of

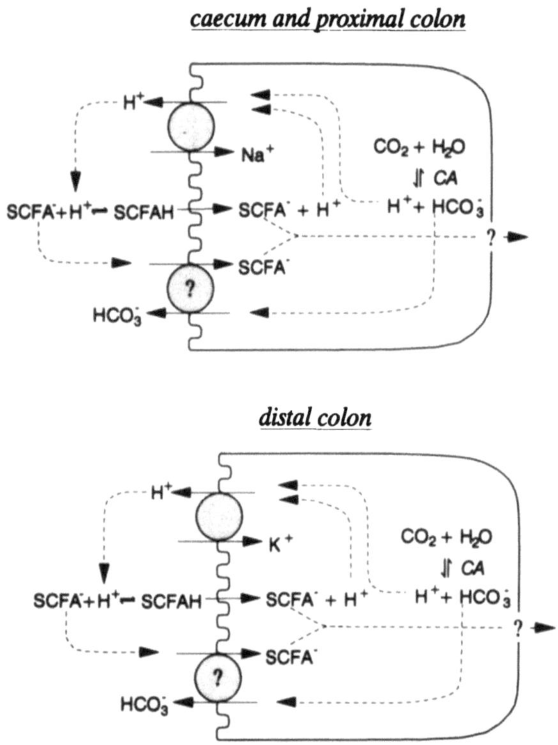

Fig. 11. Model for the apical membrane uptake of SCFA in the large intestine of guinea pig. *SCFAH* characterizes permeation in the non-ionic form and *SCFA*$^-$ transport in the ionic form

bicarbonate for absorption of SCFA anions is also supported by recent results in guinea pigs. In HCO_3- and CO_2-free solutions J^{Pr}_{ms} was reduced by 40–60% in the caecum and the proximal colon (unpubl. observ.). Ussing-chamber flux studies do not convincingly support the participation of a SCFA-Cl anion exchange mechanism; in Cl-free solution J^{Pr}_{ms} in the distal colon was diminished, no effect was observed in the caecum, and in the proximal colon J^{Pr}_{ms} was increased (unpubl. observ.). The interrelation between Cl, bicarbonate and SCFA transport certainly has to be studied in more detail in the future to elucidate a possible association between these anions.

3.4 Model for Apical Membrane Permeation of Short-Chain Fatty Acids in the Caecum, the Proximal Colon, and the Distal Colon of Guinea Pig

In the caecum and in the proximal colon SCFA may enter the enterocytes by non-ionic diffusion and in parallel through an electroneutral anion exchanger. The contribution of each of these pathways has not been determined as yet. In the distal colon of guinea pig the major portion of SCFA may pass the apical membrane in the undissociated form via non-ionic diffusion, and probably a much smaller fraction, if any, passes through an anion exchanger. Hydrogen ions provided by Na,H exchange in the caecum and in the proximal colon and by the K,H-ATPase in the distal colon are important for the protonation of SCFA anions and subsequent non-ionic diffusion (Fig. 11). The presence of a bicarbonate-SCFA exchange in the apical membrane has not been conclusively demonstrated in the guinea pig large intestine. Mechanisms involved in the passage of SCFA across the basolateral membrane are unknown so far.

Acknowledgments. Research from the authors was supported by the Deutsche Forschunggemeinschaft. Part of the work cited in this paper was performed at the Department of Physiology and Biophysics of the University of Alabama at Birmingham in collaboration with Drs. R.A. Frizzell and D.R. Halm. GR greatly appreciates the generous support of Dr. Frizzell during the course of a 2-year stay in his laboratory (supported by DFG grant Re 592/2).

References

Baird AW, Barclay WS, Blazer-Yost BL, Cuthbert AW (1987) Affinity purified immunoglobulin G transfers immediate hypersensitivity to guinea pig colonic epithelium in vitro. Gastroenterology 92: 635–642

Baron DA, Baird AW, Cuthbert AW, Margolius HS (1988) Intestinal anaphylaxis: rapid changes in mucosal ion transport and morphology. Am J Physiol 254: G307–G314

Benos DJ (1982) Amiloride: a molecular probe of sodium transport in tissues and cells. Am J Physiol 242: C131–C145

Biber J, Rechkemmer G, Bodmer M, Schröder P, Haase W, Murer H (1983) Isolation of basolateral membranes from columnar cells of the proximal colon of the guinea pig. Biochim Biophys Acta 735: 1–11

Binder HJ, Mehta P (1989) Short-chain fatty acids stimulate active sodium and chloride absorption in vitro in the rat distal colon. Gastroenterology 96: 989–996

Binder HJ, Stange G, Murer H, Stieger B, Hauri H-P (1986) Sodium-proton exchange in colon brush-border membranes. Am J Physiol 251: G382–G390

Carter MJ, Parsons DS (1970) The carbonic anhydrase of some guinea-pig tissues. Biochim Biophys Acta 206: 190–192

Carter MJ, Parsons DS (1972) The isoenzymes of carbonic anhydrase: kinetic properties with particular reference to the functions of the intestinal tract. J Physiol 220: 465–478

Claude P, Goodenough DA (1973) Fracture faces of zonulae occludentes from "tight" and "leaky" epithelia, J Cell Biol 58: 390–400

Clauss W, Dürr J, Rechkemmer G (1985) Characterization of conductive pathways in guinea pig distal colon in vitro. Am J Physiol 248: G176–G183

Cooke HJ, Wang Y-Z, Frieling T, Wood JD (1991) Neural 5-hydroxytryptamine receptors regulate chloride secretion in guinea pig distal colon. Am J Physiol 261: G833–G840

Cooper G, Schiller AL (1975) Anatomy of the guinea pig. Harvard University Press, London

Edmonds CJ, Smith T (1979) Epithelial transport pathways of rat colon determined in vivo by impulse response analysis. J Physiol 296: 471–485

Engelhardt W von, Rechkemmer G (1983a) Absorption of inorganic ions and short-chain fatty acids in the colon of mammals. In: Gilles-Baillien M, Gilles R (eds) Intestinal transport. Springer, Berlin Heidelberg New York pp 26–45

Engelhardt W von, Rechkemmer G (1983b) The physiological effects of short-chain fatty acids in the hind gut. In: Wallace G, Bell L (eds) Fibre in human and animal nutrition. R Soc N Z Bull 20: 149–155

Engelhardt W von, Rechkemmer G (1992) Segmental differences of short-chain fatty acid transport across guinea pig large intestine. Exp Physiol 77: 491–499

Engelhardt W von, Burmester M, Hansen K, Becker G, Rechkemmer G (1993) Effects of amiloride and ouabain on short chain fatty acid transport in guinea pig large intestine. J Physiol 460: 455–466

Foster ES, Hayslett JP, Binder HJ (1984) Mechanism of active potassium absorption and secretion in the rat colon. Am J Physiol 246: G611–G617

Foster ES, Sandle, GI, Hayslett JP, Binder HJ (1986) Dietary potassium modulates active potassium absorption and secretion in rat distal colon. Am J Physiol 251: G619–G626

Fromm M, Schultz SG (1981) Potassium transport across rabbit descending colon in vitro: evidence for single file diffusion through a paracellular pathway. J Membr Biol 63: 93–98

Gorelick FS, Sarras MP Jr, Jamieson JD (1982) Regional differences in lectin binding to colonic epithelium by fluorescent and electron microscopy. J Histochem Cytochem 30: 1097–1108

Graur D, Hide WA, Li W-H (1991) Is the guinea-pig a rodent? Nature 351: 649–652

Gustin MC, Goodman DBP (1981) Isolation of a brush-border membrane from rabbit descending colon epithelium. Partial characterization of a unique K-activated ATPase. J Biol Chem 256: 10651–10656

Gustin MC, Goodman DBP (1982) Characterization of the phosphorylated intermediate of the K^+-ouabain-insensitive ATPase of the rabbit colon brush-border membrane. J Biol Chem 257: 9629–9633

Halm DR, Frizzell RA (1986) Active K transport across rabbit distal colon: relation to Na absorption and Cl secretion. Am J Physiol 251: C252–C267

Halm DR, Rick R (1992) Secretion of K and Cl across colonic epithelium: cellular localization using electron microprobe analysis. Am J Physiol 31: 1392–1402

Harig JM, Knaup SM, Shoshara J, Dudeja PK, Ramaswamy K, Brasitus TA (1990) Transport of n-butyrate into human colonic luminal membrane vesicles. Gastroenterology 98: A543 (Abstr)

Hayslett JP, Binder HJ (1982) Mechanism of potassium adaptation. Am J Physiol 243: F103–F112

Ishida H, Suzuki Y (1987) Potassium secretion in the guinea pig distal colon. Jpn J Physiol 37: 33–48

Kaunitz JD, Sachs G (1986) Identification of a vanadate- sensitive potassium-dependent proton pump from rabbit colon. J Biol Chem 261: 14005–14010

Kuwahara A, Cooke HJ (1990) Tachykinin-induced anion secretion in guinea pig distal colon: role of neural and inflammatory mediators. J Pharmacol Exp Ther 252: 1–7

Kuwahara A, Radowicz-Cooke HJ (1988) Epithelial transport in guinea-pig proximal colon: influence of enteric neurones. J Physiol 395: 271–284

Kuwahara A, Bowen S, Wang J, Condon C, Cooke HJ (1987a) Epithelial responses evoked by stimulation of submucosal neurons in guinea pig distal colon. Am J Physiol 252: G667–G674

Kuwahara A, Tien X-Y, Wallace LJ, Cooke HJ (1987b) Cholinergic receptors mediating secretion in guinea pig colon. J Pharmacol Exp Ther 242: 600–606

Lönnerholm G (1977) Carbonic anhydrase in the intestinal tract of the guinea-pig. Acta Physiol Scand 99: 53–61

Luciano L, Reale E, Rechkemmer G, Engelhardt Wv (1984) Structure of zonulae occlucentes and the permeability of the epithelium to short-chain fatty acids in the proximal and the distal colon of guinea pig. J Membr Biol 82: 145–156

Luciano L, Konitz H, Reale E (1989) Localization of cholesterol in the colonic epithelium of the guinea pig: regional differences and functional implications. Cell Tissue Res 258: 339–347
Marcial MA, Carlson SL, Madara JL (1984) Partitioning of paracellular conductance along the ileal crypt-villus axis: a hypothesis based on structural analysis with detailed consideration of tight junction structure-function relationships. J Membr Biol 80: 59–70
Mascolo N, Rajendran VM, Binder HJ (1991) Mechanism of short-chain fatty acid uptake by apical membrane vesicles of rat distal colon. Gastroenterology 101: 331–338
McCulloch CR, Kuwahara A, Condon CD, Cooke HJ (1987) Neuropeptide modification of chloride secretion in guinea pig distal colon. Regul Pept 19: 35–43
McLaughlin ML, McBride DE, Perrone RD (1990) Secondary hyperaldosteronism stimulates acidification in rat distal colon. Pflügers Arch Eur J Physiol 416: 639–645
Messenger JP, Furness JB (1990) Projections of chemically-specified neurons in the guinea-pig colon. Arch Histol Cytol 53: 467–495
Mora-Galindo J (1986) Maturation of tight junctions in guinea-pig cecal epithelium. Cell Tissue Res 246: 169–175
Naftalin RJ, Pedley KC (1990) Video enhanced imaging of the fluorescent Na^+ probe SBFI indicates that colonic crypts absorb fluid by generating a hypertonic interstitial fluid. FEBS Lett 260: 187–194
Perrone RD, McBride DE (1988) Aldosterone and P_{CO_2} enhance rubidium absorption in rat distal colon. Am J Physiol 254: G898–G906
Rajendran VM, Binder HJ (1990) Characterization of Na-H exchange in apical membrane vesicles of rat colon. J Biol Chem 265: 8408–8414
Rechkemmer G (1988) Mechanisms and regulation of electrolyte transport in different large intestinal segments of the guinea pig. Habilitationsschrift, Tierärztliche Hochschule Hannover
Rechkemmer G (1989a) Adrenergic agonists stimulate electrogenic potassium secretion across guinea pig distal colon. Naunyn-Schmiedeberg's Arch Pharmacol 340: R85
Rechkemmer G (1989b) Electrogenic potassium secretion by guinea pig distal colon: effects of adrenergic agonists. FASEB J 3: A1149
Rechkemmer G (1991a) Regulation of electrolyte absorption and secretion in the large intestine. J Anim Physiol Anim Nutr 22: 101–110
Rechkemmer G (1991b) Transport of weak electrolytes. In: Field M, Frizzell RA (eds) Handbook of physiology, sect 6: the gastrointestinal system. Vol IV. Intestinal absorption and Secretion. American Physiological Society, Oxford University Press, New York, pp 371–388
Rechkemmer G, Engelhardt W von (1982) Absorptive processes in different colonic segments of the guinea pig and the effects of short-chain fatty acids. In: Kasper H, Goebell H (eds) Colon and nutrition. MTP Press, Lancaster, pp 61–69
Rechkemmer G, Engelhardt W von (1988) Concentration- and pH-dependence of short-chain fatty acid absorption in the proximal and distal colon of guinea pig (Cavia porcellus). Comp Biochem Physiol 91: 659–663
Rechkemmer G, Halm DR (1989) Aldosterone stimulates K secretion across mammalian colon independent of Na absorption. Proc Natl Acad Sci USA 86: 397–401
Rechkemmer G, Wahl M, Kuschinsky W, Engelhardt W von (1986a) pH-microclimate at the luminal surface of the intestinal mucosa of guinea pig and rat. Pflügers Arch Eur J Physiol 407: 33–40
Rechkemmer G, Halm DR, Frizzell RA (1986b) Potassium transport across guinea pig distal colon. Fed Proc 45: 748
Rechkemmer G, Krause M, Becker G, Engelhardt W von (1987a) Konzentrationen von Elektrolyten und kurzkettigen Fettsäuren im Magen-Darm-Trakt des Meerschweinchens bei Normaldiät und bei natriumarmer Diät. Dtsch Tierärztl Wochenschr 94: 12–15
Rechkemmer G, Halm DR, Bynum EP, Frizzell RA (1987b) Potassium transport by guinea pig distal colon: active absorption and active secretion. Pflügers Arch 408: R46

Rechkemmer G, Roennau K, Engelhardt W von (1988a) Fermentation of polysaccharides and absorption of short chain fatty acids in the mammalian hindgut. Comp Biochem Physiol 90A(4): 563–68
Rechkemmer G, Halm DR, Frizzell RA (1988b) Differences in electrolyte transport mechanisms in the guinea pig proximal and distal colon. Pflügers Arch Eur J Physiol 411: R80
Rönnau K, Guth D, Engelhardt W von (1989) Absorption of dissociated and undissociated short-chain fatty acids across the colonic epithelium of guinea pig. Q J Exp Physiol 74: 511–519
Russell DA, Castro GA (1989) Immunological regulation of colonic ion transport. Am J Physiol 256: G396–G403
Sakaguchi E, Becker G, Rechkemmer G, Engelhardt W von (1985) Volume, solute concentrations and production of short-chain fatty acids in the caecum and upper colon of the guinea pig. Z Tierphysiol Tierernähr Futtermittelkd 54: 276–285
Sakaguchi E, Heller R, Becker G, Engelhardt W von (1986) Retention of digesta in the gastrointestinal tract of the guinea pig. J Anim Physiol Anim Nutr 55: 44–50
Sander K (1990) Anpassung einzelner Abschnitte des Magen-Darm-Kanals des Meerschweinchens an Diäten mit unterschiedlichem Natrium- und Kaliumgehalt. Vet Med Dissertation, Hannover
Smith PL, McCabe RD (1984) Mechanism and regulation of transcellular potassium transport by the colon. Am J Physiol 250: G445–G456
Smith PL, McCabe RD (1986) Potassium secretion by rabbit descending colon: effects of adrenergic stimuli. Am J Physiol 250: G432–G439
Snipes RL (1982) Anatomy of the guinea-pig cecum. Anat Embryol 165: 97–111
Suzuki Y (1983) Acid secretion in the guinea pig colonic mucosa. Proc Int Union Physiol Sci 24th Sydney, Australia 15: 421
Suzuki Y (1984) H^+ secretion in the guinea pig colonic mucosa. In: Case RM, Lingard JA, Young JA (eds) Secretion: mechanisms and control. University Press, Manchester, pp 71–75
Suzuki Y, Kaneko K (1987) Acid secretion in isolated guinea pig colon. Am J Physiol 253: G155–G164
Suzuki Y, Kaneko K (1989) Ouabain-sensitive H^+-K^+ exchange mechanism in the apical membrane of guinea pig colon. Am J Physiol 256: G979–G988
Sweiry JH, Binder HJ (1989) Characterization of aldosterone-induced potassium secretion in rat distal colon. J Clin Invest 83: 844–851
Sweiry JH, Binder HJ (1990) Active potassium absorption in rat distal colon. J Physiol 423: 155–170
Takeguchi M. Asano S, Tabuchi Y, Takeguchi N (1990) The presence of H^+,K^+-ATPase in the crypt of rabbit distal colon demonstrated with monoclonal antibodies against gastric H^+,K^+-ATPase. Gastroenterology 99: 1339–1346
Ussing HH, Andersen B (1956) The relation between solvent drag and active transport of ions. Proc Int Congr Biochem Brussels. Academic Press, New York, pp 434–440
Wang YZ, Cooke HJ (1990) H_2 receptors mediate cyclical chloride secretion in guinea pig distal colon. Am J Physiol 258: G887–G893
Wang YZ, Cooke HJ, Su HC, Fertel R (1990) Histamine augments colonic secretion in guinea pig distal colon. Am J Physiol 258: G432–G439
Watanabe T, Suzuki T, Suzuki Y (1990) Ouabain-sensitive K^+-ATPase in epithelial cells from guinea pig distal colon. Am J Physiol 258: G506–G511
Watkins DW, Pintauro WL (1979) Luminal alkalinization by guinea-pig cecum in vitro, an electro-neutral process. Experientia 35: 356–358
Welsh MJ, Smith PL, Fromm M, Frizzell RA (1982) Crypts are the site of intestinal fluid and electrolyte secretion. Science 218: 1219–1221
Wills NK, Biagi B (1982) Active potassium transport by rabbit descending colon epithelium. J Membr Biol 64: 195–203
Wolffram S, Stingelin Y, Schneider B, Scharrer E (1985) Dietary potassium depletion stimulates potassium absorption in rat distal colon. Nutr Rep Int 32: 1099–1106

Yajima T, Suzuki T, Suzuki Y (1988) Synergism between calcium-mediated and cyclic AMP-mediated activation of chloride secretion in isolated guinea pig distal colon. Jpn J Physiol 38: 427–443

Chapter 8

Structure-Function Correlations of Colon and Distal Nephron

John P. Hayslett[1]

Contents

1 Introduction

Intestinal epithelia and epithelia lining the nephron have a special attraction for membrane physiologists because of prominent structure-function correlations, unique segmental differentiation, and highly specialized functional differences between individual cell types within segments. This review aims to compare the similarities and differences in structure-function characteristics of epithelia in the colon and collecting duct system in the mammal. Our intent is to provide support for the proposition that the terminal segments of intestine and nephron, involved in the final regulation of electrolyte composition of fecal water and urine, employ anatomical structures and transport mechanisms for the regulation of transepithelial movement of electrolytes that are basically similar. Differences in structure and the incorporation of specialized cells are best explained by the unique roles played by intestine and kidney in the absorbance of nutrients as a source of energy and in overall regulation of water and electrolyte homeostasis, respectively. Details of molecular mechanisms that subserve ion transport will only be listed to the extent that they serve our aim, since they are comprehensively reviewed in other parts of this volume and elsewhere.

[1] Department of Internal Medicine, Section of Nephrology, Yale University School of Medicine, 333 Cedar Street, New Haven, CT 06510, USA

Advances in Comparative and Environmental Physiology, Vol. 16

2 Structure-Function Correlations of the Intestine and Kidney

Similarities between colon and collecting duct system can be appreciated from their respective contributions to the function of the intestine and kidney. Absorptive processes exhibited by intestinal epithelia represent the primary pathway for assimilation of the dietary intake of nutrients, water, electrolytes and minerals into the organism, and a means to recover significant amounts of electrolyte-containing fluids added to the intestinal lumen for nutrient digestion. The intestine is not primarily concerned with homeostatic regulation of water and electrolytes. In addition to the oral intake of 1 to 1.5 l of aqueous solutions per day, 7–9 l of aqueous solutions are added to the intestinal lumen in the upper intestinal tract from salivary, gastric, biliary, pancreatic, and intestinal secretions per day. In carnivores and omnivores the bulk of water, electrolytes, carbohydrates, protein, and fats is absorbed in the small intestine. In man, for example, approximately 60% of the water and electrolyte load is absorbed isosmotically by the jejunum and 20–25% by the ileum (Binder 1983). Regarding nonelectrolyte nutrients, glucose is entirely absorbed within the duodenum and proximal portion of the ileum, and amino acids are absorbed in the proximal and middle portions of small intestine (Borgstrom et al. 1957). The epithelia in these epithelial segments have a low resistance, transepithelial (R_T) of $<30\,\Omega \cdot cm^2$ and a transepithelial potential difference (V_T) of $\pm 4\,mV$, and serve to provide bulk transfer of water and solutes (Schultz and Curran 1968). Although constructed as a single tube, the active absorptive area is increased by circumferential folds and villi that increase the surface area about 35-fold. In addition, microvilli on the apical surface of absorptive cells amplify the surface area 20-fold over that which would be presented by a cell with a flat apical plasma membrane (Brown 1962). Specific transport processes in absorptive cells that provide a means for electroneutral absorption of cations and anions, and absorption of nutrients, are reviewed elsewhere (Sullivan and Field 1991).

In man, approximately 10–15% of the fluid load present in the proximal portion of the intestinal tract is delivered to the large intestine, which includes cecum and colon. In contrast to small intestine, the large intestine is a moderately "leaky" epithelium, $R_T \sim 200-300\,\Omega \cdot cm^2$ with a V_T that may achieve levels $>20\,mV$, lumen negative, and the active absorptive surface area is amplified only moderately by microvilli on the surface of absorptive cells. Although the capacity for absorption of water and sodium is low, estimated in man to be 30–40% of the level in jejunum (Billich and Levitan 1969), transport mechanisms permit NaCl absorption against a steep concentration gradient of approximately 120 mM and active electrogenic secretion of chloride and potassium. In contrast to small intestine, the capacity to regulate levels of individual ions above or below levels in plasma provides a means to adjust the rate of excretion of electrolytes in relationship to overall electrolyte balance. From the viewpoint of efficiency the large intestine absorbs 90% of the load of water and sodium presented to it by small intestine, and can increase the rate of absorption about twofold under conditions where the load delivered is increased (Binder 1983). Since nonruminant herbivores, such as the rabbit, are often used as experimental models designed to elucidate intestinal function in man, it should be recalled that the cecum and colon undergo extensive and complex structural development for microbial catabolism of carbohydrates into short-chain fatty acids which are subsequently absorbed (Lacy 1991). There are corresponding increases in the amount of water present in the luminal compartment and in the amount of energy derived from nutrients absorbed in large intestine. Since anatomical and functional differentiation of large intestine reflects the composition of diet, direct comparisons of nonhuman animals and between experimental animals and man should be undertaken with caution. Figure 1 illustrates the anatomy of the intestinal tract in man and in two animal species, rat and rabbit, which have been extensively used in experimental studies.

In contrast to large intestine, the kidney is primarily concerned with the homeostatic regulation of water and electrolytes achieved by filtering approximately 150 l of plasma water daily, containing approximately 21 000 mmol of sodium chloride and proportionate amounts of other salts in plasma water, and nutrients with small molecular weights such as glucose, amino acids, and polypeptides. The anatomical characteristics of the mammalian nephron are shown in Fig. 2. Structure-function characteristics of the proximal tubule correspond to those of the small intestine. The proximal tubule reabsorbs approximately 70% of water and electrolytes contained in the original ultrafiltrate (Malnic et al. 1966) and nearly all of filtered nutrients (Ullrich 1976). Although not involved in the assimilation of dietary nutrients as an energy source, the reabsorption of glucose and amino acids limits losses into urine, while the turnover of polypeptides provides a mechanism for their catabolism without loss of amino acids. Amplification of the active absorptive surface area is achieved by the use of multiple small tubular structures (nephrons) in parallel, rather than a single large tubular structure, and amplification of the apical membrane of proximal tubule cells with numerous closely packed microvilli. Morphometric analysis of rabbit proximal tubule has shown that the area of these extensions is approximately 20-fold greater than the area of the tubular basement membrane and decreases gradually towards the terminal portion of the proximal tubule (Welling and Welling 1976). This epithelium has a low R_T of $<30\,\Omega \cdot cm^2$ and correspondingly low $V_T < \pm 4\,mV$ (Boulpaep 1976). As in small intestine, NaCl is absorbed by an electroneutral mechanism. Although a review of the individual transporters responsible for solute transport in small intestine and proximal tubule is beyond the scope of this analysis, detailed analyses have demonstrated the same basic molecular mechanisms in epithelia lining both structures (Burg 1986; Sullivan and Field 1991). The

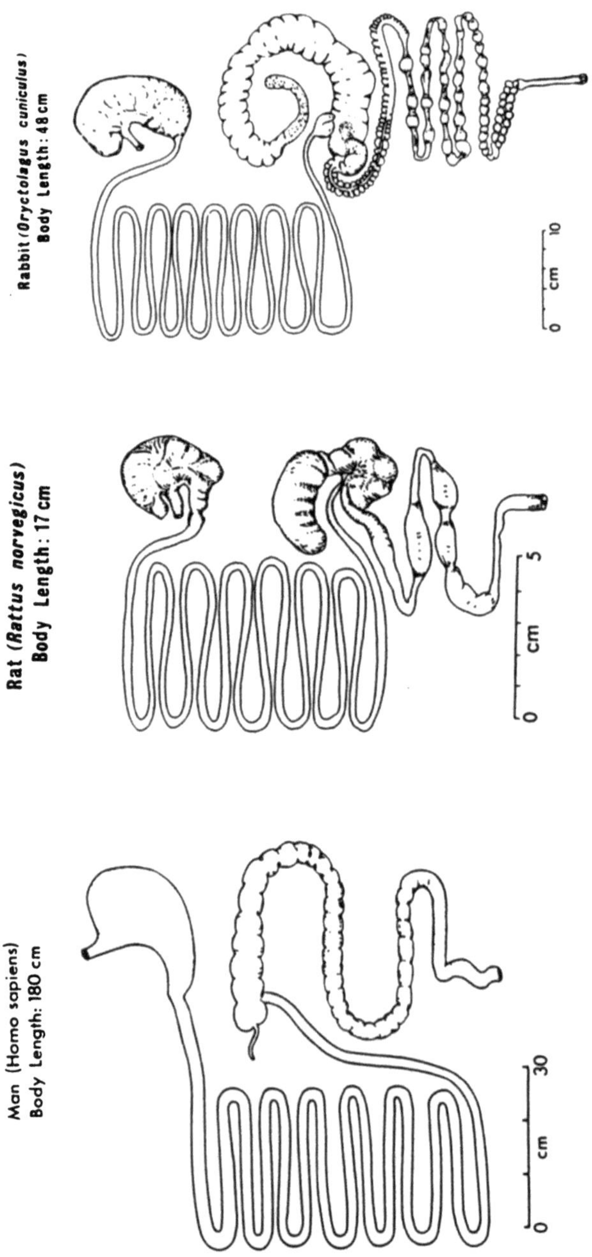

Fig. 1. The gross anatomy of the intestine in man, rat, and the rabbit. (Stevens 1977; Wrong et al. 1981)

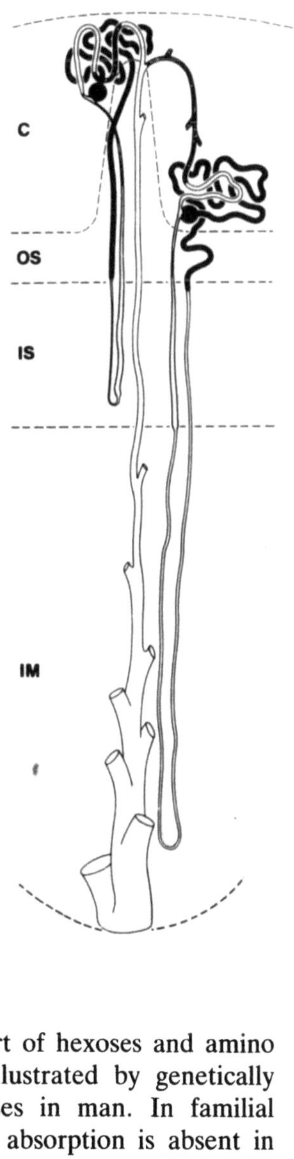

Fig. 2. This diagram shows the anatomical arrangement of a superficial nephron and deep cortical nephron in relationship to zones of the kidney, indicated as *C* cortex; *OS* and *IS* outer strip and inner strip of the outer medulla; *IM* inner medulla. Portions of the nephron derived from the blastema are shown by intermittent dark and light areas, while the portion derived from the ureteric bud is the continuous distal segment shown by clear area extending to the papilla. (Jamison and Kriz 1982)

existence of common mechanisms for the transport of hexoses and amino acids in intestine and kidney, for example, is illustrated by genetically derived abnormalities of these transport processes in man. In familial glucose-galactose malabsorption intestinal glucose absorption is absent in the homozygous individual, who also shows partial impairment in renal glucose transport (Binder 1970). In cystinuria and in Hartnap disease there is a defect in dibasic and neutral amino acid transport, respectively, in both jejunal and renal epithelia.

Approximately 10% of the original ultrafiltrate is delivered to the distal nephron for final regulation of the composition of urine. The anatomical and functional characteristics of this portion of the nephron are designed as a low capacity system with high efficiency. In contrast to proximal tubule, the absorbate is nonisosmotic and ions can be transported against steep concentration gradients such that the urine osmolality varies between 60 and 1200 mOsmol/kg in man and the fraction of filtered ions excreted into urine can be reduced to 1% or less, or, in the case of potassium, to levels greater than the filtered load. The collecting duct system has a moderately high R_T of 300–800 $\Omega \cdot cm^2$ and V_T can increase in some segments to 50 mV, lumen negative (Boulpaep 1976). As in large intestine, the structure of epithelia cells lining the distal nephron is relatively simple compared to proximal tubule, and the apical membrane surface area is only slightly amplified by microvilli (Kaissling and Kriz 1979). The anatomical development of the distal nephron is determined by the environment in which the organism lives, and not by the composition of diet. Because terrestrial life requires a capacity to efficiently conserve water, evolutionary adaptation has led to the establishment of a zone in the kidney (the medulla) that has an extracellular fluid osmolality greater than plasma. The distal nephron courses through this zone and the rate of passive water absorption is regulated by hormone-induced changes in apical membrane water permeability. The length of the medullary portion of the collecting duct is directly proportional to the maximal concentration capacity, being short in aquatic animals like the beaver and long in desert rodents (Bankir and de Bouffignac 1988). There is no parallel anatomical arrangement for the distal portion of the intestinal tract.

In summary, there are remarkable similarities in the architecture of the intestinal tract and the nephron. The proximal portions of both structures provide for isosmotic absorption of approximately 85–90% of water and electrolytes and nearly complete absorption of nutrients. The system which subserves these processes involves a large active surface area for absorption and a low resistance epithelium with electroneutral transporters. Distal portions of intestine and nephron are designed as low capacity systems with transport mechanisms for fine regulation of water and the chemical composition of the final excreta. This functional role requires epithelia with relatively high transepithelial resistances to provide a means to sustain large chemical gradients of ions across the epithelium. Differences between colon and the distal nephron correlate with the specific roles of these tissues in body function.

3 Embryonic Development of Colon and Distal Nephron

The colon is formed from portions of the embryonic midgut and hindgut in mammals. As the embryo begins to fold on itself, the primitive gut forms from partial internalization of the endoderm-lined yolk sac, with blind endings in the cephalic and caudal portions of the embryo. The midgut extends from the embryonic liver buds to the midportion of the colon, and includes the cecum and proximal colon. The hindgut differentiates into the

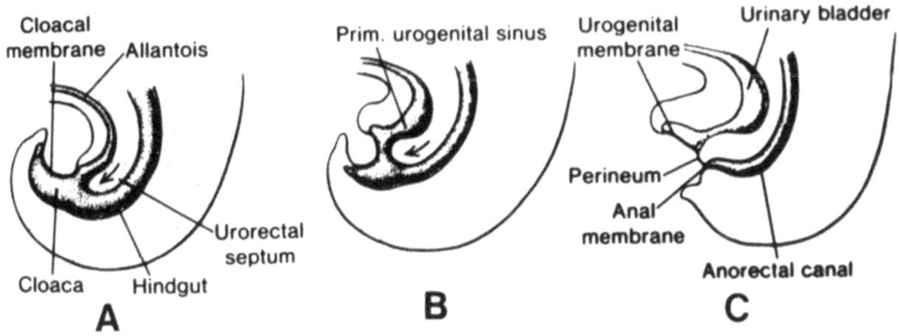

Fig. 3A–C. The embryonic development of the hindgut showing separation of distal colon and rectum from the urogenital sinus by the urogenital septum. (Lacy 1991)

distal colon, rectum, and anal canal. All parts of the gut are lined with endoderm. As the hindgut extends caudally it is compressed by the urorectal septum between the hindgut and the allantois, leading the blind end of the hindgut to form a cloaca in the embryo and lower vertebrates. In mammals the septum extends further and divides the cloaca into two parts, the anteriorly positioned primitive urogenital sinus and the bladder, and the posterior anorectal canal as shown in Fig. 3. After further development, the wedge of tissue forming the septum will become the perineal body. This course of development in the embryo indicates that cecum and proximal colon are derived from a different portion of the primitive gut than the distal colon and rectum, and that both distal colon and lower urinary tract are derived from the hindgut.

Concurrently, the ureteral bud, lined by endoderm, arises from the lowermost portion of the isolateral mesonephric duct located close to the cloaca near the termination of the hindgut. The ureteral bud begins as a branched tube surrounded by metanephric blastema. The spherical termination of the ureteral bud and the metanephric cells in the blastema by which it is surrounded will become the kidney, while the more proximal portion of the ureteral bud will become the ureter. As the ureteral bud extends cranially, in a dorsolateral direction, it divides into branches, such that the pelvis and major calyces will form within the first three to five generations of branches. Additional dichotomous branchings form the collecting tubules within the kidney which extend through the medulla and into the cortex. The blind end of the branches of the ureteral bud are called ampulla, capable of inducing the formation of nephrons from cells in the blastema as shown in Fig. 4. Eventually, after development of the glomerulus and tubule, the lumen of the tubule opens into the lumen of the collecting tubule. The ureteral bud, therefore, forms the ureter, which opens caudally into the bladder, all the major extraparenchymal cavities within the kidney, and the collecting tubules which form the distal portion of each nephron.

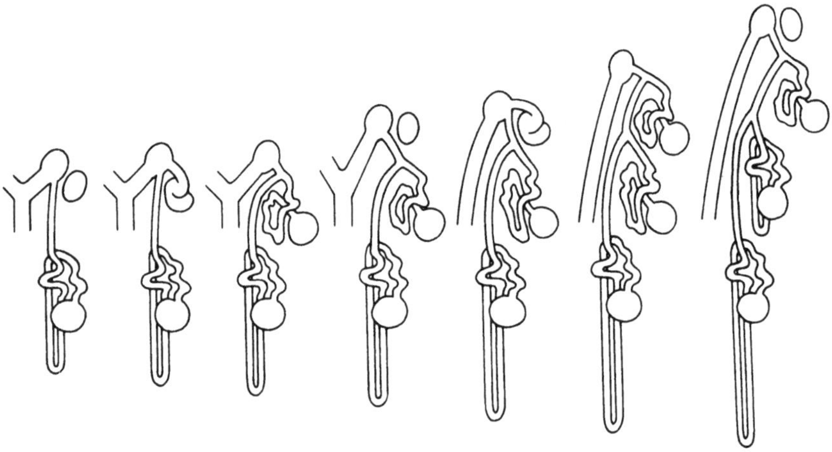

Fig. 4. Diagram of the arcade formation of multiple nephrons with a single cortical collecting tubule. Two nephrons are temporarily attached to each terminal ampulla, but the connecting piece of the older nephron shifts to communicate only with the connecting piece of the younger nephron. (Potter 1972)

The two portions of the nephron, the proximal nephron and collecting duct system, exhibit distinctive structural and functional characteristics.

4 Segmental Function and Sodium Reabsorption

Since the mammalian proximal and distal colon are derived from different portions of the embryonic gut, it is, perhaps, not surprising that the large intestine exhibits heterogeneity along its axis. In the rat, for example, V_T, lumen negative, is relatively low in the proximal colon in vivo (-6 ± 2 mV), but increases to -15 to -20 mV at about the midpoint of distal colon (Fisher et al. 1976). Correspondingly, R_T is lower in proximal colon compared to the distal segment (Foster et al. 1984). Morphological studies demonstrated marked differences in surface absorptive cells, or principal cells, between the proximal and distal colon in the rat. Cells lining the proximal colon were larger, measuring 37 ± 1 µm in height and 6 ± 0.2 µm in width, compared to dimensions of 21 ± 0.8 µm and 4 ± 0.2 µm in distal colon (Kashgarian et al. 1980). In addition, morphometric measurements showed that the ratio of basolateral membrane surface area to cell volume was greater in distal colon (1.85 ± 0.8 µm^2/µm^3) than in the proximal colon (1.48 ± 0.04). The surface density of microvilli on the luminal membrane (S_s) was also computed, and represents the ratio of villus surface to luminal surface area. This factor indicates amplification of the apical surface by microvilli. In parallel with differences in basolateral membrane area between the two segments, S_s was greater in the distal colon (33.2 ± 1.1 μm²/μm²) compared to proximal colon (19.3 ± 1.5).

The mucosal surface in colon consists of two major regions, the flat superficial region between crypts, comprised of columnar cells (principal cell) capable of NaCl absorption and responsive to aldosterone, and the crypts themselves which extend deep into the mucosa. The crypt, a structure unique to the intestine, contains undifferentiated cells, enteroendocrine cells, goblet cells, vacuolated cells, and columnar cells (principal cells). The undifferentiated cells are confined to the lower one-third of the crypt and serve as stem cells for the other cell types, which migrate towards the lumen surface. Regarding intestinal ion transport by cells lining the crypt, there is evidence that at least one type of cell, probably the early principal cell, is capable of active chloride secretion. The mechanism of this specialized function involves cell entry of chloride into the cell across the basolateral cell membrane via a Na:K:2Cl transporter and passive diffusion through apical chloride channels into luminal fluid (Cartright et al. 1985). This process is mediated by cAMP, which serves to activate chloride channels, is regulated by vasoactive intestinal peptide, and probably other factors, and provides a means to add a NaCl-rich solution to the bulk of luminal contents. Interestingly, the capacity of principal cells for cell division is lost when cells are differentiated to absorptive cells on the mucosal surface.

The three-compartment model for sodium absorption requires a mechanism for cell entry of sodium across the apical membrane, down an electrochemical gradient, and a mechanism for cell extrusion across the basolateral membrane by the Na:K exchanger. The mode of entry across the apical membrane could involve passive diffusion via a sodium channel, and, thus, result in electrogenic sodium transport, or an electroneutral, secondary, active mechanism involving either Na:K:2Cl or Na:Cl co-transport, or the Na:H antiport exchanger. Transepithelial absorption of chloride could occur by passive diffusion through the paracellular pathway and/or by a transcellular pathway. Studies in the rat indicate that the primary mode of NaCl absorption is electroneutral under basal conditions in proximal and distal colon. In both segments there is also evidence for low level electrogenic sodium transport, since the short-circuit current (Isc), which represents the net current due to flow of ions across the epithelium, is reduced to zero after removal of sodium from apical bathing solution.

In distal colon the mechanism of electroneutral NaCl absorption involves Na:H and Cl:HCO_3 (or Cl:OH) exchangers coupled by intracellular pH. Evidence to support this mechanism includes: (1) inhibition of NaCl absorption by acetazolamide and 1 mM amiloride, a concentration that inhibits the Na:H exchanger in other epithelia (Binder and Rawlins 1973; Binder et al. 1987); (2) dependence of NaCl absorption on the presence of both ions in luminal fluid, but independence from the presence of K in luminal fluid (Binder et al. 1987); and (3) inhibition or stimulation of NaCl transport under conditions of systemic alkalosis and acidosis, respectively

(Charney and Feldman 1984). Recently, the Na:H exchanger has been demonstrated in membrane vesicles derived from proximal and distal apical membranes of rat colon (Binder et al. 1986; Foster et al. 1986b). A Cl:OH exchanger has been identified in apical membrane from rat distal colon (Rajendran et al. 1991). Under basal conditions, electroneutral NaCl transport in proximal colon appears to result from active sodium transport through a transcellular route via an apical Na:H exchanger, and passive chloride diffusion through the paracellular pathway (Foster et al. 1986b).

Rabbit colon has been extensively studied and an electroneutral mechanism for NaCl absorption stimulated by epinephine was demonstrated in proximal colon (Sellin and DeSoignie 1984). Since sodium absorption was inhibited by mM concentrations of amiloride (Sellin and DeSoignie 1984) and intracellular pH was decreased by mM concentrations of amiloride and by sodium removal from the apical bath solution (Ahn et al. 1985), it is likely the sodium absorption depends on a Na:H exchanger. In distal colon, in contrast, Isc was roughly equivalent to J_{net}^{Na}, suggesting that an electrogenic mechanism accounts for sodium absorption (Frizzell et al. 1976).

Segments of human colon have been studied in in vitro experiments. In flux studies, under voltage-clamp conditions, electroneutral NaCl absorption was demonstrated in both proximal and distal colon in parallel with electrogenic sodium absorption (Sellin and DeSoignie 1987). Electrophysiological experiments showed that Isc was similar in both segments. The Isc in distal colon, but not in proximal colon, was amiloride-sensitive (Sandle and McClone 1987). Comparison studies with rat colon demonstrated that the voltage across the basolateral membrane was similar in both species (~40 mV), while the voltage across apical membrane in human colon was one-third lower than in rat, due apparently to higher conductance.

Although the cecum shares the same embryonic origin with proximal colon, it is of interest that the mode of sodium transport has been shown to differ between these segments of the colon, at least in flux experiments in the rabbit, which demonstrated that net sodium absorption was higher in the cecum and that sodium transport was due to an electrogenic mechanism (Sellin et al. 1988). In contrast to distal colon, however, electrogenic sodium transport in cecum was resistant to the inhibiting action of amiloride, but was inhibited by the potent amiloride analog phenamil, suggesting a difference in the affinity of the amiloride-binding site on the sodium channel among different intestinal segments. The mode of NaCl absorption in rat cecum appears to be by an electroneutral mechanism, as in rat proximal colon (Escobar et al. 1990).

The difference in the mechanisms for sodium handling between proximal and distal colon is amplified during stimulation by aldosterone. Specific, high-affinity receptors for aldosterone have been demonstrated in proximal and distal colon in the rat (Schulman et al. 1986). Chronic hyperaldosteronism, induced by a Na-free diet, stimulated electroneutral NaCl transport in rat proximal colon, while in distal colon aldosterone stimulated electrogenic amiloride-sensitive sodium absorption and inhibited electroneutral NaCl transport (Foster et al. 1983). Under this condition electroneutral NaCl transport in proximal colon appears to involve Na:H and Cl:HCO_3 antiporters acting in parallel (Foster et al. 1986a). Other types of cell expression induced by aldosterone, however, were similar in both segments; chronic hyperaldosteronism resulted in an increase in the number of copies of Na:K ATPase in basolateral cell membrane and amplification of the area of basolateral cell membrane in both intestinal segments (Kashgarian et al. 1980).

Both the proximal and distal colon possess glucocorticoid receptors (Bastl et al. 1989a), in addition to the mineralocorticoid receptor. Recent studies in the rat show that activation of glucocorticoid receptors with low-dose glucocorticoids stimulated electroneutral NaCl absorption in proximal and distal colon (Bastl et al. 1989a). Under in vivo conditions this mode of transport was not inhibited by 0.1 mM amiloride or the specific Na channel blocker benzamil, or by the mineralocorticoid receptor antagonist spironolactone, but was inhibited by the Na-H antiport inhibitor 5-N-ethyl-N-isopropylamiloride. The synthetic glucocorticoid RU26988, which does not bind to the mineralocorticoid receptor, resulted in similar changes in NaCl transport (Bastl 1987). However, when dexamethasone was administered at a high dose that occupied greater than 80% of aldosterone receptors in the distal colon, sodium conductive absorption was stimulated due to crossover binding to the mineralocorticoid receptor (Bastl et al. 1989b). Taken together, these data indicate that glucocorticoids regulate electroneutral NaCl absorption in the entire large intestine, probably by stimulating apical membrane Na:H, Cl:OH exchangers.

It has not been determined whether the cellular actions of aldosterone and glucocorticoids are complementary of counterregulatory. Based on observations made in rat colon, Bastl suggests that glucocorticoids regulate basal NaCl absorption by an electroneutral process (Bastl et al. 1989a). Since aldosterone levels are low in the euvolemic subject, and probably below the threshold for mineralocorticoid activation, the role of aldosterone in the regulation of sodium transport is probably negligible. During volume depletion, however, the rise in serum levels of aldosterone by as much as 100-fold is sufficient to activate the mineralocorticoid receptor. In this condition aldosterone stimulates electroneutral NaCl absorption in proximal colon, but activates, as noted above, electrogenic sodium absorption and inhibits electroneutral NaCl transport in distal colon. In this scheme, the action of aldosterone to stimulate electrogenic sodium absorption in the most distal portion of the colon may provide a more rapid and efficient means to defend the homeostasis of sodium and water balance than could otherwise be provided by mechanisms which prevail umder basal conditions.

In this comparative analysis of the colon and collecting duct system, it is instructive to review the cellular mechanisms for electrogenic sodium transport employed by these tissues. Studies on rat distal colon during chronic hyperaldosteronism demonstrated a sevenfold increase in V_T and a 52% decrease in R_T (Sandle et al. 1984). The rise in V_T reflected an increase in the voltage of the basolateral membrane (V_b) and a fall in apical membrane voltage (V_a). Resolution of R_T into its separate membrane components showed that the fall in R_T was entirely due to a 66% decrease in apical membrane resistance, caused by activation of amiloride-sensitive sodium conductance. Similar findings were reported under basal conditions in rabbit distal colon (Wills et al. 1979). These findings indicate that electrogenic sodium transport is modulated by factors that control the rate of diffusive movement of Na ions into the cell across the apical membrane. Since studies in rabbit distal colon showed that intracellular sodium levels were unchanged during large increases in electrogenic transport (Turnheim et al. 1987), there is evidence that the rate of cell sodium extrusion by the basolateral Na:K exchanger varies in parallel with changes in the rate of apical membrane conductance for sodium.

The collecting duct system in the mammalian kidney extends from the late part of the distal tubule to the ducts of Bellini. Collecting tubules begin in the cortex not far from the kidney surface and join with other tubules (an average of 5–10) as they descend through the cortex before transversing the outer medulla as unbranched tubules (Potter 1972). On entering the inner medulla, the collecting tubules undergo successive fusions until they form the final tubule, called the duct of Bellini. During their descent towards the papilla, there is a gradual increase in the outer and inner diameters of tubules. In rat, rabbit, and man approximately 1200–2000 nephrons drain into a single duct of Bellini. There are two cell types in the collecting duct system, the principal cell, capable of absorptive and secretory processes involving Na, K, Cl, Ca, PO_4, and other ions, and intercalated cells which secrete H or HCO_3 into luminal fluid, depending on whether the proton pump is located on the apical or basolateral membrane (Steinmetz 1986). Transition from the distal convoluted tubule to the initial collecting tubule is distinct morphologically in some species, such as rabbit, but gradual in others, such as the rat, by the replacement of an apical membrane with dense, short microvilli to a cellular configuration with few apical microvilli; in addition, there is a marked reduction in lateral cell membrane interdigitations (Jamison and Kriz 1982). Detailed descriptions of the principal cell along the course of the collecting duct system are found elsewhere (Madsen et al. 1988). Intercalated cells are characterized structurally by numerous mitochondria and polysomes and prominent microplicae on the luminal surface, and are found in the initial and cortical collecting tubule and outer medullary collecting tubule (Jamison and Kriz 1982).

Analysis of ion transport in this portion of the nephron is confounded by the small size of individual tubules and their relative inaccessibility to direct study in vivo by techniques such as micropuncture except for the initial collecting tubule present on the kidney surface. Our understanding of the function of the cortical and medullary portions of the collecting duct system, therefore, is derived primarily from in vitro experiments using the isolated perfused tubule technique in the rabbit, an herbivore, and, more recently, in the rat, an omnivore. Insights into the inner medullary collecting duct in vivo, however, have been provided by microcatherization studies in which the catheter, inserted into the duct of Bellini, can be advanced to the corticomedullary border. Since neither technique permits voltage clamping, the mode of ion transport in the nephron cannot readily be determined, in contrast to intestinal tissue where the larger size allows analysis in voltage-clamp chambers.

Taken together, these experimental approaches have demonstrated continuous absorption of sodium along the axis of the collecting duct system in the rat and rabbit. In the basal state approximately 5–8% of the filtered load of sodium is delivered to the initial collecting tubule and 2–3% to the medullary collecting tubule (Giebisch et al. 1964; Sonnenberg 1977). The fractional excretion of sodium in final urine is 0.1 to 0.5% of the amount in the initial ultrafiltrate. The collecting duct system, therefore, has a relatively low capacity for sodium absorption, compared to proximal tubule, and exhibits a capacity to transport sodium against a steep electrochemical gradient, similar to mammalian colon. The profile of transepithelial potential differences along the axis of the collecting duct system is heterogeneous. In the rat, in vivo, V_T is about 30 mV, lumen negative, in the initial collecting tubule (Hirsch et al. 1984) and near zero (±2 mV) along the inner medullary collecting tubule (Hayslett et al. 1980). In rabbit and rat cortical collecting tubule the luminal potential difference varies from +10 to about −15 mV, depending on diet and acid-base balance (Gross et al. 1975).

Since the concentration of sodium in tubular fluid is below plasma levels along the entire axis of the collecting duct system, it is clear that sodium absorption is mediated by an active mechanism. In the steady, or near steady state, the maximum gradient for sodium is 0.32 in the rat initial collecting tubule (Hierholzer et al. 1965) and 0.40 in rabbit cortical collecting tubule (Grantham et al. 1970). Because the sodium concentration in medullary interstitial fluid can exceed 200 mM in the rat, and probably also in man, and can be reduced to levels below 5 mM in final urine, the maximal transepithelial gradient for sodium absorption is greatest along the inner medullary collecting tubule. Because voltage-clamp experiments have not been performed in experiments on segments of the collecting duct system, the mode of sodium transport has not been directly assessed. Insights into whether movement of sodium is regulated by electrogenic or electroneutral mechanisms can be inferred, however, from the inhibitory action of amiloride in μM concentrations and correlates of net sodium transport and V_T. Based on these indirect criteria, it seems likely that sodium absorption in rat initial collecting tubule is electroneutral since V_T is unaffected by adrenalectomy (Hierholzer et al. 1965) or administration of aldosterone (Hirsch et al. 1984), although sodium absorption is reduced by adrenalectomy (Hierholzer et al. 1965). In rat cortical collecting tubule the basal level of absorption was too low to evaluate, but desoxycorticosterone stimulated sodium absorption by both amiloride-sensitive and amiloride-insensitive mechanisms (Tomito et al. 1986). These results were interpreted as showing two modes of sodium transport, electrogenic and electroneutral. Sodium transport in rabbit cortical collecting tubule is regarded to involve only an electrogenic mechanism under basal conditions, while in the rabbit outer medullary collecting duct sodium movement is governed solely by a passive mechanism (Stokes 1982). Lastly, sodium absorption in rat inner medullary collecting tubule appears to involve an electroneutral process since V_T persists at a level of $\pm 2\,mV$ despite variations in sodium absorption (Hayslett et al. 1980).

Studies in rat and rabbit nephrons demonstrated a distinct pattern of [^{3}H] aldosterone binding, with a high concentration of binding sites (receptors) and specificity only in divisions of the collecting duct system, including initial collecting tubule, cortical collecting tubule, and outer medullary collecting tubule (Doucet and Katz 1981). Specific binding was negligible in more proximal portions of the nephron, but the inner medullary collecting tubule was not examined because extensive branching precluded analysis. Recently, aldosterone receptors have been identified in inner medullary collecting tubule with a monoclonal antibody to the aldosterone receptor (Lombès et al. 1990). Analysis of tubular transport with a free-flow micropuncture and microcatherization technique in the rat in vivo, and with in vitro isolated perfused nephrons in the rabbit, demonstrated that all segments of the collecting duct system, like the proximal and distal segments of colon, are responsive to the action of aldosterone which induces increases in sodium, potassium, and/or proton transport (Wiederholt et al. 1966; Ullrich and Papavassilisu 1979; Field et al. 1984; Hirsch et al. 1984). As noted above, mineralocorticoids appear to induce electroneutral absorption in the initial collecting tubule and inner medullary collecting tubules, in contrast to the cortical collecting tubule where electrogenic sodium absorption is stimulated. Taken together, these reports indicate that nephron segments derived from the ureteral bud are target sites for aldosterone, in contrast to nephron segments derived from the blastema which lack specific hormone receptors.

As in colon, the collecting duct system contains glucocorticoid-responsive epithelia. The profile of high affinity binding sites parallels that for aldosterone in the sequential segments of the collecting duct system (Lee et al. 1983). Because reported experiments have employed pharmacological doses or pharmacological concentrations of synthetic glucocorticoids, in which crossover binding to the mineralocorticoid receptor cannot be excluded, the physiological role of endogenous glucocorticoids in the regulation of sodium transport has not been determined. Further studies with low physiological concentrations of endogenous hormones or "pure" synthetic glucocorticoids which lack an affinity for the mineralocorticoid receptor are required to determine whether glucocorticoids regulate basal sodium absorption in euvolemic animals as proposed in the large intestine.

The molecular basis for electroneutral sodium absorption in the collecting duct system has not been elucidated. The lack of information in this area probably reflects a dominant focus on electrogenic mechanisms by renal physiologists and the restrictions placed on experimental studies by the small tubular structures. Cell impalement studies on rabbit cortical collecting duct, however, have demonstrated that aldosterone stimulated electrogenic sodium absorption by the same mechanism described in distal colon. The action of aldosterone in principal cells is characterized by a marked increase in V_T, due to depolarization of the apical cell membrane and hyperpolarization of the basolateral membrane (Sansom and O'Neil 1985). Aldosterone modulates apical sodium conductance over a wide range by activating amiloride-sensitive sodium channels, and thus modulates a wide-range control of sodium entry into the cell and, hence, the rate of net sodium absorption.

An overview of structure-function characteristics of colon and collecting duct system in the mammal suggests a common design for the regulation of sodium absorption. Both systems employ an epithelium with a moderately low conductance that permits the generation of steep transepithelial gradients for sodium, and relatively noncomplex cells with only moderate amplification of the apical membrane. Evidence shows that both electroneutral and electrogenic mechanisms are utilized for sodium transport in the large intestine and distal nephron and both tissues are target sites for the action of aldosterone. Analysis of the mechanisms for sodium absorption along the axes of colon and collecting duct, however, reveals substantial differences; a mechanism for electrogenic sodium transport is present in the distal portion of colon, while in the medullary collecting duct the transport mechanism is more heterogeneous and available evidence is most consistent with electroneutral transport in inner medullary collecting tubule. Although any explanation for the functional differences of these systems remains speculative, it seems likely that functional properties of the inner collecting tubule are imposed by the architecture of the nephron. In contrast to the distal colon, the inner medullary collecting tubule is bathed in a hypertonic solution with steep corticomedullary gradients for the major cations in extracellular fluids, while the concentration of sodium in luminal fluid can fall to levels that equal intracellular levels. Under these conditions it is possible that an apical membrane with a high sodium conductance would be less effective for sodium conservation, since the chemical gradient for sodium across the apical cell membrane would be markedly reduced.

5 Segmental Function and Potassium Transport

In the basal state potassium is secreted by colonic mucosa in vivo. Because of luminal negativity, it seemed possible initially that net flux was passive, driven by the electrical gradient through conductive paracellular pathways. Studies conducted in vivo in rat colon, however, demonstrated active potassium secretion in proximal colon against an electrochemical gradient of ~32 ± 1.3 mV when net water and electrolyte movement approached zero. In distal colon, but not in proximal colon, active absorption against a gradient of −17 ± 2.2 mV was demonstrated under the same conditions (Kliger et al. 1981). These phenomena were subsequently confirmed in voltage-clamp experiments (Foster et al. 1984). Further studies demonstrated that rat distal colon was capable of net potassium absorption when V_T was reduced by limiting the rate of sodium absorption (Hayslett et al. 1982).

Identification of active secretory and absorptive mechanisms in rat colon indicated a transcellular pathway for transmucosal movement of potassium. It seemed likely that for potassium secretion to occur potassium entered the principal cell, at least in part, by the Na:K exchanger, and was extruded by a passive mechanism across the basolateral and apical membranes via K channels. The fraction of potassium which exited across the apical membrane and thereby accounted for transepithelial movement was dependent on the conductance of the apical membrane for potassium and the electrochemical force acting on potassium ions across that cell surface. Cell impalement studies demonstrated that free potassium ions were above their equilibrium potential in both rat (Wills and Biagi 1982) and rabbit distal colon (Sandle et al. 1985). It seemed likely, therefore, that a primary active or secondary active mechanism across apical membrane in distal colon accounted for transepithelial potassium absorption. In rat colon the potassium entry mechanism was shown to be electroneutral and was chloride- and sodium-independent. It was proposed that this mechanism was best explained by an electroneutral K:H exchanger, and evidence was cited suggesting that the exchanger could be K:H ATPase, similar to the exchanger described in gastric parietal cells (Foster et al. 1984). Subsequently, an oubain-insensitive potassium entry mechanism was reported in apical membrane of rabbit distal colon that was consistent with the same molecular mechanism (Wills and Biagi 1982).

Segmental differences between proximal and distal portions of colon are amplified under conditions where potassium secretion is stimulated. During chronic secondary hyperaldosteronism, induced by a high potassium diet, the active secretory transport potential in vivo was increased in proximal colon from 32 to 47 ± 0.9 mV and active absorption was transformed to active secretion against a potential of 54 ± 2.7 mV (Kliger et al. 1981). These hormone-induced changes in transcellular movement of potassium were corroborated by in vitro voltage-clamp experiments performed on colon tissue obtained from animals with hyperaldosteronism, induced by a low sodium diet or by a high potassium diet (Foster et al. 1984). In addition, it was found that net potassium secretion was unaltered after sodium absorption was abolished by the presence of amiloride in the apical solution

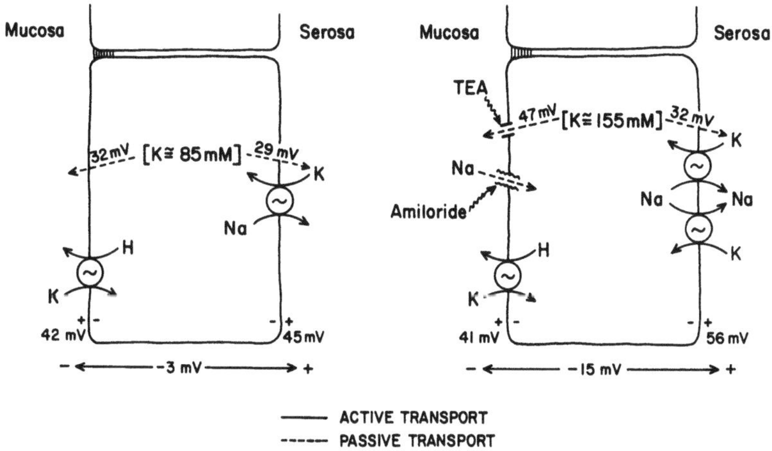

Fig. 5. Proposed model for potassium transport in distal colon in controls (*left*) and potassium-loaded (*right*) animals. In these experiments, the activities of potassium and sodium in the bathing solution were 5.7 and 103 mM, respectively. —, Active transport; --- passive transport. (Sandle et al. 1985)

(Foster et al. 1984). Moreover, under this condition the short-circuit was equal to net potassium secretion, indicating that potassium secretion was electrogenic.

The cellular mechanism responsible for sustained increases in potassium secretion in potassium-loaded animals was examined in rat distal colon under in vitro conditions (Sandle et al. 1985). Compared to control, potassium loading resulted in a fivefold increase in V_T, lumen negative, a 15-fold increase in short-circuit current and a 52% decrease in R_T, as shown in Fig. 5. The rise in V_T was associated with an increase in V_b from 45 ± 2 to 56 ± 2 mV, in the absence of a change in V_a compared to the control value of 42 ± 2 mV. Using potassium-sensitive microelectrodes, the cellular activety of potassium rose to 153 ± 12 mM, compared to a control value of 85 ± 4 mM, in proportion to the increase in V_b, such that the potassium equilibrium potential (E_k) across the basolateral membrane was unchanged. In further experiments the action of potassium loading on apical membrane conductance was analyzed by testing the effects of the sodium channel blocker amiloride and the potassium channel blocker tetraethylammonium on the ratio of R_a/R_b. These studies indicated that potassium loading induced activated sodium and potassium channels in apical membrane. Taken together, these observations suggested that chronic potassium loading resulted in the increased rate of potassium secretion by increasing the rate of cell uptake of potassium by increasing the number of Na:K exchangers in basolateral membrane and the rate of passive efflux of cellular potassium across the apical membrane by inducing activated apical potassium channels. Moreover, the rise in the electrochemical gradient for potassium ions across the apical membrane from 32 to 47 mV augmented the driving force favoring potassium efflux.

As in the study of sodium absorption, transport of potassium by the collecting duct system has been analyzed by different experimental approaches. Free-flow micropuncture studies demonstrated that under basal and most experimental conditions, 90 to 95% of filtered potassium is reabsorbed by the proximal tubule and loop of Hendle, resulting in the delivery of 5 to 10% of the filtered load to the early distal convolution (Malnic et al. 1964). The amount of potassium excreted in final urine is, therefore, determined by the collecting duct system, and ranges from less than 5% to more than 100%, depending on the physiological condition. All segments of the collecting duct system exhibit the capacity for net potassium secretion, but net absorption has been demonstrated only in the inner medullary collecting tubule (Backman and Hayslett 1982). The role of the various segments of the collecting duct system in regulation of potassium balance can be appreciated by analyzing the rates and direction of net potassium movement under conditions where an increase in excretion rate is stimulated or potassium retention is induced. During chronic potassium loading, net secretion is markedly stimulated in the initial collecting tubule (Wright et al. 1971), cortical collecting tubule (Fine et al. 1979), and inner medullary collecting tubule (Schon et al. 1981). During potassium deprivation, net potassium secretion falls to zero in the initial collecting tubule (Stanton et al. 1981) and cortical collecting tubule (Schwartz and Burg 1978), resulting, in the rat, in the delivery of about 3% of filtered potassium to the inner medullary collecting duct, where a mechanism of potassium absorption causes the absorption of 90% of the load delivered to this segment (Backman and Hayslett 1982).

The mechanism responsible for potassium secretion in cortical segments of the collecting duct system is similar to that in the colon. Potassium enters the cell by the Na:K exchanger and diffuses across the apical membrane into apical fluid when the apical membrane is conductive for potassium. The rate of transport across the apical membrane is dependent on apical membrane conductance and the electrochemical driving force. In rabbit cortical collecting tubule mineralocorticoids cause a marked increase in apical membrane conductance (O'Neil 1990). In contrast to the finding in rat distal colon (Sandle et al. 1985), a recent study reports that cellular potassium activity did not increase in rabbit cortical collecting duct after treatment with pharmacological doses of deoxycorticosterone, although V_b increased (Sansom et al. 1989). Since $E_K < V_T$ these data suggested the existence of a passive driving force favoring diffusion of potassium into the cell across the basolateral membrane. This finding is not, however, in accord with measurements of cellular potassium by electron-probe analysis that found an increase in total potassium in the principal cell of the rat cortical collecting tubule after exposure to aldosterone (Beck et al. 1987). Because the cell potassium concentration is about ~90 mM in papillary collecting duct cells, it seems likely that potassium secretion by the inner medullary collecting tubule into tubular fluid, containing high levels of potassium, requires a nonconductive transporter in the apical cell membrane. The molecular mechanism for potassium cell entry across the apical membrane to account for net potassium absorption has not been identified. However, in rabbit outer medullary collecting tubule the potent inhibitor of gastric H-K-ATPase, omeprazole, abolished net potassium absorption and proton secretion, suggesting that H-K-ATPase regulates potassium absorption (Wingo 1989). In addition, monoclonal antibodies against gastric H-K-ATPase stained the cytoplasma of intercalated cells in outer medullary collecting tubule (Wingo et al. 1990).

In summary, there is a remarkable similarity between colon and collecting duct in regard to potassium handling. Epithelia extending along the axis of both structures exhibit a capacity for potassium secretion and can adapt to provide high levels of secretion when body stores of potassium are increased. Conversely, when body stores are reduced, both structures exhibit changes in transport which serve to conserve potassium, involving elimination of secretion in their proximal parts and net absorption in the terminal portions of colon and collecting duct system. It should be noted that in health, the intestinal excretion of potassium plays a minor role in maintaining potassium balance. Approximately, 10% of the oral intake of potassium is excreted in fecal water (Hayes et al. 1967), and when the colon is excised the kidney can easily compensate for the absence of the colon's contribution to total potassium excretion. In renal failure, however, the rate of intestinal potassium excretion can account for as much as 35% of dietary potassium (Hayes et al. 1967)

6 Functional Differences Between Colon and Collecting Duct System

The foregoing discussion illustrates the parallel mechanisms for transport of sodium and potassium in the terminal portions of intestine and nephron. As noted, these structure-function relations serve to prevent the external loss of salt solutions added to the intestinal lumen to promote digestion, and to permit the kidney to regulate the total amount and concentration of solutes in body fluids within a narrow range. The anatomical arrangement in the kidney which uses multiple small tubules rather than a single tube, provides an appropriate larger membrane surface area to transport larger amounts of electrolytes transported by the collecting duct system compared to colon. For example, approximately 90 mEq of sodium and 20 mEq of chloride are absorbed daily by the colon in man. In contrast, 1000 mEq of sodium are absorbed by the renal collecting duct system, assuming that approximately 5% of the filtered load is delivered to the initial collecting tubule.

Similar differences in absolute potassium secretion by these structures can be estimated.

Functional differences between colon and collecting duct system probably relate to the unique roles these structures play in subserving the needs of the organism which require specialized adaptation. For example, a system to regulate water balance characterizes the kidney and includes steep corticomedullary gradients of solutes and, in some cases, recycling systems between different nephron segments in the medulla for urea, ammonia, and potassium metabolism. A similar structure-function relationship does not exist in large intestine. The specialized crypt cell in colon with a capacity to secrete chloride, presumably to promote the movement of solid waste material along the large intestine, has no counterpart in the kidney. Conversely, highly specialized cells for proton secretion are found in the collecting duct system that permit a means to effectively excrete 50–75 mEq of metabolic acid daily in man; net acid excretion by intestine does not occur since luminal fluid in colon and stool water is rich in bicarbonate. The presence of similar segmental differentiation and molecular mechanisms for transepithelial movement of salts in the terminal portion of the intestine and distal portion of the nephron suggests that successful systems were retained during evolution and, perhaps, reflects similar embryonic origins.

Besides the esthetic features of these structure-function relationships, the parallel systems serve a useful role for investigators of epithelial function. Historically, the identification of cotransporters and antiporters in either small intestine or proximal tubule led to rapid and usually successful searches for the same transport protein in the alternate system. In studies on the mode of transport, active vs passive or electrogenic vs electroneutral, and correlations between biochemical events and transport, intestinal tissue, because of its larger size, has permitted insights that were not readily achieved in studies on segments of the nephron. These insights, however, serve as an important basis for designing experiments to elucidate similar processes in the kidney. For example, when we initially demonstrated that increases in the excretory load of potassium per nephron were associated with an increase in the activity of Na-K-ATPase in homogenates of renal tissue (Silva et al. 1973), it was important to determine whether changes in enzyme levels occurred in the same cells that were capable of potassium secretion. Since techniques were not available at the time to estimate Na-K-ATPase activity at the level of individual nephron segments, we turned to the mammalian colon as another major tissue capable of potassium secretion. Using an in vivo technique to estimate net potassium transport, it was possible to correlate enzyme levels and flux rates of potassium directly in the same epithelium (Fisher et al. 1976), and in further studies determined that potassium loading stimulated active potassium secretion (Kliger et al. 1981). Subsequent advances in technology permitted others to confirm the phenomena in segments of the collecting duct system (Doucet and Katz 1980). Current interest in signal transduction mechanisms favors cultured cells as models for experimentation. Because cultures of absorptive intestinal cells are not available, studies in cultured renal epithelial cells provide model systems to determine mechanisms which are likely to be common to both organs.

Clearly, the intestine is not a nephron nor is the nephron a miniature intestine concerned with the assimilation of nutrients as a source of energy for the organism. It seems likely, however that a broad prospective of the overlapping similarities, as well as the differences, in the transport characteristics of these epithelial structures will be useful in future studies aimed at understanding the molecular and genetic mechanisms operative in the regulation of electrolyte transport.

Acknowledgments. This work was supported in part by a grant from the National Institutes of Health AM 18061.

References

Ahn J, Chang EB, Field M (1985) Phorbol ester inhibition of Na-K exchange in rabbit proximal colon. Am J Physiol 249: C527–C530

Backman KA, Hayslett JP (1982) Role of the medullary collecting duct in potassium conservation. Pflügers Arch Eur J Physiol 396: 297–300

Bankir L, deRouffignac C (1988) Urinary concentrating ability: insights from comparative anatomy. Am J Physiol 249: R643–666

Bastl CP (1987) Regulation of cation transport by low doses of glucocorticoids in in vivo adrenalectomized rat colon. J Clin Invest 80: 348–359

Bastl CP, Barnett CA, Schmidt TJ, Litwack G (1989a) Glucocorticoid stimulation of sodium absorption in colon epithelium is mediated by corticosteroid IB receptor. J Biol Chem 259: 1186–1195

Bastl CP, Schulman G, Cragoe EJ (1989b) Low-dose glucocorticoids stimulate electroneutral NaCl absorption in rat colon. Am J Physiol 257: F1027–F1038

Beck FX, Dörge A, Rick R, Schramm M, Thurau K (1987) Effect of potassium adaptation in the distribution of potassium, sodium and chloride across the apical membrane of renal tubular cells. Pflügers Arch Eur J Physiol 409: 477–485

Billich CO, Levitan R (1969) Effects of sodium concentration and osmolarity on water and electrolyte absorption from the intact human colon. J Clin Invest 48: 1336–1347

Binder HJ (1970) A comparison of intestinal and renal transport systems. Am J Clin Nutr 23: 330–335

Binder HJ (1983) Absorption and secretion of water and electrolytes by small and large intestine. In: Sleisenger MH, Fordtran JS (eds) Gastrointestinal disease 3rd edn. Edition Saunders, Philadelphia, pp 133–155

Binder HJ, Rawlins CL (1973) Electrolyte transport across isolated large intestine mucosa. Am J Physiol 225: 1232–1239

Binder HJ, Stange G, Murer H, Stieger B, Jauri J-P (1986) Sodium-proton exchanges in colon brush-border membranes. Am J Physiol 251: G382–G390

Binder HJ, Foster ES, Budinger ME, Hayslett JP (1987) Mechanism of electroneutral sodium chloride absorption in distal colon of the rat. Gastroenterology 93: 449–455

Borgstrom B, Dahlgvist A, Luordh G, Sjovall J (1957) Studies of intestinal digestion and absorption in the human. J Clin Invest 36: 1521–1536

Boulpaep EL (1976) Electrical phenomena in the nephron. Kidney Int 9: 88–102

Brown AL Jr (1962) Microvilli of the human jejunal epithelium cell. J Cell Biol 12: 623–627
Burg MB (1986) Renal handling of sodium, chloride, water, amino acids and glucose. In: Brenner BM, Rector FC (eds) The kidney. Saunders, Philadelphia, pp 145–176
Cartright CA, McRoberts JA, Mandel KG, Dharmsathaphorn K (1985) Synergistic action of cyclic adenosine monophosphate- and calcium-mediated chloride secretion in a colonic epithelial cell line. J Clin Invest 76: 1837–1842
Charney A, Feldman G (1984) Systemic acid-base disorders and intestinal electrolyte transport. Am J Physiol 247: G1–G12
Doucet A, Katz AI (1980) Renal potassium adaptation: Na-K-ATPase activity along the nephron after chronic potassium loading. Am J Physiol 238: F380–F386
Doucet A, Katz AI (1981) Mineralocorticord receptors along the nephron: [^{3}H] aldosterone binding in rabbit tubules. Am J Physiol 241: F605–F611
Escobar E, Ibarra C, Todesco E, Parise M (1990) Water and ion handling in the rat cecum. Am J Physiol 259: G786–G791
Field MJ, Stanton BA, Giebisch GH (1984) Differential acute effect of aldosterone, dexamethasone and hyperkalemia on distal tubule potassium secretion in the rat kidney. J Clin Invest 74: 1792–1802
Fine LG, Yanagawa N, Schultze RG, Tuck M, Trizna W (1979) Functional profile of the isolated uremic nephron. J Clin Invest 64: 1033–1042
Fisher KA, Binder HJ, Hayslett JP (1976) Potassium secretion by colonic mucosal cells after potassium adaptation. Am J Physiol 281: 987–994
Foster ES, Zimmerman TW, Hayslett JP, Binder HJ (1983) Corticosteroid alteration of active electrolyte transport in rat distal colon. Am J Physiol 245: G668–G675
Foster ES, Hayslett JP, Binder HJ (1984) Mechanism of active potassium absorption and secretion in the rat colon. Am J Physiol 246: G611–G617
Foster ES, Budinger ME, Hayslett JP, Binder HJ (1986a) Ion transport in proximal colon of the rat. J Clin Invest 77: 228–235
Foster ES, Dudeja PK, Brasitus TA (1986b) Na^+-H^+ exchange in rat colon brush-border membrane vesicles. Am J Physiol 250: G781–G787
Frizzell RA, Koch MJ, Schultz SG (1976) Ion transport by rabbit colon. I. Active and passive components. J Membr Biol 27: 297–316
Giebisch G, Klose RM, Windhager EE (1964) Micropuncture study of hypertonic sodium chloride loading in the rat. Am J Physiol 206: 687–693
Grantham JJ, Burg MB, Orloff J (1970) The nature of transtubular Na and K transport in isolated rabbit collecting tubules. J Clin Invest 49: 1815–1826
Gross JB, Imai M, Kokko J (1975) A functional comparison of the cortical collecting tubule and the distal convoluted tubule. J Clin Invest 55: 1284–1294
Hayes CP Jr, McLeod M, Robinson R (1967) An extrarenal mechanism for the maintenance of potassium balance in severe chronic renal failure. Trans Assoc Am Phys 80: 207–216
Hayslett JP, Backman KA, Schon DA (1980) Electrical properties of the medullary collecting duct in the rat. Am J Physiol 239: F258–F264
Hayslett JP, Halevy J, Pace PE, Binder HJ (1982) Demonstration of net potassium absorption in mammalian colon. Am J Physiol 242: G209–G214
Hierholzer K, Wiederholt M, Holzgreve N, Giebisch G, Klose RM, Windhager EE (1965) Micropuncture study of renal transtubular concentration gradients of sodium and potassium in adrenalectomized rats. Pflügers Arch 285: 193–210
Hirsch D, Kashgarian M, Boulpaep EL, Hayslett JP (1984) Role of aldosterone in the mechanism of potassium adaptation in the initial collecting tubule. Kidney Int Eur J Physiol 26: 798–807
Jamison RL, Kriz W (1982) Urinary concentrations mechanism: structure and function. Oxford University Press, New York
Kaissling B, Kriz W (1979) Structural analysis of the rabbit kidney. Adv Anat Embryol Cell Biol 56: 1–123
Kashgarian M, Taylor CR, Binder HJ, Hayslett JP (1980) Amplification of cell membrane surfaces in potassium adaptation. Lab Invest 42: 580–581

Kliger AS, Binder HJ, Bastl CP, Hayslett JP (1981) Demonstration of active potassium transport in the mammalian colon. J Clin Invest 67: 1189–1196
Lacy ER (1991) Functional morphology of the large intestine. In: Schultz SG (ed) The gastrointestinal tract, sect 6. Handbook of physiology. American Physiological Society, Bethesda, Maryland, pp 121–194
Lee SMK, Chekal MA, Katz AI (1983) Corticosterone binding along the rat nephron. Am J Physiol 244: F504–F509
Lombès M, Farman N, Oblin ME, Baulieu EE, Bonvalet JP, Erlanger BF, Gasc JM (1990) Immunohistochemical localization of renal mineralocorticoid receptor by using an anti-idiotypic antibody that is an internal image of aldosterone. Proc Natl Acad Sci USA 87: 1086–1088
Madsen KM, Clapp WL, Verlander JW (1988) Structure and function of the inner medullary collecting duct. Kidney Int 34: 441–454
Malnic G, Klose RM, Giebisch G (1964) Micropuncture study of renal potassiun excretion in the rat. Am J Physiol 206: 674–686
Malnic G, Klose RM, Giebisch G (1966) Micropuncture study of distal tubular potassium and sodium transport in rat kidney. Am J Physiol 211: 548–559
O'Neil RG (1990) Aldosterone regulation of sodium and potassium transport in the cortical collecting duct. Semin Nephrol 10: 365–374
Potter EL (1972) Normal and abnormal development of the kidney. Year Book Medical Publ, Chicago
Rajendran VM, Oesterlin N, Binder HJ (1991) pH regulated Cl uptake in apical membrane vesicles (AMV) of rat distal colon. Gastroenterology 100: A700
Sandle GI, McClone F (1987) Segmental variability of membrane conductances in rat and human colonic epithelia. Pflügers Arch Eur J Physiol 410: 173–180
Sandle GI, Hayslett JP, Binder HJ (1984) Effect of chronic hyperaldosteronism on the electrophysiology of rat distal colon. Pflügers Arch Eur J Physiol 401: 22–26
Sandle GI, Foster ES, Lewis SA, Binder HJ, Hayslett JP (1985) The electrical basis for enhanced potassium secretion in rat distal colon during dietary potassium loading. Pflügers Arch Eur J Physiol 403: 433–439
Sansom SC, O'Neil RG (1985) Mineralocorticoid regulation of apical cell membrane Na^+ and K^+ transport of the cortical collecting duct. Am J Physiol 248: F858–F868
Sansom SC, Agulian S, Mirto S, Illig V, Giebisch G (1989) K activity in CCD principal cells from normal and DOCA-treated rabbits. Am J Physiol 256: F136–F142
Schon DR, Backman KA, Hayslett JP (1981) Role of the medullary collecting duct in potassium excretion in potassium-adapted animal. Kidney Int 20: 655–662
Schulman G, Miller-Diener A, Litwack G, Bastl CP (1986) Characterization of the rat colonic aldosterone receptor and its activation process. J Biol Chem 261: 12102–12108
Schultz SG, Curran PF (1968) Intestinal absorption of sodium chloride and water. In: Code CF (ed) Handbook of physiology, sect 6. Alimentary canal. American Physiological Society, Bethesda, Maryland
Schwartz GJ, Burg MB (1978) Cation transport by the cortical collecting tubule in vitro. Am J Physiol 253: F576–F585
Sellin JH, DeSoignie RC (1984) Rabbit proximal colon: a distinct transport epithelium. Am J Physiol 26: G603–G610,
Sellin JH, DeSoignie RC (1987) Ion transport in human colon in vitro. Gastroenterology 93: 441–448
Sellin JH, Oyarzabal H, Cragoe EF (1988) Electrogenic sodium absorption in rabbit cecum in vitro. J Clin Invest 81: 1275–1283
Silva P, Hayslett JP, Epstein FH (1973) The role of Na-K-activated adenosine triphosphate in potassium adaptation. J Clin Invest 52: 2665–2671
Sonnenberg H (1977) Effect of adrenalectomy on medullary collecting-duct function in rats before and during blood volume expansion. Pflügers Arch Eur J Physiol 368: 55–62
Stanton BA, Biemesderfer D, Wade JB, Giebisch G (1981) Structural and functional study of the rat distal nephron: effects of potassium adaptation and depletion. Kidney Int 19: 36–48

Steinmetz PR (1986) Cellular organization of urinary acidification. Am J Physiol 251: F173–F187

Stevens CE (1977) Comparative physiology of the digestive system. In: Swenson MJ (ed) Duke's physiology of domestic animals, 9th edn. Cornell University Press, Ithaca, NY

Stokes JB (1982) Na and K transport across the cortical and outer medullary collecting tubule of the rabbit: evidence for diffusion across the outer medullary portion. Am J Physiol 242: F514–F520

Sullivan SK, Field M (1991) Ion transport across mammalian small intestine. In: Schultz SG (ed) The gastrointestinal tract, sect 6. Handbook of physiology. American Physiological Society, Bethesda, Maryland, pp 287–301

Tomito K, Pisano JJ, Burg MB, Knepper MA (1986) Effects of vasopressin and bradykinin on anion transport by the rat cortical collecting duct. J Clin Invest 77: 136–141

Turnheim K, Hudson RL, Schultz SG (1987) Cell Na^+ activities and transcellular Na^+ absorption by descending colon from animal and Na^+-deprived rabbits. Pflügers Arch Eur J Physiol 410: 279–283

Ullrich KJ (1976) Renal tubular mechanisms of organic solute transport. Kidney Int 9: 134–148

Ullrich KJ, Papavassilisu F (1979) Sodium reabsorption in the papillary collecting duct of rats. Effect of adrenalectomy, low Na^+ diet, acetazolamide, HCO_3^- free solutions and of amiloride. Pflügers Arch Eur J Physiol 379: 49–52

Welling LE, Welling DJ (1976) Shape of epithelial cells and intercellular channels in the rabbit proximal nephron. Kidney Int 9: 385–394

Wiederholt M, Stolte H, Brecht JP, Hierholzer K (1966) Micropuncture research on the effect of aldosterone, cortisone and dexamethasone on renal sodium resorption in adrenalectomized rats. Pflügers Arch Eur J Physiol 292: 316–333

Wills NK, Biagi B (1982) Active potassium transport by rabbit descending colon epithelium. J Membr Biol 64: 195–203

Wills NK, Lewis SA, Eaton DC (1979) Active and passive properties of rabbit discending colon. A microelectrode and nystatin study. J Membr Biol 45: 81–108

Wingo CS (1989) Active proton secretions and potassium absorption in the rabbit outer medullary collecting duct. Functional evidence for proton-potassiun-activated adenosine triphosphatase. J Clin Invest 84: 361–365

Wingo CS, Madsen KM, Smolka A, Tisher CC (1990) H-K-ATPase immunoreactivity in cortical and outer medullary collecting duct. Kidney Int 38: 985–990

Wright F, Strieder N, Fowler N, Giebisch G (1971) Potassium secretion by distal tubule after potassium adaptation. Am J Physiol 221: 437–448

Wrong OM, Edmonds CJ, Chadwick VS (1981) The large intestine: its role in mammalian nutrition and homeostasis. Wiley, New York

Chapter 9

Fluid and Electrolyte Transport in Human Colon in Health and Disease

Geoffrey I. Sandle[1]

Contents

1 Introduction

The human colon plays an important role in salt and water homoeostasis. In health, 1.5–2 l of sodium- and chloride-rich small intestinal fluid enters the colon daily, but ~90% of this is absorbed (Phillips and Giller 1973; Debongnie and Phillips 1978). The colon also possesses a large reserve, absorptive capacity, and is capable of absorbing up to 5 l of fluid per 24 h (Phillips and Giller 1973; Debongnie and Phillips 1978). Diarrhoea will ensue when this reserve capacity is overwhelmed (e.g., in patients with Asiatic cholera, where small intestinal fluid secretion is massive and acute), or when colonic mucosal disease is sufficiently severe and extensive to impair fluid and electrolyte transport processes in the colonic epithelium

[1] Department of Medicine, University of Manchester School of Medicine, Hope Hospital, Salford M6 8HD, UK

Advances in Comparative and Environmental Physiology, Vol. 16

(e.g., in ulcerative colitis and Crohn's disease of the colon). Partial resection of the colon for localized disease may also result in a significant loss of absorptive capacity, particularly if the proximal (ascending) segment is removed.

Despite the obvious importance of the human colon in maintaining salt and water balance, it is only recently that we have begun to acquire a clearer picture of the array of electrolyte transport processes that operate throughout this organ.

2 Insights from in Vivo Studies

Our understanding of fluid and electrolyte transport in the human colon has developed during the past three decades, beginning with studies performed in vivo. Although in vivo assessments of colonic transport processes are less detailed than those performed in vitro, they allow water and electrolyte transport to be studied in the presence of normal mucosal blood flow, neurohumoral factors, and electrochemical gradients. Thus, Levitan et al. (1962) demonstrated net absorption of water, sodium and chloride, and net secretion of potassium and bicarbonate, during constant perfusion of the entire normal human colon with isotonic saline. Furthermore, unidirectional (but not net) water fluxes, net sodium and chloride absorption, and net potassium and bicarbonate secretion, were greater in the right (ileocaecal valve to splenic flexure) than in the left (splenic flexure to rectum) colon. Other in vivo studies have confirmed that sodium absorption is greatest in the caecum and decreases progressively in the transverse colon, descending colon, and rectum (Devroede et al. 1971). In broad agreement with this pattern of decreasing sodium and chloride absorption (Levitan et al. 1962; Devroede et al. 1971), water absorption shows little variability throughout the colon, but is negligible or absent in the rectum (Levitan et al. 1962; Devroede and Phillips 1970; McNeil 1984). Colonic perfusion studies (Devroede and Phillips 1969) have also shown that (1) sodium and chloride absorption occurs against large concentration gradients; (2) the absorptions of sodium and water correlate more strongly than those of chloride and water; (3) sodium and chloride absorption continues in the absence of net water transport; and (4) potassium and bicarbonate are absorbed or secreted depending on whether the intraluminal concentrations of these ions are above or below critical levels (15 and 25 mmol/l for potassium and bicarbonate respectively). Taken together, these observations suggested that absorption of sodium and chloride (rather than water) was the main function of the human colon, and provided an early indication of segmental differences in electrolyte transport.

Tritiated water, which is absorbed in proportion to the surface area of the perfused segment, has recently been used to provide a more detailed assessment of the contribution of the proximal (caecum to descending colon) and distal (sigmoid colon and rectum) segments to the water and electrolyte transport capabilities of the entire colon (Schiller et al. 1988). These studies indicated that the distal segment accounts for ~20% of total colonic surface area, but only 5–7% of the sodium, chloride, and water absorbed by the entire colon, i.e., the rates of sodium, chloride, and water absorption in the distal colon are about one quarter of those in a similar area of proximal colon. However, the rates of potassium and bicarbonate secretion in the proximal and distal colon were roughly proportional to the areas of the two segments, suggesting that pancolonic secretion of these ions was relatively uniform. The clinical importance of the dominant role of the proximal colon (and the ileum) in salt and water absorption is well illustrated by those patients who develop refractory diarrhoea following combined terminal ileal and proximal colonic resection, which cannot be explained by bile acid and/or fat malabsorption. Such patients have been shown to have 23–31% reductions in sodium, chloride, and water absorption compared with normal subjects during whole gut (pylorus-rectum) perfusion with a balanced plasma-like electrolyte solution, differences which become even more marked under conditions that require sodium and chloride to be absorbed against concentration gradients (Arrambide et al. 1989). Since the distal segments of the large intestine (particularly the rectum) have limited capacities for sodium, chloride, and water absorption, it can be seen that removal of the high-capacity proximal colon will contribute significantly to excessive faecal water and electrolyte losses.

It should be emphasized that in vivo studies have also provided important clues regarding possible *mechanisms* of electrolyte transport in human large intestine. For example, Edmonds and Godfrey (1970) measured transmucosal electrical potential difference (pd) and faecal fluid electrolyte concentrations in the rectum and sigmoid colon of normal subjects in the basal state and at intervals after the intravenous injection of one or two 0.5-mg doses of aldosterone. They found (1) high transmucosal pd values (range 14–61 mV, mean 35 mV lumen-negative, corrected for skin pd), consistent with active (electrogenic) sodium absorption; (2) concentrations of potassium in faecal fluid which were significantly greater than expected from passive diffusion, suggesting the presence of an active potassium secretory process; (3) concentrations of chloride in faecal fluid consistent with passive chloride movement into the lumen along favourable electrochemical gradients; and (4) aldosterone significantly increased pd and decreased faecal sodium within 4 h, suggesting that the mineralocorticoid stimulated electrogenic sodium absorption in the distal colon. Segmental perfusion studies have shown that instillation of 100 μmol/l amiloride (a concentration of the diuretic which blocks sodium channels *and* inhibits Na-H exchange at the apical membrane) into the proximal colon (i.e., caecum to descending colon, equivalent to 80% of the entire colonic surface area) significantly decreased net sodium absorption by 31% from 21 mmol/h to 14.4 mmol/h without changing transmucosal pd (Schiller et al. 1988). In contrast, in the distal colon (i.e., sigmoid to rectum, equivalent to 20% of the entire colonic surface area), amiloride had no significant effect on the much lower rate of net sodium absorption (1.2 mmol/h to 0.9 mmol/h), although transmucosal pd decreased significantly from −46 to −30 mV (Schiller et al. 1988). Rectal perfusion of isotonic saline in six normal subjects also revealed that intraluminal amiloride decreased transmucosal pd by 50–100% in a dose-dependent manner, and complete depolarization of the mucosa occurred in two subjects at amiloride concentrations of 4 and 40 μmol/l respectively (Mohamed et al. 1978). Taken together, these studies suggest that (1) sodium absorption is faster in the proximal than in the distal colon, irrespective of segmental differences in mucosal surface area; (2) most of the sodium absorbed in vivo reflects amiloride-insensitive electrogenic or electroneutral transport; and (3) amiloride-sensitive electrogenic sodium transport accounts for less than 30% of total sodium absorption in vivo, and occurs predominantly in the sigmoid colon and rectum. Despite regional variations in its amiloride sensitivity, the presence of a lumen-negative transmucosal pd throughout the colon (which appears to mainly reflect electrogenic sodium transport in both the proximal and distal segments) provides a strong driving force for the passive absorption of chloride. Recent perfusion studies suggest that passive chloride transport accounts for ~75% of the total chloride absorbed by the human colon, the remainder (~25%) reflecting Cl-HCO_3 exchange (Davis et al. 1983).

3 Insights from in Vitro Studies

Despite their ability to reveal segmental differences in water and electrolyte transport, in vivo studies have played a somewhat limited role in establishing the precise nature of ion transport processes in human colonic epithelia. However, electrophysiological and ion flux measurements performed in isolated sheets of human colonic mucosa have provided clearer (but by no means complete) insights into the mechanisms involved. It is important to note that reported basal values of transepithelial pd, short-circuit current (Isc), and ion fluxes (particularly those of sodium) vary widely between different in vitro studies. These variations probably reflect factors such as pre-operative sodium balance, type of pre-operative bowel preparation, the degree of tissue hypoxia during colonic resection, and the speed with which mucosal specimens are mounted in chambers and bathed in oxygenated electrolyte solutions (Grady et al. 1970), rather than the segment of colon resected (Hubel et al. 1987; Sandle and McGlone 1987; Sellin and De Soignie 1987; Sandle 1989) or the degree of handling of the specimen prior to mounting in the chamber (Grady et al. 1970).

3.1 Sodium Absorption

Initial studies of sodium transport across human colonic mucosa under short-circuit conditions (transepithelial pd clamped to zero) revealed (1) no variability in pd between different segments; (2) a net absorptive flux of sodium (J_{Na}^{net}); and (3) an Isc/J_{Na}^{net} ratio of 0.68–0.8, indicating that active sodium transport accounted for most of the spontaneous transepithelial pd and Isc (Grady et al. 1970; Archampong et al. 1972). In more carefully controlled studies, Hawker et al. (1978) used rapidly mounted mucosa from the left (distal) colon with high values of pd, and found (1) an Isc/J_{net}^{net} ratio of ~1; (2) values of pd and Isc which were almost entirely dependent on the presence of sodium in the mucosal (but not serosal) bathing solution; and (3) that the serosal addition of 1 mmol/l ouabain abolished J_{Na}^{net}, pd, and Isc. These data suggested that sodium absorption in the human colon (at least in the distal segment) was mediated entirely by an ouabain-sensitive electrogenic process, as seen in rabbit descending colon (Frizzell et al. 1976).

Subsequent electrophysiological and ion flux experiments have shown that sodium transport mechanisms in human colon are in fact more complex than suggested by the above-mentioned studies. For example, Rask-Madsen and Hjelt (1977) showed that the mucosal addition of 1 mmol/l amiloride to distal colonic (sigmoid and rectal) epithelium decreased pd, Isc and total electrical conductance (G_t) by 61, 76 and 38% respectively, while J_{Na}^{net} decreased by only 26%. The presence of a large amiloride-insensitive component of J_{Na}^{net} (3.4 μmol cm^{-2} h^{-1}) strongly suggested that mechanisms other than amiloride-sensitive electrogenic transport contribute to overall sodium absorption in human distal colon.

Additional insights into the nature and distribution of sodium transport processes in human colon have come from purely electrophysiological studies. Thus, although basal values of transepithelial pd, Isc and G_t were similar in isolated sheets of proximal (ascending) and distal (descending colon-proximal rectum) colonic epithelium, the mucosal addition of 0.1 mmol/l amiloride decreased these parameters by 73–94, 77–93, and 15% respectively in the distal colon but had little or no effect on the electrical properties of the proximal colon (Sandle et al. 1986a; Sandle and McGlone 1987). Using intracellular microelectrodes, the effects of amiloride in the distal (but not the proximal) colon were shown to reflect hyperpolarization of the apical membrane and a two-fold increase in the apical/basolateral membrane resistance ratio, changes consistent with amiloride inhibition of a substantial apical sodium conductive pathway in this region of the large bowel (Sandle et al. 1986a; Sandle and McGlone 1987). Although amiloride-sensitive electrogenic sodium transport appeared to account for most if not all of the Isc in the distal colon, these studies did not establish the ionic basis for Isc in the proximal colon or whether electroneutral sodium transport processes contributed significantly to sodium absorption in different segments of the human large bowel. However, additional studies have shown that Isc values in proximal and distal colon decreased to almost zero under sodium-free conditions but were unchanged under chloride-free conditions (Sellin and De Soignie 1987; Sandle 1989), indicating that sodium-dependent electrogenic transport processes (assumed to reflect predominantly electrogenic *sodium* transport) accounted for Isc in both colonic segments. Ion flux measurements under voltage-clamp conditions have further shown that J_{Na}^{net} was equal to Isc and net chloride transport was zero in human caecum (Hubel et al. 1987), while J_{Na}^{net} was greater than Isc and net chloride absorption was substantial in human proximal (ascending) colon (Sellin and De Soignie 1987). These studies also indicated that the transverse colon and distal colon were similar to the proximal colon in that J_{Na}^{net} was greater than Isc and there was appreciable net chloride absorption. However, 0.1 mmol/l amiloride decreased J_{Na}^{net} and Isc in the distal colon by 49 and 70% respectively, whereas the sodium channel blocker had virtually no effect on J_{Na}^{net} and Isc in the proximal colon and intermediate effects in the transverse colon; in addition, J_{Na}^{net} in both the proximal and distal colon appeared to contain a significant component (17–40%) which was amiloride-insensitive and chloride-independent (Sellin and De Soignie 1987).

Taken together, these studies have begun to provide a clearer view of how various sodium transport processes are distributed throughout the human colon. In the proximal colon (i.e., that segment from the ileo-caecal valve to the hepatic flexure), sodium absorption in the caecum reflects purely amiloride-insensitive, electrogenic sodium transport, while sodium absorption in the ascending colon reflects both chloride-dependent, electroneutral sodium transport *and* amiloride-insensitive, electrogenic sodium transport, with a small amiloride-sensitive, electrogenic component. A combination of electroneutral and electrogenic processes also mediates sodium absorption in the transverse and distal (splenic flexure to rectum) segments, but Isc and J_{Na}^{net} become progressively more amiloride-sensitive moving distally along the colon. Thus, human colon possesses several sodium absorptive processes: (1) electrogenic sodium absorption, involving apical sodium channels which show marked segmental variability in their sensitivity to amiloride; (2) chloride-dependent electroneutral sodium absorption, which may reflect apical Na-Cl cotransport or Na-H : Cl-HCO_3 dual exchange (there is presently no evidence for apical Na-K-2Cl cotransport), and (3) chloride-independent, amiloride-insensitive sodium absorption, which may reflect a combination of apical Na-H exchange and amiloride-insensitive electrogenic sodium transport.

As regards electrogenic sodium transport in human colon, two further observations suggest that there are fundamental differences in the apical sodium entry step between the distal and proximal segments. First, although substitution of sulphate ions for chloride ions in the solutions bathing distal colon has been shown to have no effect on total Isc, its amiloride-sensitive component was stimulated by 56%, whereas the amiloride-sensitive Isc in proximal colon was unchanged (Sandle 1989). The basis for the effect of sulphate ions in human distal colon is not clear, but the same manoeuvre has been shown to increase amiloride-sensitive (i.e, total) Isc in rabbit descending colon to a degree which was inversely related to basal Isc in the presence of chloride ions (Turnheim et al. 1977). This effect may reflect the ability of sulphate ions to alter the surface charge of the apical membrane and hence the gating of amiloride-sensitive sodium channels (Turnheim et al. 1977), and/or to increase the amiloride sensitivity of a population of otherwise amiloride-insensitive sodium channels. Second, similar to its effects in rabbit descending colon (Frizzell and Schultz 1978), incubation of human distal colon with 1 μmol/l aldosterone for 5 h has been shown to increase total Isc by 52%, entirely by stimulating its amiloride-sensitive component (Sandle 1989). In contrast, aldosterone had no effect on total or amiloride-sensitive Isc in human proximal colon (Sandle 1989), despite the presence of mineralocorticoid receptors in this segment (Rafestin-Oblin et al. 1984). The time course of aldosterone's effect in the distal colon was consistent with enhanced receptor-mediated synthesis of sodium channel protein, or the activation of "latent" apical sodium channels as seen in amphibian urinary bladder (Palmer and Edelman 1981; Garty and Edelman 1983), or a combination of both possibilities. These observations do not exclude a role for aldosterone in the short- or long-term stimulation of electroneutral sodium absorption in either segment of human colon, as seen in rat proximal colon during chronic hyperaldosteronism induced by dietary sodium depletion (Foster et al. 1986).

3.2 Chloride Absorption

In vivo perfusion studies suggest that about 75% of the chloride absorbed by human colon reflects passive chloride movement along the favourable electrical gradient generated by electrogenic sodium absorption (Davis et al. 1983), and studies in isolated human colon suggest that the permeability of the paracellular shunt pathway to chloride is similar in the proximal and distal segments (Sandle and McGlone 1987). Potential-dependent chloride absorption may normally be the *only* chloride absorptive mechanism in human caecum, as net chloride transport across caecal mucosa is zero under voltage-clamp conditions (Hubel et al. 1987). Although about 25% of the total chloride absorbed by human colon in vivo appears to involve a Cl-HCO_3 exchange mechanism (Davis et al. 1983), the question of whether Cl-HCO_3 exchange represents the *sole* active chloride absorptive process remains controversial. Studies in voltage-clamped human distal colon have shown that active chloride absorption reflects an electroneutral, sodium-independent process consistent with a Cl-HCO_3 exchange (Hawker et al. 1978). In contrast, in both human proximal and distal colon, removal of sodium has been shown to decrease net chloride absorption to zero mainly by decreasing the unidirectional chloride flux from mucosa to serosa (Sellin and De Soignie 1987). Although this response is consistent with coupled Na-Cl uptake at the apical membrane, the pattern of changes in ion fluxes produced by theophylline in the transverse and distal colon suggests that this secretagogue stimulates net chloride secretion by triggering an electrogenic chloride secretory process rather than by exerting its well-known inhibitory effect on Na-Cl cotransport (Sellin and De Soignie 1987). The data of Hubel et al. (1987) also suggest that the caecum lacks a theophylline-sensitive Na-Cl cotransport process, while the transverse colon and sigmoid (distal) colon possess both Na-Cl cotransport and Cl-HCO_3 exchange mechanisms, the latter becoming more marked moving distally through the colon. Thus, in vitro studies to date suggest that active chloride absorption in human colon (excluding the caecum) is mediated by two electroneutral processes, i.e., sodium-dependent chloride absorption (which may reflect Na-Cl cotransport or dual Na-H:Cl-HCO_3 exchanges) and Cl-HCO_3 exchange.

3.3 Potassium Secretion

Although it is well recognized that the human colon secretes potassium, little is known of the underlying mechanisms. It is likely that colonic potassium secretion in vivo reflects, at least in part, passive movement of potassium from plasma to lumen secondary to the high, lumen-negative pd. However, ^{43}K tracer studies have suggested that the passive component may be relatively small (Kermode and Edmonds 1980), and intraluminal amiloride has been shown to have no effect on net potassium secretion in the proximal rectum despite decreasing transmucosal pd by 30% (Sandle et al. 1986b). Microelectrode studies also indicate that paracellular pathways in human proximal and distal colonic epithelia possess no significant potassium selectivity (Sandle and McGlone 1987).

Active potassium secretion occurs in human colon under voltage-clamp conditions, and is mediated by a sodium-dependent process (Archampong et al. 1972; Hawker et al. 1978). The factors involved in potassium uptake at the basolateral membrane and potassium exit at the apical membrane have yet to be fully characterized, but it is likely that active potassium secretion in human colon reflects membrane transport processes which are similar to those identified in rabbit and rat colon (Binder and Sandle 1987). Thus, basolateral potassium uptake is mediated by ouabain-sensitive Na,K-ATPase (and possibly also by Na-K-2Cl cotransport), and apical potassium exit occurs through a potassium conductive pathway. Although there is strong evidence for the existence of apical potassium channels (under basal conditions and particularly in the presence of hyperaldosteronism) in rabbit and rat colon (Wills et al. 1982; Sweiry and Binder 1989), their presence in the apical membrane of human colon remains to be confirmed. However, microelectrode studies suggest that an appreciable amiloride-insensitive conductance exists in the apical membrane of human distal colon which may reflect, at least in part, a potassium conductive pathway (Sandle and

McGlone 1987). Noise analysis studies in human distal colon have also revealed evidence of apical potassium channels which bear some similarities to those found in rabbit distal colon (Wills et al. 1984). Thus on balance, it appears that active potassium secretion in human colon (at least in the distal segment) involves membrane processes which are similar to those identified in other mammalian species. It remains to be established whether human distal colonic epithelium also possesses an active potassium *absorptive* process as seen in rat and rabbit distal colon, where apical potassium uptake is mediated by an electroneutral, sodium-independent, and partially chloride-independent process which appears to reflect K-ATPase activity (Gustin and Goodman 1981; Wills and Biagi 1982; Foster et al. 1984; Halm and Frizzell 1986; Sweiry and Binder 1990).

3.4 Bicarbonate Secretion

Relatively little is known of bicarbonate secretory processes in human colon. In vivo perfusion studies indicate that net bicarbonate secretion (1) occurs when intraluminal bicarbonate concentrations are less than 25 mmol/l (Devroede and Phillips 1969); (2) is dependent on the presence of intraluminal chloride (Davis et al. 1983); and (3) is relatively uniform throughout the colon (Schiller et al. 1988). However, flux measurements across voltage-clamped human colon bathed in bicarbonate-containing solutions have revealed values of "residual" flux (J^R) which increase progressively moving distally along the colon, and are generally assumed to reflect (either wholly or in part) an active bicarbonate secretory process (Hawker et al. 1978; Hubel et al. 1987; Sellin and De Soignie 1987). Active bicarbonate secretion appears to reflect electroneutral Cl-HCO_3 exchange (particularly in the distal colon) (Hawker et al. 1978; Hubel et al. 1987), although it remains to be established whether human colon also possesses an electrogenic bicarbonate secretory process, as seen in renal tubular epithelia (Alpern 1985; Akiba et al. 1986).

4 Abnormalities of Colonic Fluid and Electrolyte Transport

4.1 Ulcerative Colitis and Crohn's Disease

Diarrhoea is the main symptom in ulcerative colitis and Crohn's disease of the colon and reflects, at least in part, reduced water absorption or water secretion and decreases in net sodium and chloride absorption (Duthie et al. 1964; Head et al. 1969; Harris and Shields 1970; Archampong et al. 1972; Edmonds and Pilcher 1973; Rask-Madsen 1973). Colonic inflammation is associated with a fall in the lumen-negative transmucosal pd (Rask-Madsen 1973; Sandle et al. 1986c), which reflects a marked decrease in electrogenic

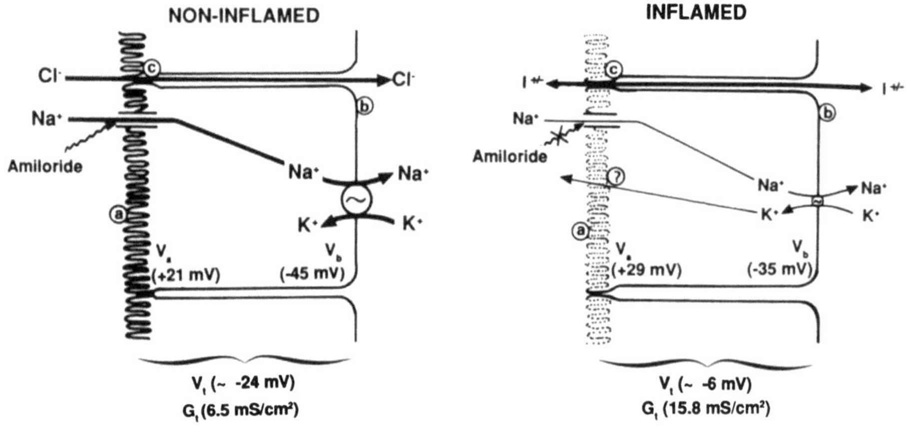

Fig. 1. Proposed model of defective ion transport in inflamed human distal colon based on in vitro studies (Hawker et al. 1980; Sandle et al. 1990). Non-inflamed distal colon is characterized by (i) apical Na^+ uptake via amiloride-sensitive Na^+ channels; (ii) substantial electrogenic Na^+-K^+ pump activity ($Na^+:K^+$ ratio assumed to be 3:2); (iii) a high lumen-negative transepithelial potential difference (V_t), as the basolateral membrane (*b*) voltage (V_b; negative with respect to serosa) is greater than the apical membrane (*a*) voltage (V_a; positive with respect to cell interior); and (iv) passive Cl^- movement from lumen to plasma, which occurs via the paracellular pathway (*c*) secondary to the favourable transepithelial electrical gradient. In contrast, inflamed distal colon is characterized by (i) negligible or absent amiloride-sensitive apical Na^+ uptake, which may reflect a defect in apical Na^+ channels per se, or the significant reduction in basolateral Na^+-K^+ pump activity, or a combination of both possibilities; (ii) loss of the normal cell membrane voltage profile with a reduction in V_t; and (iii) a marked increase in transepithelial conductance (G_t) which may reflect increases in the Cl^- and K^+ permeabilities of the paracellular pathwaay ($I^{+/-}$) and the apical membrane. No account is taken of possible defects in other apical Na^+ uptake processes or the basolateral K^+ shunt conductance

sodium absorption (Hawker et al. 1980). Although the inflammatory process increases both the overall "leakiness" and the electrical conductance of the colonic epithelium (Edmonds and Pilcher 1973; Rask-Madsen 1973; Hawker et al. 1980; Sandle et al. 1990), sodium flux measurements suggest that the decrease in net sodium absorption reflects a defect in electrogenic sodium transport rather than back-diffusion of transported sodium (Hawker et al. 1980). Microelectrode studies (Sandle et al. 1990) have revealed that several factors contribute to the abnormal electrical properties and the defective electrogenic transport of sodium seen in ulcerative colitis and Crohn's disease of the colon: (1) loss of the normal cell membrane voltage profile (basolateral voltage ≫ apical voltage); (2) non-specific increases in the conductances of the apical membrane and paracellular shunt pathway; (3) virtual loss of the amiloride-sensitive apical sodium entry step; and (4) a 76% decrease in the activity of the basolateral Na-K-ATPase-mediated sodium-potassium pump (Fig. 1). Loss of the lumen-negative pd from inflamed colon also removes the electrical driving force for the passive absorption of chloride (Rask-Madsen 1973; Sandle et al. 1986c). Whether mucosal inflammation impairs electroneutral sodium transport is unclear, but the combined loss of electrogenic sodium absorption and the predominant (passive) mechanism for chloride absorption appears sufficient to explain the marked decrease in water absorption which occurs in active colitis. It should be noted that prostaglandins are present in excessive amounts in colitic mucosa (Sharon et al. 1978), and have been shown to reduce sodium and chloride absorption and stimulate electrogenic chloride secretion in intact colonic epithelia and monolayers of cultured colonocytes (Racusen and Binder 1980; Weymer et al. 1985). However, prostaglandins may play only a minor role in the pathogenesis of diarrhoea in colitis, since flubiprofen (a prostaglandin synthetase inhibitor) has been shown to have no effect on abnormal rectal water and electrolyte transport in colitic patients, despite its ability to inhibit mucosal prostaglandin production (Rampton and Sladen 1984).

Although basal sodium and water transport may be grossly defective in ulcerative and Crohn's colitis, the inflamed mucosa remains responsive to systemically administered glucocorticoid agonists, which form the mainstay of treatment in acute colitis. Thus, net sodium absorption has been shown to be similar across normal distal colon and distal colon from prednisolone-treated colitic patients, but markedly reduced in distal colon from non-prednisolone-treated colitic patients (Hawker et al. 1980). Rectal dialysis studies have shown that intravenous therapeutic doses of hydrocortisone (100 mg) and methylprednisolone (40 mg) stimulate similar increases in net sodium and water absorption and transepithelial pd after 5 h in normal subjects and patients with active colitis (Sandle et al. 1986c). Aldosterone stimulates electrogenic sodium absorption in normal human distal colon over a similar period (Sandle 1989), and high-dose dexamethasone treatment (which activates mineralocorticoid and glucocorticoid receptors) stimulates electrogenic sodium absorption in rat distal colon (Foster et al. 1983; Binder et al. 1989). Thus it seems likely that the transport effects of glucocorticoids in inflamed distal colon reflect a combination of crossover binding to mineralocorticoid receptors and glucocorticoid receptor activation. The ability of systemic glucocorticoids to decrease diarrhoea in colitic patients probably reflects direct stimulation of colonic sodium and water transport, in addition to their anti-inflammatory effects.

4.2 Microscopic Colitis

In microscopic colitis there is no radiological or endoscopic evidence of mucosal disease, but diarrhoea is associated with a diffuse and variable mucosal inflammation and reactive changes in the surface epithelium (Kingham et al. 1982; Bo-Linn et al. 1985). Colonic perfusion studies suggest that water, sodium and chloride absorption, and Cl-HCO_3 exchange are decreased in this disease (Bo-Linn et al. 1985). These transport changes are similar to those seen in ulcerative colitis (Harris and Shields 1970). However, whereas transmucosal pd appears to be normal and mucosal permeability is decreased in patients with microscopic colitis (Bo-Linn et al. 1985), transmucosal pd is low and mucosal permeability is increased in patients with ulcerative colitis (Edmonds and Pilcher 1973; Rask-Madsen 1973; Hawker et al. 1980; Sandle et al. 1990). The cause of these differences between the two diseases is unclear, but they may reflect breaches in the surface epithelium in ulcerative colitis.

4.3 Collagenous Colitis

The hallmark of collagenous colitis is a continuous subepithelial collagenous layer 25–60 μm in thickness. This disease is a rare cause of secretory diarrhoea, i.e., diarrhoea persisting in the fasting state. Saline perfusion of the entire colon in one patient revealed net secretion of sodium, chloride and water and a rise in colorectal transmucosal pd, suggesting the presence of an active (electrogenic) chloride secretory process (Rask-Madsen et al. 1983). Chloride secretion was associated with increased intraluminal concentrations of prostaglandin E_2, which presumably originated from the mucosa. However, it remains unclear whether prostaglandins play a central role in the pathogenesis of diarrhoea in collagenous colitis, since treatment with indomethacin (a prostaglandin synthetase inhibitor) had no effect on stool volume or frequency (Rask-Madsen et al., 1983).

4.4 Congenital Chloridorrhoea

This rare disorder is inherited as an autosomal recessive trait and is seen worldwide, the largest number of reported cases coming from Finland (for review, see Holmberg 1986). Watery diarrhoea in neonates and infants results in dehydration, hypochloraemia, hyponatraemia, hypokalaemia, and metabolic alkalosis. Stool chloride concentrations are high (~150 mmol/l) and exceed the sum of the sodium and potassium concentrations, while stool bicarbonate concentrations and pH are low (Holmberg 1986). Perfusion studies have demonstrated abnormal mucosal chloride transport in the ileum and colon, but water and electrolyte transport is normal in the duodenum and jejunum (Turnberg 1971; Bieberdorf et al. 1972; Holmberg et al. 1975). The results of these studies suggest that the $Cl-HCO_3$ exchange mechanism normally present in the ileum and colon is absent or defective (i.e., operating in a "reversed" manner), which leads to a low intraluminal pH, possible down-regulation of Na-H exchange, and the production of watery, acidic stools rich in chloride and sodium.

4.5 Cholera

In cholera, watery diarrhoea and the resulting dehydration are severe and potentially fatal. Secretion of water and electrolytes originates from the jejunal and ileal mucosa (Banwell et al. 1970) following the interaction of a *Vibrio cholerae*-derived enterotoxin with an apical membrane GM_1-ganglioside receptor, which triggers the activation of adenylate cyclase and a rise in intracellular cAMP concentration (Fishman 1980). Until recently, it had been unclear whether colonic fluid and electrolyte absorptive processes were also impaired by cholera enterotoxin, as *Vibrio cholerae* infection produced negligible changes in water transport in canine colon (Sack et al. 1966; Carpenter et al. 1968). However, the enterotoxin has been shown to initiate net sodium and water secretion in rat caecum (Donowitz and Binder 1976), and colonic perfusion studies in patients with acute cholera have demonstrated minimal water secretion (Speelman et al. 1986) rather than the net water absorption seen in normal subjects (Levitan et al. 1962). Nevertheless, these patients continued to absorb sodium and chloride from their colons at variable rates, although there was marked potassium secretion in each case (Speelman et al. 1986). In those patients studied during the convalescent phase, the abnormalities in net water, sodium, and chloride transport persisted, while potassium secretion reversed to potassium absorption (Speelman et al. 1986). It therefore appears that while the brunt of cholera enterotoxin's effects occurs in the jejunum and ileum, a reduction in the colon's capacity for fluid absorption may contribute to fluid and electrolyte losses in some patients.

4.6 Shigellosis

Shigella infection has been shown to elicit small intestinal fluid secretion in monkeys and rabbits (Keusch et al. 1972; Rout et al. 1975), but the colon is the main site of water and electrolyte secretion in this disease. Studies involving perfusion of all but the most proximal 20 cm of colon in patients with acute shigella colitis have demonstrated minimal net water secretion associated with decreased net sodium and chloride absorption, rather than net water absorption and the higher rates of sodium and chloride absorption seen in normal subjects (Butler et al. 1986). These studies also revealed that net water absorption was impaired to a greater extent in patients with extensive colitis than in patients with disease limited to the rectosigmoid colon. When studied again during the convalescent phase, net water and chloride absorption had increased significantly while net sodium absorption had also improved (Butler et al. 1986). Although colonic absorptive function appears to be impaired to similar extents in acute shigella colitis and cholera, shigella has little or no effect on small intestinal transport, and patients with shigellosis have normal ileocaecal flow rates and consequently much lower daily stool outputs than patients with cholera (Butler et al. 1986).

4.7 Carcinoid Syndrome

Diarrhoea, which may be intermittent, is a common feature of the carcinoid syndrome when the primary tumour in the distal small intestine or appendix has metastasized to the liver. Perfusion studies in patients with the carcinoid syndrome have demonstrated chloride secretion and impaired sodium absorption or sodium secretion in the small intestine (Donowitz and Binder 1975; Davis et al. 1980; Antonen et al. 1982), and reduced rates of sodium and water absorption in the colon (Davis et al. 1980). Several lines of evidence suggest that the changes in small and large intestinal transport seen in the carcinoid syndrome are mediated by serotonin. Increased concentrations of serotonin in the plasma and 5-hydroxyindole acetic acid in the urine are diagnostic features in these patients (Rambaud et al. 1986), and serotonin elicits chloride secretion or reduces sodium absorption in rat colon (Zimmerman and Binder 1984) and rabbit ileum (Donowitz et al. 1980). It also seems likely that diarrhoea in the carcinoid syndrome reflects a degree of intestinal motor dysfunction, since the administration of serotonin to healthy subjects induces a pattern of small and large intestinal dysmotility which is similar to that seen in patients with diarrhoea associated with carcinoid syndrome (Smith et al. 1965; Misiewicz et al. 1966). It should be noted that carcinoid tumours may secrete other neurhumoural agents (e.g. bradykinin, calcitonin, substance P) which are capable of acting as intestinal secretagogues, or which may enhance the effects of serotonin.

4.8 Watery Diarrhoea Syndrome

Watery diarrhoea of varying severity may occur in patients with islet cell tumours of the pancreas, phaeochromocytoma, ganglioneuroma or ganglioblastoma (Verner and Morrison 1958; Modigliani and Bernier 1982). Faecal water and electrolyte losses reflect a true secretory diarrhoea which may cause dehydration, renal impairment, hypokalaemia, and metabolic acidosis. Vasoactive intestinal polypeptide (VIP) is the most common peptide secreted by tumours associated with the watery diarrhoea syndrome, even when a variety of peptides are secreted by a single tumour (Long et al. 1981; Modigliani and Bernier 1982; Bloom et al. 1983). Although studies in normal subjects and patients with the watery diarrhoea syndrome indicate that VIP-induced fluid and electrolyte secretion occurs in the jejunum (Rambaud et al. 1975; Krejs et al. 1977, 1980; Davis et al. 1981), the response of the colon to VIP remains unclear. In rat colon, VIP stimulates fluid and electrolyte secretion in vivo and electrogenic chloride secretion in vitro, effects which are probably mediated by cAMP (Racusen and Binder 1977). VIP has also been shown to stimulated adenylate cyclase activity in human colonic mucosa (Simon and Kather 1980), and decrease colonic sodium and water absorption in normal subjects (Krejs 1982). However, it is by no means certain that the colon contributes to fluid and electrolyte losses in patients with the watery diarrhoea syndrome (Rambaud et al. 1975, 1981), and a degree of secondary hyperaldosteronism may actually enhance colonic sodium and water absorption. Indeed, in some patients with the watery diarrhoea syndrome, VIP may not be the principal secretory agent (Schmitt et al. 1975), as peptide histidine methionine (Bloom et al. 1983), pancreatic polypeptide (Long et al. 1981), and gastrin (Bloom et al. 1980) are often secreted excessively, and these peptides may stimulate water and electrolyte secretion predominantly in the small intestine.

4.9 Bile Acid and Fatty Acid-Induced Diarrhoea

A series of clinical and laboratory studies have established that bile acid-induced colonic fluid and electrolyte secretion contribute significantly to the diarrhoea that occurs in patients with disease of the ileal mucosa or in those who have undergone ileal resection (Hofmann and Poley 1969; Mekhjian et al. 1971; Chadwick et al. 1979). This has formed the basis for the use of the anion binding resin cholestyramine in the treatment of patients with bile acid-induced diarrhoea. The observations that fatty acids and hydroxy fatty acids (produced from fatty acids by enteric bacteria) also stimulate water and electrolyte secretion in rat colon (Bright-Asare and Binder 1973) and inhibit water and electrolyte absorption in human colon (Ammon and Phillips 1973), suggest that the diarrhoeal component of steatorrhoea is at least partly due to direct effects of fatty acids on colonic transport function. However, the cellular basis for the colonic secretory effects of bile and fatty acids remains unclear. In experimental animals, both types of laxative increase colonic mucosal permeability and the paracellular transport of small solutes (Gaginella et al. 1977; Chadwick et al. 1979; Freel et al. 1983), which probably accounts for the enhanced absorption of dietary oxalate and the increased incidences of hyperoxaluria and oxalate stone formation in patients with ileal disease (Dobbins and Binder 1976). It should be emphasized that an increase in mucosal permeability per se may be insufficient to produce fluid secretion, as there is evidence that a combination of increased mucosal permeability *and* the hydrostatic pressure gradient normally present across colonic mucosa is required to achieve net fluid secretion in vivo (Wanitschke et al. 1977; Karbach and Wanitschke 1984). In addition, there are data which suggest that diarrhoea induced by bile and fatty acids reflects stimulation of transcellular (active) chloride secretion, as well as increased fluid and electrolyte movement into the lumen secondary to changes in permeability. Experimental observations suggest that the induction of active chloride secretion by bile and fatty acids involves several mechanisms which may be interrelated, viz: stimulation of adenylate cyclase activity and an increase in intracellular cAMP concentration (Binder et al.

1975), stimulation of protein kinase C activity by bile acids (Craven et al. 1987), induction of membrane calcium channels by both bile and fatty acids (Maenz et al. 1982), stimulation of mucosal prostaglandin production (Beubler and Juan 1979), and activation of enteric neurones with the release of neurotransmitters capable of eliciting intestinal secretion (Karlstrom et al. 1983).

4.10 Chronic Uraemia

The concept of "potassium tolerance" originally developed from the experimental observation that rats fed high potassium diets survived acute oral potassium loads which were fatal in control animals (Thatcher and Radike 1947). Subsequent studies have shown that "potassium tolerance" reflects adaptive changes in potassium transport in both renal *and* colonic epithelia (Hayslett and Binder 1982). Recent evidence suggests that adaptive changes in colonic potassium transport also play an important role in maintaining potassium balance in patients with chronic renal failure. Despite normal dietary potassium intakes, these patients often remain normokalaemic for long periods owing to compensatory increases in the potassium secretory capacity of surviving nephrons (Hayslett and Binder 1982), and increases in faecal potassium excretion (Hayes et al. 1967; Wilson et al. 1968). Metabolic balance studies have demonstrated lower faecal sodium concentrations and higher faecal potassium concentrations in uraemic patients than in normal controls (Wilson et al. 1968). These changes reversed during treatment with spironolactone (a mineralocorticoid receptor antagonist), suggesting that enhanced faecal potassium excretion in chronic uraemia was, at least to some extent, a reflection of aldosterone-mediated colonic potassium adaptation (Wilson et al. 1968). However, recent studies have demonstrated enhanced rectal potassium secretion in uraemic patients not requiring dialysis (ND), patients receiving chronic ambulatory peritoneal dialysis (CAPD), and those undergoing chronic haemodialysis (CHD) (Sandle et al. 1986b, 1987). Increased rectal potassium secretion in each of these groups occurred in the absence of secondary hyperaldosteronism, and without increases in the rate of sodium absorption, transmucosal pd, or plasma potassium concentration (at least in the ND and CAPD patients). Rectal potassium secretion in CHD patients was significantly greater before dialysis (i.e., when hyperkalaemic) than post-dialysis (i.e., normokalaemic), which indicates that distal colon in uraemic patients is sufficiently adapted to respond to increases in plasma potassium concentration during the inter-dialysis period (Sandle et al. 1987). It therefore seems likely that enhanced colonic potassium secretion contributes significantly to the maintenance of normal total body potassium contents in patients with end-stage uraemia, since no more than 50% of the near-normal dietary potassium intake could be accounted for by losses during peritoneal dialysis or haemodialysis (Sandle et al. 1987).

References

Akiba T, Alpern RJ, Eveloff J, Calamina J, Warnock DG (1986) Electrogenic sodium/bicarbonate co-transport in rabbit renal cortical basolateral membrane vesicles. J Clin Invest 78: 1472–1478

Alpern RJ (1985) Mechanism of basolateral membrane $H^+/OH^-/HCO_3^-$ transport in the rat proximal convoluted tubule: a sodium-coupled electrogenic process. J Gen Physiol 86: 613–636

Ammon HV, Phillips SF (1973) Inhibition of colonic water and electrolyte absorption by fatty acids in man. Gastroenterology 65: 744–749

Antonen S, Hansen MGJ, Bukhave K, Rask-Madsen J (1982) Influence of a new selective 5-HT_2 receptor antagonist (ketanserin) on jejunal PGE_2 release and ion secretion due to malignant carcinoid syndrome. Gut 23: A887

Archampong EQ, Harris J, Clark CG (1972) The absorption and secretion of water and electrolytes across the healthy and the diseased human colonic mucosa measured in vitro. Gut 13: 880–886

Arrambide KA, Santa Ana CA, Schiller LR, Little KH, Santangelo WC, Fordtran JS (1989) Loss of absorptive capacity for sodium chloride as a cause of diarrhoea following partial ileal and right colonic resection. Dig Dis Sci 34: 193–201

Banwell JG, Pierce NJ, Mitra RC, Brigham KL, Caranasos GJ, Keimowitz RI, Fedson DS, Thomas J, Gorbach SL, Sack RB, Modal A (1970) Intestinal fluid and electrolyte transport in human cholera. J Clin Invest 49: 183–195

Beubler E, Juan H (1979) Effect of ricinoleic acid and other laxatives on net water flux and prostaglandin E release by the rat colon. J Pharm Pharmacol 31: 681–685

Bieberdorf FA, Gorden P, Fordtran JS (1972) Pathogenesis of congenital alkalosis with diarrhea. J Clin Invest 51: 1958–1968

Binder HJ, Sandle GI (1987) Electrolyte absorption and secretion in the mammalian colon. In: Johnson LR (ed) Physiology of the gastrointestinal tract, 2nd edn. Raven Press, New York, pp 1389–1418

Binder HJ, Filburn C, Volpe B (1975) Bile salt alteration of colonic electrolyte transport: role of cyclic adenosine monophosphate. Gastroenterology 68: 503–508

Binder HJ, McGlone F, Sandle GI (1989) Effects of corticosteroid hormones on the electrophysiology of rat distal colon: implications for Na^+ and K^+ transport. J Physiol 410: 425–441

Bloom SR, Long RG, Bryant MG, Mitchell SJ, Polak JM (1980) Clinical, biochemical and pathological studies on 62 VIPomas. Gastroenterology 78: 1143 (abstract)

Bloom SR, Christofides ND, Delamarter J, Buell G, Kawashima E, Polak JM (1983) Diarrhoea in VIPoma patients associated with cosecretion of a second active peptide (peptide histidine isoleucine) explained by single coding gene. Lancet 2: 1162–1165

Bo-Linn GW, Vendrell DD, Lee E, Fordtran JS (1985) An evaluation of the significance of microscopic colitis in patients with chronic diarrhea. J Clin Invest 75: 1559–1569

Bright-Asare P, Binder HJ (1973) Stimulation of colonic secretion of water and electrolytes by hydroxy fatty acids. Gastroenterology 64: 81–88

Butler T, Speelman P, Kabir I, Banwell J (1986) Colonic dysfunction during shigellosis. J Infect Dis 154: 817–824

Carpenter CCJ, Sack RB, Feeley JC, Steenburg RW (1968) Site and characteristics of electrolyte loss and effect of intraluminal glucose in experimental canine cholera. J Clin Invest 47: 1210–1220

Chadwick VS, Gaginella TS, Carlson GL, Debongnie JC, Phillips SF, Hofmann AF (1979) Effect of molecular structure on bile acid-induced alterations in absorptive function, permeability, and morphology in the perfused rabbit colon. J Lab Clin Med 94: 661–674

Craven PA, Pjanstiel J, DeRubertis FR (1987) Role of activation of protein kinase C in the stimulation of colonic epithelial proliferation and reactive oxygen formation by bile acids. J Clin Invest 79: 532–541

Davis GR, Camp RC, Raskin P, Krejs GJ (1980) Effect of somatostatin infusion on jejunal water and electrolyte transport in a patient with secretory diarrhea due to malignant carcinoid syndrome. Gastroenterology 78: 346–349
Davis GR, Santa Ana CA, Morawski SG, Fordtran JS (1981) Effect of vasoactive intestinal polypeptide on active and passive transport in the human jejunum. J Clin Invest 87: 1687–1694
Davis GR, Morawski SG, Santa Ana CA, Fordtran JS (1983) Evaluation of chloride/bicarbonate exchange in the human colon in vivo. J Clin Invest 71: 201–207
Debongnie JC, Phillips SF (1978) Capacity of the human colon to absorb fluid. Gastroenterology 74: 698–703
Devroede GJ, Phillips SF (1969) Conservation of sodium, chloride, and water by the human colon. Gastroenterology 56: 101–109
Devroede GJ, Phillips SF (1970) Failure of the human rectum to absorb electrolytes and water. Gut 11: 438–442
Devroede GJ, Phillips SF, Code CF, Lind JF (1971) Regional differences in rates of insorption of sodium and water from the human large intestine. Can J Physiol Pharmacol 49: 1023–1029
Dobbins JW, Binder HJ (1976) Effect of bile acids and fatty acids on the colonic absorption of oxalate. Gastroenterology 70: 1096–1100
Donowitz M, Binder HJ (1975) Jejunal fluid and electrolyte secretion in carcinoid syndrome. Am J Dig Dis 20: 1115–1122
Donowitz M, Binder HJ (1976) Effect of enterotoxins of *Vibrio cholerae*, *Escherichia coli*, and *Shigella dysenteriae* type 1 on fluid and electrolyte transport in the colon. J Infect Dis 134: 135–143
Donowitz M, Asarkif N, Pike G (1980) Calcium dependence of serotonin induced changes in rabbit ileal electrolyte transport. J Clin Invest 66: 341–352
Duthie HL, Watts JM, de Dombal FT, Goligher JC (1964) Serum electrolytes and colonic transfer of water and electrolytes in chronic ulcerative colitis. Gastroenterology 47: 525–530
Edmonds CJ, Godfrey RC (1970) Measurement of electrical potentials of the human rectum and pelvic colon in normal and aldosterone-treated patients. Gut 11: 330–337
Edmonds CJ, Pilcher D (1973) Electrical potential difference and sodium and potassium fluxes across rectal mucosa in ulcerative colitis. Gut 14: 784–789
Fishman PH (1980) Mechanism of action of cholera toxin: events on the cell surface. In: Field M, Fordtran JS, Schultz SG (eds) Secretory diarrhea. American Physiological Society, Maryland, pp 85–106
Foster ES, Zimmerman TW, Hayslett JP, Binder HJ (1983) Corticosteroid alteration of active electrolyte transport in rat distal colon. Am J Physiol 245: G668–G675
Foster ES, Hayslett JP, Binder HJ (1984) Mechanism of active potassium absorption and secretion in the rat colon. Am J Physiol 246: G611–G617
Foster ES, Budinger ME, Hayslett JP, Binder HJ (1986) Ion transport in proximal colon of the rat. Sodium depletion stimulates neutral sodium chloride absorption. J Clin Invest 77: 228–235
Freel R, Hatch M, Earnest D, Goldner A (1983) Role of tight-junctional pathways in bile salt-induced increases in colonic permeability. Am J Physiol 245: G816–G823
Frizzell RA, Schultz SG (1978) Effect of aldosterone on ion transport by rabbit colon in vitro. J Membr Biol 39: 1–26
Frizzell RA, Koch MJ, Schultz SG (1976) Ion transport by rabbit colon. 1. Active and passive components. J Membr Biol 27: 297–316
Gaginella TS, Chadwick VS, Debongnie JC, Lewis JC, Phillips SF (1977) Perfusion of rabbit colon with ricinoleic acid: dose-related mucosal injury, fluid secretion, and increased permeability. Gastroenterology 73: 95–101
Garty H, Edelman IS (1983) Amiloride-sensitive trypsinization of apical sodium channels. Analysis of hormonal regulation of sodium transport in toad bladder. J Gen Physiol 81: 785–803
Grady GF, Duhamel RC, Moore EW (1970) Active transport of sodium by human colon in vitro. Gastroenterology 59: 583–588

Gustin MC, Goodman DBP (1981) Isolation of brush border membrane from the rabbit descending colon epithelium: partial characterization of a unique potassium-activated ATPase. J Biol Chem 256: 10651–10656
Halm DR, Frizzell RA (1986) Active K transport across rabbit distal colon: relation to Na absorption and Cl secretion. Am J Physiol 251: C252–C267
Harris J, Shields R (1970) Absorption and secretion of water and electrolytes by the intact human colon in diffuse untreated proctocolitis. Gut 11: 27–33
Hawker PC, Mashiter KE, Turnberg LA (1978) Mechanisms of transport of Na, Cl, and K in the human colon. Gastroenterology 74: 1241–1247
Hawker PC, McKay JS, Turnberg LA (1980) Electrolyte transport properties of colonic epithelium in patients with inflammatory bowel disease. Gastroenterology 79: 508–511
Hayes CP, McLeod ME, Robinson RR (1967) An extrarenal mechanism for the maintenance of potassium balance in severe chronic renal failure. Trans Assoc Am Phys 80: 207–216
Hayslett JP, Binder HJ (1982) Mechanism of potassium adaptation. Am J Physiol 243: F103–F112
Head LH, Heaton JW, Kivel RM (1969) Absorption of water and electrolytes in Crohn's disease of the colon. Gastroenterology 56: 571–579
Hofmann AF, Poley JR (1969) Cholestyramine treatment of diarrhea associated with ileal resection. N Engl J Med 281: 397–402
Holmberg C (1986) Congenital chloride diarrhoea. Clin Gastroenterol 15: 583–602
Holmberg C, Perheentupa J, Launiala K (1975) Colonic electrolyte transport in health and in congenital chloride diarrhea. J Clin Invest 56: 302–310
Hubel KA, Renquist K, Shirazi S (1987) Ion transport in human cecum, transverse colon, and sigmoid colon in vitro. Gastroenterology 92: 501–507
Karbach U, Wanitschke R (1984) Influence of serosal hydrostatic pressure on net water and electrolyte transport across the isolated rat colonic mucosa exposed to different secretagogues. Naunyn-Schmiedeberg's Arch Pharmacol 327: 336–341
Karlstrom L, Cassuto J, Jodal M, Lundgren O (1983) The importance of the enteric nervous system for the bile salt induced secretion in the small intestine of the rat. Scand J Gastroenterol 18: 117–123
Kermode JC, Edmonds CJ (1980) Pathways of transepithelial potassium movement in the epithelium of distal colon in man. Clin Sci 59: 29–39
Keusch GT, Grady GF, Mata LJ, McIver J (1972) The pathogenesis of shigella diarrhea. I. Entertoxin production by *Shigella dysenteriae* I. J Clin Invest 51: 1212–1218
Kingham JGC, Levison DA, Ball JA, Dawson AM (1982) Microscopic colitis – a cause of chronic watery diarrhoea. Br Med J 285: 1601–1604
Krejs GJ (1982) Effect of VIP infusion on water and electrolyte transport in the human intestine. In: Said SI (ed) Vasoactive intestinal peptide. Raven Press, New York, pp 193–200
Krejs GJ, Walsh JM, Morawski B, Fordtran JS (1977) Intractable diarrhea: intestinal perfusion studies and plasma VIP concentrations in patients with pancreatic cholera syndrome and surreptitious ingestion of laxatives and diuretics. Am J Dig Dis 22: 280–292
Krejs GJ, Fordtran JS, Bloom SR, Fahrenkrug J, Schaffalitzky de Muckadell OB, Fischer JE, Humphrey CS, O'Dorisio TM, Said SI, Walsh JH, Shulkes AA (1980) Effect of VIP infusion on water and ion transport in the jejunum. Gastroenterology 78: 722–727
Levitan R, Fordtran JS, Burrows BA, Ingelfinger FJ (1962) Water and salt absorption in the human colon. J Clin Invest 41: 1754–1759
Long RG, Bryant MG, Mitchell SJ, Adrian TE, Polak JM, Bloom SR (1981) Clinicopathological study of pancreatic and ganglioneuroblastoma tumours secreting vasoactive intestinal polypeptide (VIPomas). Br Med J 282: 1767–1771
Maenz DD, Forsyth GW (1982) Ricinoleate and deoxycholate are calcium ionophores in jejunal brush border vesicles. J Membr Biol 70: 125–133
McNeil NI (1984) Differences in electrolyte handling through the human large intestine. In: Skadhauge E, Heintze K (eds) Intestinal absorption and secretion. Falk Symposium 36. MTP Press, Lancaster, pp 111–116

Mekhjian HS, Phillips SF, Hofamm AF (1971) Colonic secretion of water and electrolytes induced by bile acids: perfusion studies in man. J Clin Invest 50: 1569–1577
Misiewicz JJ, Waller SL, Eisner M (1966) Motor responses of human gastrointestinal tract to 5-hydroxytryptamine in vivo and in vitro. Gut 7: 208–211
Modigliani R, Bernier JJ (1982) Physiopathology of hormonal diarrhoea. In: Chadwick VS, Phillips SF (eds) Gastroenterology 2: small intestine. Butterworths, London, pp 265–279
Mohamed F, McSorley P, Warren DJ (1978) A comparison of methods for measurement of rectal potential difference in man: effects of rectal infusion of amiloride. Clin Sci Mol Med 55: 231–234
Palmer LG, Edelman IS (1981) Control of apical sodium permeability in the toad urinary bladder by aldosterone. Ann NY Acad Sci 372: 1–14
Phillips SF, Giller J (1973) The contribution of the colon to electrolyte and water conservation in man. J Lab Clin Med 81: 733–746
Racusen LC, Binder HJ (1977) Alteration of large intestinal electrolyte transport by vasoactive intestinal polypeptide in the rat. Gastroenterology 73: 790–796
Racusen LC, Binder HJ (1980) Effect of prostaglandin on ion transport across isolated colonic mucosa. Dig Dis Sci 25: 900–904
Rafestin-Oblin HE, Lombes M, Michel JB, Michaud A, Claire M (1984) Mineralocorticoid receptors in the epithelial cells of human colon and ileum. J Steroid Biochem 20: 311–315
Rambaud JC, Modigliani R, Matuchansky C, Bloom SR, Passayre D, Bernier JJ (1975) Pancreatic cholera: studies on tumoural secretions and pathophysiology of diarrhea. Gastroenterology 69: 110–122
Rambaud JC, Modigliani R, Matuchansky C (1981) Hormones and diarrhea. Clin Res Rev 1: 23–32
Rambaud JC, Hautefeuille M, Ruskone A, Jacquenod P (1986) Diarrhea due to circulating agents. Clin Gastroenterol 15: 603–629
Rampton DS, Sladen GE (1984) The relationship between rectal mucosal prostaglandin production and water and electrolyte transport in ulcerative colitis. Digestion 30: 13–22
Rask-Madsen J (1973) Simultaneous measurement of electrical polarization and electrolyte transport by the entire normal and inflamed human colon during in vivo perfusion. Scand J Gastroenterol 8: 327–336
Rask-Madsen J, Hjelt K (1977) Effect of amiloride on electrical activity and electrolyte transport in human colon. Scand J Gastroenterol 12: 1–6
Rask-Madsen J, Grove O, Hansen MGJ, Bukhave K, Scient C, Henrik-Nielsen R (1983) Colonic transport of water and electrolytes in a patient with secretory diarrhea due to collagenous colitis. Dig Dis Sci 28: 1141–1146
Rout WR, Formal SB, Giannella RA, Dammin GJ (1975) Pathophysiology of shigella diarrhea in the rhesus monkey: intestinal transport, morphological and biochemical studies. Gastroenterology 68: 270–278
Sack RB, Carpenter CCJ, Steenburg RW, Pierce NF (1966) Experimental cholera. A canine model. Lancet 2: 206–207
Sandle GI (1989) Segmental heterogeneity of basal and aldosterone-induced electrogenic Na transport in human colon. Pflügers Arch Eur J Physiol 414: 706–712
Sandle GI, McGlone F (1987) Segmental variability of membrane conductances in rat and human colonic epithelia. Implications for Na, K and Cl transport. Pflügers Arch Eur J Physiol 410: 173–180
Sandle GI, Wills NK, Alles W, Binder HJ (1986a) Electrophysiology of the human colon: evidence of segmental heterogeneity. Gut 27: 999–1005
Sandle GI, Gaiger E, Tapster S, Goodship THJ (1986b) Enhanced rectal potassium secretion in chronic renal insufficiency: evidence for large intestinal potassium adaptation in man. Clin Sci 71: 393–401
Sandle GI, Hayslett JP, Binder HJ (1986c) Effect of glucocorticoids on rectal transport in normal subjects and patients with ulcerative colitis. Gut 27: 309–316

Sandle GI, Gaiger E, Tapster S, Goodship THJ (1987) Evidence for large intestinal control of potassium homoeostasis in uraemic patients undergoing long-term dialysis. Clin Sci 73: 247–252

Sandle GI, Higgs N, Crowe P, Marsh MN, Venkatesan S, Peters TJ (1990) Cellular basis for defective electrolyte transport in inflamed human colon. Gastroenterology 99: 97–105

Schiller LR, Santa Ana CA, Morawski SG, Fordtran JS (1988) Effect of amiloride on sodium transport in the proximal, distal and entire human colon in vivo. Dig Dis Sci 33: 969–976

Schmitt MG, Soergel K, Hensley GT, Chey WY (1975) Watery diarrhea associated with pancreatic islet cell carcinoma. Gastroenterology 69: 209–216

Sellin JH, De Soignie R (1987) Ion transport in human colon in vitro. Gastroenterology 93: 441–448

Sharon P, Ligumsky M, Rachmilewitz D, Zor U (1978) Role of prostaglandins in ulcerative colitis: enhanced production during active disease and inhibition by sulfasalazine. Gastroenterology 75: 638–649

Simon B, Kather H (1980) Human colonic adenylate cyclase. Stimulation of enzyme activity by vasoactive peptide and various prostaglandins via distinct receptor sites. Digestion 20: 62–67

Smith AW, Preshaw RM, Sircus W (1965) Clinical features in patients with carcinoid syndrome. Gastroenterology 48: 738–744

Speelman P, Butler T, Kabir I, Ali A, Banwell J (1986) Colonic dysfunction during cholera infection. Gastroenterology 91: 1164–1170

Sweiry JH, Binder HJ (1989) Characterization of aldosterone-induced potassium secretion in rat distal colon. J Clin Invest 83: 844–851

Sweiry JH, Binder HJ (1990) Active potassium absorption in rat distal colon. J Physiol 423: 155–170

Thatcher JC, Radike AW (1947) Tolerance to potassium intoxication in the albino rat. Am J Physiol 151: 138–146

Turnberg LA (1971) Abnormalities in intestinal electrolyte transport in congenital chloride diarrhoea. Gut 12: 544–551

Turnheim K, Frizzell RA, Schultz SG (1977) Effect of anions on amiloride-sensitive, active sodium transport across rabbit colon, in vitro. J Membr Biol 37: 63–84

Verner JB, Morrison AB (1958) Islet cell tumor and a syndrome of refractory watery diarrhea and hypokalemia. Am J Med 25: 374–380

Wanitschke R, Nell G, Rummel W (1977) Influence of hydrostatic pressure gradients on net transfer of sodium and water across isolated rat colonic mucosa. Naunyn-Schmiedeberg's Arch Pharmacol 297: 191–194

Weymer A, Huott P, Liu W, McRoberts JA, Dharmsathaphorn K (1985) Chloride secretory mechanism induced by prostaglandin E_1 in a colonic epithelial cell line. J Clin Invest 76: 1828–1836

Wills NK, Biagi B (1982) Active potassium transport by rabbit descending colon epithelium. J Membr Biol 64: 195–203

Wills NK, Zeiske W, Van Driessche W (1982) Noise analysis reveals K^+ channel conductance fluctuations in the apical membrane of rabbit colon. J Membr Biol 69: 187–197

Wills NK, Alles WP, Sandle GI, Binder HJ (1984) Apical membrane properties and amiloride binding kinetics of the human descending colon. Am J Physiol 247: G749–G757

Wilson DR, Ing TS, Metcalfe-Gibson A, Wrong OM (1968) The chemical composition of faeces in uraemia, as revealed by in vivo faecal dialysis. Clin Sci 35: 197–209

Zimmerman TW, Binder HJ (1984) Serotonin-induced alteration of colonic electrolyte transport in the rat. Gastroenterology 86: 310–317

Chapter 10

Chloride Secretion by the Colonic Epithelial Cell Line T_{84}: Mechanisms and Regulation[1]

Kim E. Barrett[2]

Contents

1 Introduction

The gastrointestinal tract has as its primary function the digestion and absorption of nutrients. However, as a necessary concomitant of this function, the intestine also serves as a site of fluid and salt absorption and

[1] This chapter is dedicated to the memory of Kiertisin Dharmsathaphorn, M.D., the investigator who was first to recognize the potential of using the T_{84} cell line as a model for the study of intestinal secretion. His untimely death in 1990 has left a significant void among the ranks of researchers in gastroenterology.

[2] Department of Medicine, University of California, San Diego, School of Medicine UCSD Medical Center 8414, 225 Dickinson Street, San Diego, CA 92103-8414, USA

Advances in Comparative and Environmental Physiology, Vol. 16

secretion (Barrett and Dharmsathaphorn 1991). In addition to the approximate volume of 1 l provided by daily oral intake, the intestine and the organs that drain into it provide about 8 l more. Obviously, this considerable volume must be largely reabsorbed if rapid dehydration is to be avoided. Conversely, a change in the balance between active absorption and secretion, such that secretion predominates, will rapidly lead to clinical diarrhea. Thus, the clinical implications of intestinal fluid secretion, while not contributing significantly to normal fluid homeostasis, have driven the desire of many physiologists to understand how intestinal salt and water transport is controlled.

Many of the barriers that blocked our understanding of intestinal secretion are related to the complexity of the gastrointestinal mucosa. Firstly, at the level of the epithelium, both secretory and absorptive epithelial cells are present, meaning that any secretory processes that occur can be partly masked by simultaneous absorption. Second, intestinal secretion is controlled in an integrated fashion, with many mucosal cell types such as nerves, endocrine cells and immunocytes contributing to regulation of the epithelium. Third, native intestinal tissue is available in limited supply (particularly from humans) and may not be sufficiently viable ex vivo to permit detailed biochemical or mechanistic studies.

The study of intestinal transport mechanisms was greatly facilitated in 1984. Dharmsathaphorn and coworkers (1984) reported that a human intestinal cell line, T_{84}, could be grown in culture such that it displayed the differentiated properties of crypt epithelial cells and performed vectorial electrolyte transport. The T_{84} cell line, derived from the lung metastasis of a human colonic carcinoma, has enabled the dissection of the precise electrolyte transport pathways that make up the intestinal chloride secretory mechanism. Further, having a well-described chloride channel, the cell line contributed to the recent discovery of the gene that is defective in the majority of cases of cystic fibrosis (Riordan et al. 1989).

It is the aim of the present chapter to review some features of the chloride secretory mechanism in T_{84} cells, and how it is regulated. This chapter will also focus on experiments which demonstrate the utility of the tissue culture model for dissecting pathways of cellular interaction in the control of intestinal chloride secretion. It will not be the goal of this chapter to review practical aspects of studies with T_{84} cells. The reader is referred to a recent summary of this area by Dharmsathaphorn and Madara (1990) for such information.

2 Electrolyte Transport Pathways

Ion-transporting epithelia are characterized by their development of functional and morphological polarity. Specific ion transport pathways are inserted in the membranes of the epithelial cells in an asymmetric fashion and thereby constitute electrolyte transport mechanisms which transport a given species unidirectionally (Barrett and Dharmsathaphorn 1991). The transport pathways that make up the chloride secretory mechanism in T_{84} cells have been identified, and will be discussed in some detail in this section. It should be noted that there is no evidence that T_{84} cells perform any absorptive functions. In this regard, they are therefore thought to be functionally analogous to the epithelial cells in the intestinal crypts.

2.1 Basolateral Transport Pathways

2.1.1 The Na^+,K^+ ATPase

The active transport of ions implies that cellular energy must be expended. In T_{84} cells, as in many other epithelia, this requirement for a driving force is supplied by the activity of a basolaterally located Na^+,K^+ ATPase. The involvement of this pump was first suggested when it was observed that ouabain, an established inhibitor of this enzyme, was able to block chloride secretion across T_{84} cells induced by a variety of stimuli. This drug was later shown to inhibit the uptake of $^{86}Rb^+$ (a tracer for potassium) into T_{84} cells preloaded with Na^+, with this ouabain-sensitive Rb^+ uptake being localized to the basolateral aspect of the cells (Dharmsathaphorn et al. 1985). The stoichiometry of the pump has not been formally demonstrated, though preliminary findings suggest that it is electrogenic (J.A. McRoberts, pers. comm.), making it likely that this ATPase has the same stoichiometry as that of the enzyme identified in other intestinal tissues, transporting three sodium ions out of the cell cytoplasm for every two potassium ions brought in, with the accompanying hydrolysis of ATP (Kirk et al. 1980; Halm and Dawson 1983).

2.1.2 The Na^+, K^+, $2Cl^-$ Cotransporter

The chloride uptake step in the T_{84} cell chloride secretory mechanism is provided for by a cotransporter or symport. Stimulated increases of chloride secretion across T_{84} cell monolayers are readily inhibitable by loop diuretics, with the inhibition dependent on provision of the drugs to the basolateral aspect of the cells (Dharmsathaphorn et al. 1985). In the presence of ouabain, bumetanide-sensitive uptake of any of $^{86}Rb^+$, $^{22}Na^+$, or $^{36}Cl^-$ requires the simultaneous presence of the other two ions. Experiments of this type were also used to demonstrate the stoichiometry of the co-transporter (two chloride ions transported for each sodium plus potassium ion); the uptake pathway is thus electrically neutral. Although the model for chloride secretion in T_{84} cells is based on this transporter being activated secondarily in the transport mechanism and being in and of itself a passive process, it is crucial to ongoing secretion. Addition of bumetanide during stimulated chloride secretion will promptly terminate the response.

2.1.3 Basolateral K^+ Channel(s)

The activity of the cotransporter described above suggests that chloride secretion will be accompanied by a rapid increase in cellular potassium activity. Obviously, this potassium must be effluxed from the cell in some way to prevent depolarization. Indeed, it has long been known that chloride secretion in a variety of epithelial tissues is accompanied by an increase in basolateral membrane potassium conductance (Shorofsky et al. 1982; Welsh et al. 1982). Using radioactive tracers, it was demonstrated that the active secretion of chloride across T_{84} cells was accompanied by a small increase in the transport of potassium, but that this flux was no more than 5% of that observed for chloride (Mandel et al. 1986b). It was therefore concluded that the bulk of the potassium that is cotransported across the basolateral membrane with chloride is recycled across that same membrane via potassium channels.

$^{86}Rb^+$ efflux from preloaded cells has been used to demonstrate that chloride secretion is accompanied by potassium recycling. The rate of efflux of Rb^+ into the basolateral chamber is one to two orders of magnitude greater than that into the apical chamber, and can be further stimulated by agents which induce chloride secretion (see below), while efflux of $^{86}Rb^+$ across the apical membrane remains unchanged (McRoberts et al. 1985). The potassium conductance on the basolateral membrane that is stimulated is likely to be a channel, because the established potassium channel blocker, barium, inhibits both $^{86}Rb^+$ efflux and chloride secretion occurring in response to vasoactive intestinal polypeptide (VIP) (McRoberts et al. 1985; Mandel et al. 1986b). It may prove to be technically difficult to definitively prove that this conductance is, in fact, a channel via patch-clamp techniques, due to its basolateral location.

There may be more than one class of potassium channel on the basolateral membrane of T_{84} cells involved in the chloride secretory responses to agents which employ different second messenger pathways. Maximally effective doses of VIP (a cAMP-dependent stimulus) and the ionophore A23187 (a calcium-dependent stimulus) had effects on ^{86}Rb efflux that were additive when the agents were given simultaneously (McRoberts et al. 1985). Further, calcium-related chloride secretion, such as that induced by histamine or carbachol, is much less susceptible to inhibition by barium than is cAMP-stimulated secretion (Mandel et al. 1986b; Wasserman et al. 1988). The conclusion has been drawn that differing intracellular effectors can either activate separate classes of potassium channels, or that protein kinase-dependent phosphorylations (or other signal transduction pathways specific for the various intracellular effectors) induce conforma-

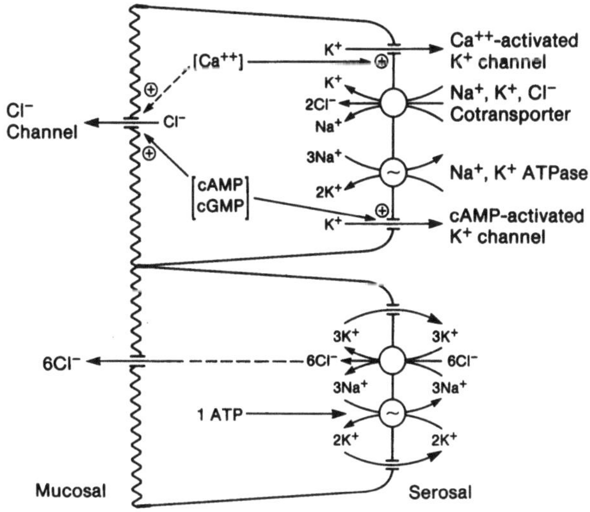

Fig. 1. Model of ion transport pathways involved in the T_{84} cell chloride secretory mechanism. The *upper* cell depicts the proposed nature and regulation of the individual transport pathways, while the *lower* cell indicates the predicted stoichiometry of the mechanism. (McRoberts and Barrett 1989)

tional changes in the potassium channel that modulate its functional properties.

2.2 Apical Transport Pathways

2.2.1 The Chloride Channel

The final component of the chloride secretory model, as shown in Fig. 1, is an apical chloride channel. The open status of this channel is thought to be a primary site for the regulation of overall chloride secretion, particularly for cAMP-dependent agonists (as will be discussed in more detail below, calcium-dependent stimuli may also exert effects on potassium channels to stimulate secretion). Thus, when chloride and/or potassium channel opening is induced by the binding of a secretagogue, there are changes in the cellular activities of Cl^- and K^+ which secondarily influence the transport rate of the cotransporter. Since Cl^- cannot be recycled across the basolateral membrane, the net result is transcellular chloride secretion, with net movement of chloride ions from the serosal to the mucosal side of the cells.

As for the basolateral potassium channel described above, evidence is mounting that several different chloride channels are present in T_{84} cells, and at least some of these may be involved in secretory responses to diverse stimuli. The apical location of the secretory chloride channel makes it amenable to examination by patch-clamp techniques, both in the cell-attached and excised-patch configurations. Further, several investigators have now reported studies in the whole cell patch-clamp mode, where the activities of all the channels in a given cell can be monitored.

T_{84} cell chloride channels have been studied most extensively by Frizzell's group. Working with the whole cell configuration, these workers have described three chloride conductances. The first is a volume-sensitive channel that is activated during cell swelling in response to hyperosmotic challenge (Worrell et al. 1989). The localization of this channel is unknown, but it is postulated to reside in the basolateral membrane. It is thought to participate in the regulatory volume decrease response, but is unlikely to play a role in chloride secretion. Second, there is an apical chloride conductance which is activated by cAMP (Cliff and Frizzell 1990). This channel behaves in an ohmic fashion and is not time-dependent. These properties distinguish it from another agonist-sensitive channel, that stimulated by calcium (Cliff and Frizzell 1990). This third chloride conductance is voltage-sensitive and outwardly rectifying, and displays a time-dependent kinetic component. More recently, Lin et al. (1991) have reported that there is also a cGMP-stimulated chloride conductance in T_{84} cells which again has distinctive electrophysiological and kinetic properties. Interestingly, this channel is relatively insensitive to the inhibitory effect of the chloride channel blocker, DNDS. However, this last channel has only been reported to date in studies with excised patches, where observed channel properties may differ quite markedly from those seen in the whole cell mode (Cliff and Frizzell 1990). It remains to be determined whether these various conductances are reflective of separate molecular entities, or whether diverse second messenger systems induce different regulatory states in a single channel.

There has been considerable discussion about the relationship of the recently discovered cystic fibrosis gene product, known as the cystic fibrosis transmembrane conductance regulator (CFTR), to chloride channels in a variety of chloride-secreting epithelia (Ringe and Petsko 1990). T_{84} cells express high levels of CFTR, which is a substrate for both protein kinase A and protein kinase C in this cell type (Cohn et al. 1991). Further, overexpression of CFTR in T_{84} cells, induced by transfection with a vaccinia-T7 hybrid vector containing the CFTR sequence, led to an increase in cAMP-mediated chloride permeability (Anderson et al. 1991). It is currently controversial whether CFTR is itself the cAMP-dependent chloride channel, or whether it acts as an essential regulator of this conductance. In any event, it does not appear that CFTR is the calcium-regulated chloride channel in T_{84} cells, or that a normal CFTR is required for calcium-mediated chloride secretion, because cells from cystic fibrosis patients that contain a mutant form of CFTR can respond normally to stimuli such as ionomycin (Drumm et al. 1990; Wagner et al. 1991).

2.3 Summary

The essential features of the model for chloride secretion in T_{84} cells are shown in Fig. 1. Following agonist stimulation, chloride flows out of the cell via apical chloride channels. Chloride uptake is via the basolateral cotransporter, whose activity is secondarily driven by the energy-requiring basolateral ATPase. Cotransported potassium is recycled through basolateral channels, which may also be subject to regulation by agonists. In total, six equivalents of chloride are transported per molecule of ATP hydrolyzed.

3 Intracellular Regulation of Secretion

3.1 Second Messenger Mechanisms Activating Secretion

As in other secretory cell types, a number of second messengers have been identified in T_{84} cells which link the initial stimulus of hormone binding to the physiological response of chloride secretion (McRoberts and Barrett 1989). This laboratory and others have identified three broad classes of stimuli in T_{84} cells that act through distinct second messenger pathways; cAMP, cGMP, and calcium-related. In addition, there are other secretagogues which appear to diverge from this three-messenger paradigm. A summary of agents known to stimulate chloride secretion across T_{84} cells is given in Table 1, together with their purported second messengers.

3.1.1 cAMP-Related Stimuli

As demonstrated for intact intestinal tissue, an important second messenger that stimulates T_{84} chloride secretion is cAMP. A variety of physiologically or pathophysiologically relevant stimuli, including VIP and cholera toxin, as well as experimental agents, such as forskolin and theophylline, have effects on T_{84} cell chloride secretion which are closely correlated with their ability to increase levels of cAMP within the cells (Mandel et al. 1986a; Barrett et al. 1990). These agents have relatively slow, but sustained effects on chloride secretion, with the time course of action following the time course of cAMP accumulation. In addition, prostaglandin E_1 (PGE_1) induces both chloride secretion and cAMP accumulation in T_{84} cells (Weymer et al. 1985), but interestingly, this mediator induces significantly higher levels of chloride secretion for a given degree of cAMP elevation than do the other agents named (K.E. Barrett, unpubl. observ.). This last finding raises the possibility that some stimuli are able to increase levels in a specific pool of cAMP that is more effectively linked to stimulation of secretion, or that additional second messengers amplify the observed response.

cAMP exerts its effects in cells via the stimulation of a cAMP-dependent protein kinase (protein kinase A), with consequent increases in protein

Table 1. Substances causing chloride secretion across T_{84} cell monolayers

Substance	Dose[a]	Sidedness[b]	Second messenger
VIP	10^{-8} M	s	cAMP
Forskolin	3×10^{-6} M	s	cAMP
Cholera toxin	3 µg/ml	m	cAMP
PGE_1	10^{-6} M	s	cAMP
PGD_2	10^{-4} M	n.d.[c]	n.d.
ST_a	2.5×10^{-7} M	m	cGMP
Carbachol	10^{-4} M	s	Calcium
Histamine	10^{-4} M	s	Calcium
Bile acids	10^{-4} M	s,m	Calcium
Ionophores	10^{-5} M	s,m	Calcium
Adenosine	10^{-4} M	s,m	n.d.
Mast cell lysate	–	s,m	n.d.
Anaphylactic supernatant	–	s,m	n.d.
Neutrophil-derived secretagogue	–	m	n.d.
H_2O_2[d]	10^{-3} M	n.d.	cAMP
Bradykinin[d]	10^{-6} M	n.d.	cAMP

[a] Maximally effective dose.
[b] Side of addition causing effect. s, Serosal; m, mucosal.
[c] n.d., Not determined.
[d] These substances require fibroblasts to be juxtaposed to the T_{84} cells for significant secretory effects, and probably act through secondary release of prostaglandins.

phosphorylation and presumably changes in, for example, the function of channels and/or regulatory proteins. As described above, CFTR is known to be a substrate for protein kinase A (Gregory et al. 1990; Cohn et al. 1992). Similarly, a number of additional soluble substrates of protein kinase A have been identified in T_{84} cells stimulated with VIP, and their phosphorylation corresponds temporally with the chloride secretory response induced by this hormone (Cohn 1987).

More direct evidence for the involvement of protein kinase A in the secretory response has been obtained from whole cell patch-clamp experiments. Here, perfusion of the cytosol of T_{84} cells with a pseudosubstrate peptide inhibitor of protein kinase A fully inhibited chloride conductance responses to VIP, without affecting the response to ionomycin (Fitz and Cohn 1991). Further, Rogers et al. (1990) have reported that T_{84} cells transfected with a mutant form of the regulatory subunit of protein kinase A, and stably expressing this defective protein, are unable to secrete chloride in response to forskolin, VIP, or PGE_2, while retaining their ability to respond to calcium ionophores. Experiments of this type illustrate the direct link between protein kinase A and chloride secretory activity.

3.1.2 cGMP-Related Stimuli

In general, cGMP seems to be a second messenger that is employed less frequently than cAMP to stimulate chloride secretion in T_{84} cells. Unlike the situation in many other cell types, where signals transduced by cAMP and cGMP have opposite effects (Berridge 1975), in the intestinal epithelium both cAMP and cGMP appear to have prosecretory actions. cGMP accumulation is induced in T_{84} cells by the heat-stable enterotoxin of *E. coli*, ST_a, and levels of the messenger induced by this stimulus are closely correlated in both a time- and dose-dependent fashion to the effects of the toxin on chloride secretion (Huott et al. 1988).

In many respects, chloride secretion induced by cGMP in T_{84} cells closely resembles that induced by cAMP. Thus, there is a relatively slow onset of the effect (maximal within 10–15 min after stimulus addition), but the response is thereafter sustained for prolonged periods. Chloride secretion induced by ST_a is rapidly reversed by transport pathway inhibitors such as ouabain and bumetanide, and notably, has a similar sensitivity to inhibition by the potassium channel blocker, barium, as does cAMP-related secretion (Huott et al. 1988).

Presumably, ST_a induces chloride secretion in T_{84} cells by activation of a cGMP-dependent protein kinase, but this remains to be formally demonstrated. The effects of cGMP within the cell may well be related to the opening of a recently postulated cGMP-activated chloride channel, as described above (Lin et al. 1991). However, cAMP-dependent and cGMP-dependent stimuli must involve some common pathways for the stimulation of chloride secretion, since addition of a combination of these two types of stimuli to T_{84} cells results in a chloride secretory response that is less than additive (Huott et al. 1988). Both cAMP- and cGMP-dependent stimuli do, however, interact synergistically with calcium-related secretagogues (see below).

3.1.3 Calcium-Related Stimuli

Agents thought to induce T_{84} cell chloride secretion by elevating levels of free cytosolic calcium ($[Ca^{2+}]_i$) include carbachol (Dharmsathaphorn and Pandol 1986), histamine (Wasserman et al. 1988), bile acids such as taurodeoxycholic acid (Dharmsathaphorn et al. 1989b), and calcium ionophores (Dharmsathaphorn et al. 1989a). Each of these stimuli has relatively transient effects on secretion; particularly for the receptor-mediated agonists, such as carbachol and histamine, effects are maximal approximately 2 min after stimulus addition and decline rapidly thereafter. For these latter agents, the transient changes in chloride secretion correspond to transient changes in $[Ca^{2+}]_i$. Another notable general feature of calcium-related chloride secretion across T_{84} cells is its relative insensitivity to inhibition by barium (Mandel et al. 1986b; Wasserman et al. 1988).

Calcium-related secretion from T_{84} cells may actually represent the interaction of several intracellular effectors (Dharmsathaphorn et al. 1989a). When the ability of a number of stimuli in this class was compared in terms of their ability to increase $[Ca^{2+}]_i$ and chloride secretion, there were marked discrepancies. For example, histamine and carbachol were potent chloride secretagogues while inducing modest increases in $[Ca^{2+}]_i$. The reverse scenario was true for the calcium ionophores ionomycin and 4-BrA23187. These findings led to the hypothesis that stimuli of this type also mobilize other intracellular second messengers which modify the response to a given elevation in $[Ca^{2+}]_i$. In fact, it is well appreciated that calcium mobilization in response to carbachol and histamine, and to some extent even to calcium ionophores, involves dramatic changes in membrane phospholipid metabolism. Stimulation of T_{84} cells results in the hydrolysis of phosphatidylinositol bisphosphate, yielding inositol trisphosphate (IP_3) which is responsible for releasing calcium from intracellular stores, and diacylglycerol, an activator of protein kinase C. Many other lipid messengers, including IP_4, IP_5, and phosphatidic acid, are also generated in the cells following stimulation (Kachintorn et al. 1990a; Vajanaphanich et al. 1990). It had therefore been difficult to definitively ascribe a role for calcium per se in the observed secretory responses.

In an attempt to directly ascertain the role of calcium in controlling chloride secretion, our laboratory has recently used the agent thapsigargin to increase $[Ca^{2+}]_i$ in T_{84} cells. Because this drug affects calcium levels by blocking reuptake into intracellular stores, it induces a progressive increase in $[Ca^{2+}]_i$ which is independent of any effects on membrane phospholipid turnover. In this setting, thapsigargin induced chloride secretion with a time course identical to its effects on $[Ca^{2+}]_i$ (Kachintorn et al. 1990b). Both responses were blocked by an intracellular calcium chelator, BAPTA. We have concluded that calcium alone is probably sufficient to lead to chloride and/or potassium channel opening, but that other messengers released concurrently during agonist stimulation may be important determinants of the magnitude and/or kinetics of the observed response. A similar conclusion was reached by Wong et al. (1990), who showed that increases in basolateral potassium conductance could be accounted for by changes in $[Ca^{2+}]_i$ in T_{84} cells, but that carbachol induced a greater increase in potassium transport than did ionomycin for a given degree of elevation in $[Ca^{2+}]_i$.

The ability of calcium to induce chloride secretion in T_{84} cells does not appear to be related in a simple fashion to the activity of either of the two known classes of protein kinases activated by the cation, protein kinase C (PKC), and the calcium/calmodulin-dependent protein kinase (CaMK). Although chloride secretion invoked in T_{84} cells by carbachol was associated with the phosphorylation of PKC substrates (Cohn 1990), secretion cannot be inhibited by a number of PKC inhibitors (Kachintorn et al. 1991). In fact, with some of these drugs, a significant enhancement of stimulated secretion was observed. Further, short periods of pretreatment (5–10 min) of T_{84} cells with PKC activators such as phorbol esters or water-soluble analogs of diacylglycerol enhances secretory responses to carbachol, histamine, or ionophores while longer preincubations (>15 min) suppress both the calcium signal and chloride secretion induced by carbachol and histamine. Similarly, a CaMK inhibitor, W_{13}, enhanced rather than suppressed the effect of carbachol on chloride transport and $[Ca^{2+}]_i$ (Vajanaphanich et al. 1991b).

Calcium-related secretion may also be modified by other lipid metabolites generated following stimulation. Phosphatidic acid, which is increased in T_{84} cells stimulated with carbachol, significantly enhances both calcium responses and chloride secretory responses to this agent if given exogenously (Vajanaphanich et al. 1990). Although phosphatidic acid is a weak activator of PKC (Vajanaphanich et al., in press), its effects on chloride secretion do not resemble those of phorbol esters or diacylglycerol analogs. However, one problem that is apparent in interpreting data in this area is the existence of multiple isoforms of PKC (Nishizuka 1988). These disparate enzymes may have different subcellular localizations, different substrate affinities, and variable sensitivities to activation and inhibition by known agents.

The ability of T_{84} cells to respond to calcium-dependent stimuli appears to be related, at least in part, to phospholipid synthetic pathways. For example, treatment of T_{84} cells with W_{13}, which enhances their responsiveness to carbachol, increases the proportion of phosphatidylinositol bisphosphate in membrane phospholipids and thereby presumably enhances IP_3 generation (Vajanaphanich et al. 1991a). Prolonged (>15 min) pretreatment with the phorbol ester PMA reduces phosphatidylinositol labeling. These findings may be reflective of feedback mechanisms that play an important role physiologically to limit the extent of secretion.

Our current working model for calcium-related chloride secretion in T_{84} cells is summarized in Fig. 2. We propose that calcium, acting either directly or via a protein kinase which is insensitive to known PKC and CaMK inhibitors, can induce chloride secretion by opening chloride and/or potassium channels. The activation of PKC which follows as a inevitable consequence of membrane phospholipid turnover and calcium mobilization serves to limit or terminate both the secretory response and the calcium signal. The mediator of this termination response is currently unknown.

3.1.4 Other Possible Messengers

Adenosine agonists can induce chloride secretion from T_{84} cells at doses where they have no measurable effects on cAMP, cGMP, or $[Ca^{2+}]_i$ (K.E. Barrett et al. 1990). Similarly, as alluded to above, PGE_1 appears to stimulate a secretory response that is out of proportion to its ability to elevate cAMP. Observations of this type have led to the speculation that additional, as yet unidentified, second messengers may be able to modulate secretory events.

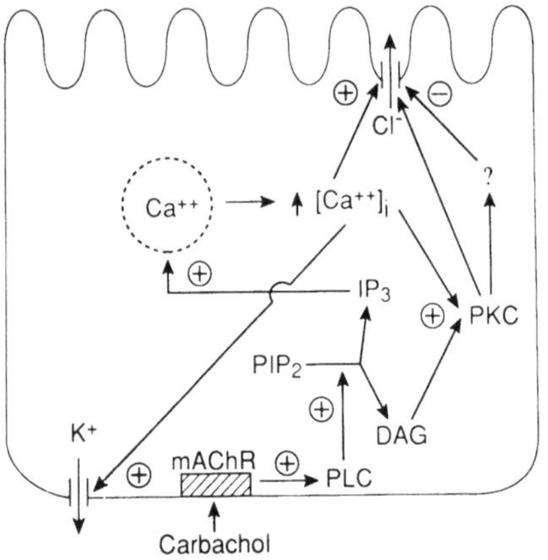

Fig. 2. Hypothetical model for regulation of calcium-regulated chloride secretion, as exemplified by the response of T_{84} cells to carbachol. Binding of agonist to basolateral receptors activates phospholipase C (*PLC*) (via G-proteins) and stimulates hydrolysis of phophatidylinositol bisphosphate (*PIP_2*). Released inositol trisphosphate (*IP_3*) mobilizes calcium from intracellular stores, and increases the intracellular calcium concentration ([*Ca^{++}*]). Calcium then activates potassium and/or chloride channels. The other product of PIP_2 hydrolysis, diacylglycerol (*DAG*), acts with calcium to stimulate protein kinase C (*PKC*) activity. This enzyme, acting either directly or via as yet unidentified intermediate products (?), down-regulates chloride and/or (not shown) potassium channels. PKC may also reduce availability of PIP_2. For further details, see text

A role for phospholipid metabolism in the response of T_{84} cells to adenosine agonists such as NECA is suggested by the observation that the secretory response is inhibited by a diglyceride lipase inhibitor, RG80267 (Barrett 1989b). Further, NECA induces the release of radioactivity from T_{84} cells preloaded with ^{3}H-arachidonic acid with a dose dependency that approximates that for its effect on chloride secretion. It is possible, therefore, that arachidonic acid or one of its metabolites acts within the cells as a second messenger. In keeping with this hypothesis, exogenous arachidonic acid has a small but significant sustained effect on chloride secretion. The effect is specific for arachidonic acid and is not reproduced by equivalent concentrations of a number of other saturated and unsaturated fatty acids. Synergism is observed between the responses to arachidonic acid and carbachol, whereas the combination of arachidonic acid plus VIP resulted in a less than additive response. In this repect, the response to arachidonic acid resembles that observed with NECA, where a similar pattern of interactions was seen (Barrett and Bigby, in press).

If arachidonic acid is to be considered as a messenger for secretion, it will be necessary to explain data showing that doses of NECA which do not measurably increase cAMP do however increase protein kinase A activity in T_{84} cells (K.E. Barrett et al. 1990). It is also of interest that diglyceride lipase inhibition can also reduce chloride secretory responses to some cyclic nucleotide-dependent stimuli, such as VIP, forskolin, and ST_a (Barrett 1990).

3.2 Second Messenger Mechanisms Terminating Secretion

For a full understanding of the mechanisms underlying T_{84} cell chloride secretion, it will be necessary to learn not only which messengers initiate the secretory response, but also how the response is terminated. As noted above, chloride secretory responses to calcium-dependent stimuli are transient in nature. Further, the transient nature of the response cannot always be simply explained by the correspondingly transient nature of the calcium signal. This implies that additional second messengers are generated within the cell which are inhibitory. We have recently shown that when T_{84} cells are treated first with carbachol, then stimulated with thapsigargin, there is no chloride secretory response to thapsigargin, even though the increase in $[Ca^{2+}]_i$ occurring in response to the second agent is unchanged (Kachintorn et al., in press).

Interestingly, pretreatment of T_{84} cells with histamine does not affect their ability to display a chloride secretory response to thapsigargin. We have therefore compared phospholipid metabolism in cells treated with carbachol or histamine to identify candidate inhibitory messengers. At 20 min after addition of carbachol, a time when inhibition of thapsigargin-stimulated secretion is observed, levels of IP_4, increased by carbachol, are still markedly elevated, whereas histamine-induced alterations in the level of this messenger have returned to baseline. Further experiments will be required to test the hypothesis that IP_4 acts to reduce subsequent calcium-mediated chloride section following carbachol stimulation. It will also be of interest to learn how the activity of PKC and/or CaMK impacts on IP_4 levels. It does, however, appear that the effect of the putative inhibitory messenger is specific for calcium-dependent secretion, since the ability of T_{84} cells to respond to VIP or ST_a is not reduced by prior carbachol stimulation.

3.3 Interactions Between Signaling Pathways

One area where the use of the T_{84} cell line has contributed greatly to the understanding of transport physiology has been in its use to identify how hormones acting through different second messenger mechanisms may interact to promote secretion. It had long been appreciated that hormones may act synergistically to promote secretion elsewhere in the gastrointestinal tract (for example, the interaction of histamine and acetylcholine to induce gastric acid secretion), but similar synergistic phenomena had not been readily identified for intestinal chloride secretion. This undoubtedly reflected the fact that multiple endogenous hormones were present when intact tissues were used as the experimental model.

Using the T_{84} cells, it can be readily demonstrated that the combination of a stimulus that acts through a cyclic nucleotide-dependent pathway with a calcium-dependent agent has effects on secretion that are significantly larger than would be predicted by simple addition of the responses to each agent added in isolation (Cartwright et al. 1985). The mechanism of this synergism is postulated to lie at the level of the various transport pathways activated. cAMP appears to preferentially activate chloride channels, while calcium may be a better stimulus of a basolateral potassium conductance. It has been hypothesized that when both pathways are activated simultaneously, the rate-limiting steps for both types of secretion are effectively abolished with a resulting potentiated response. Synergism does not appear to arise at the level of second messenger generation. T_{84} cells treated with the combination of VIP and a calcium ionophore accumulate equivalent amounts of cAMP as do cells stimulated with VIP alone (Cartwright et al. 1985). Similarly, cells treated with carbachol plus NECA display an identical calcium response to those treated with carbachol alone (K.E. Barrett et al. 1990).

The existence of synergism has significant implications for the control of intestinal secretion in vivo, where multiple hormones may be encountered simultaneously on a frequent basis. This may be of particular relevance in settings such as intestinal inflammation, where mediators from immunocytes, nerves, and endocrine cells could interact to induce clinical diarrhea (Barrett and Dharmsathaphorn 1991).

4 Extracellular Regulation of Secretion

4.1 Interactions of T_{84} Cells with Other Cell Types

The T_{84} cell model, being reductionist in nature, provides an ideal setting to examine the precise pathways of cell-cell interactions which may form the basis for integrated extracellular control of secretion in vivo. Such studies are essentially impossible to perform either in vivo or in intact tissue, because it is very difficult to define the precise contributions of resident cell types and mediators. I will discuss here studies being performed in a number of laboratories, including our own, that illustrate the strength of the model in this regard.

4.1.1 Interactions with Mast Cells

Diarrhea is a well-known sequelae of mast cell activation in the gastrointestinal tract. For example, secretory diarrhea is a common symptom of generalized anaphylaxis, food allergy, and systemic mastocytosis. In theory, there are at least three ways in which mast cell mediators could induce intestinal secretion. First, they could stimulate the secondary production of secretagogues, such as prostaglandins, from other mucosal elements.

Second, they could interact directly with epithelial cells to stimulate secretion. Third, they might somehow alter the sensitivity of the epithelium to normal levels of hormones and neurotransmitters. We have used the T_{84} model to examine the latter two of these possibilities (Barrett 1991).

A number of mast cell mediators, including histamine (Wasserman et al. 1988), adenosine (K.E. Barrett et al. 1990), and prostaglandin D_2 (Barrett 1989a), are able to induce chloride secretion directly when added to T_{84} cells. Similarly, both lysates of mast cells and supernatants derived following immunological activation can cause marked secretion (Barrett 1991). Histamine and adenosine appear to contribute to the effect of these mixed mediator preparations, because secretion is partially inhibited by H_1 antihistamines and adenosine receptor antagonism. There is evidence, however, that the mediators in these preparations interact synergistically, since the measured concentrations of both histamine and adenosine are below those that would be required to induced secretion were the substances to be added in isolation.

In an inflammatory setting, it is likely that epithelial cells would be exposed to mast cell mediators on an ongoing basis. We have therefore examined the effect of chronically treating T_{84} cells with a mast cell lysate preparation for 7 days in culture (Barrett 1991). Cells cultured in this fashion display a normal transepithelial resistance and potential difference, as well as basal chloride secretion. However, their ability to respond to certain stimuli of chloride secretion is significantly enhanced. Thus VIP, carbachol, and ST_a induce higher levels of chloride secretion in treated than control cells. At least for VIP, the effect is also seen at the level of cAMP generation. The response is not a generalized one, because responses to NECA and PGE_1 were unaffected. The active constituent in the lysate preparation is apparently of low molecular weight, because activity is removed by dialysis. Using the same experimental design, we have also studied the effect of chronic exposure to an anaphylactic supernatant from mast cells. The most prominent effect of this preparation was to induce a marked decrease in T_{84} transepithelial resistance, although the cells remained viable as attested to by their ability to respond to stimulation with chloride secretion (K.E. Barrett, unpubl. observ.).

In summary, studies of mast cell-T_{84} cell interactions (summarized in Fig. 3) have allowed us to determine that mast cell-related intestinal secretion in vivo is likely to be mediated, at least in part, by direct interactions of mast cell mediators with the intestinal epithelium. Both transcellular and paracellular pathways for solute transport are likely to be modulated in clinical settings where ongoing mast cell activation occurs.

4.1.2 Interactions with Neutrophils

Another inflammatory cell type that may have important interactions with the epithelium in inflammatory bowel disease is the neutrophil. Madara and

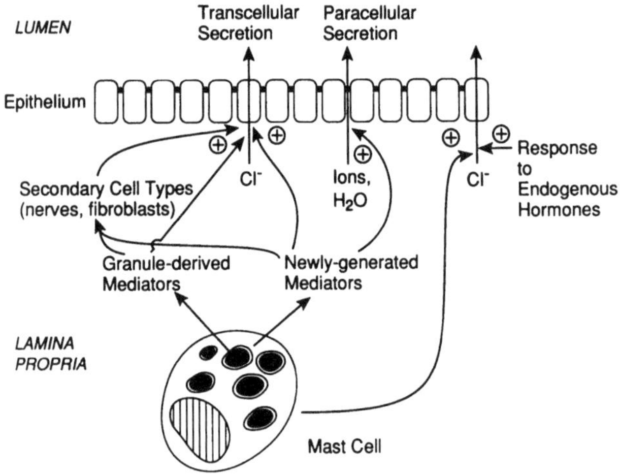

Fig. 3. Proposed pathways of mast cell-epithelial interaction, as defined by experiments with T_{84} cells and comparison of these studies with those in intact tissue. Both granule-associated and newly generated mediators may increase chloride secretion directly, or by acting on secondary cell types. Mast cell mediators, particularly newly generated ones, may affect the permeability of the paracellular pathway for solute transport. Finally, mast cell mediators may up-regulate the responsiveness of the epithelium to endogenous hormones and neurotransmitters

coworkers have utilized the T_{84} cell to develop a model of cellular interactions in the formation of crypt abscesses, a prominant feature of ulcerative colitis (Nash et al. 1987, 1991). Neutrophils can be shown to migrate specifically through T_{84} cells, a process which involves the cellular adhesion complex CD18/CD11b (Parkos et al. 1991). Further, activated neutrophils elaborate a low molecular weight mediator, as yet unidentified, which is a potent stimulus of T_{84} cell chloride secretion, but only when supplied to the apical aspect of the cells (Nash et al. 1991). Thus, were neutrophils to be stimulated either within a crypt abscess or within the gut lumen, they might contribute to the induction of inflammatory diarrhea.

Neutrophils may release other mediators with relevance for chloride secretion. For example, it has been demonstrated that treatment of intact intestinal tissue with neutrophil stimuli such as chemotactic peptides results in chloride secretion, and that this response may be dependent on the release of reactive oxygen species by the neutrophils (Bern et al. 1989; T.A Barrett et al. 1990). However, the ability of the epithelium to respond to such species may be dependent on the participation of secondary cell types, because when oxidants such as hydrogen peroxide are supplied to T_{84} monolayers, their effect on chloride secretion is modest at best and requires high doses (Berschneider and Powell 1989).

4.1.3 Interactions with Mesenchymal Cells

At one time, the mesenchymal elements of the intestinal mucosa, such as fibroblasts, were considered simply to provide structural support for the tissue. This view is giving way to that of a more active role for these mesenchymal cell types, which contends that they interact with the epithelium to control its growth, differentiation and, ultimately, function.

The ability of fibroblasts to modulate T_{84} cell chloride secretion has been examined by constructing "sandwiches" of filter-grown T_{84} cells and intestinal fibroblasts, mounted back-to-back in Ussing chambers. These studies have shown that T_{84} responses to a variety of stimuli, including histamine, reactive oxygen species, and bradykinin can be enhanced if fibroblasts are present (Berschneider and Powell 1989). In some cases, these stimuli have no measurable effect on secretion unless the fibroblasts are included. The ability of fibroblasts to mediate these responses is likely to be due to their ability to synthesize and release prostaglandins in response to the named agents. Moreover, the ability of fibroblasts to transduce these secretory effects may be subject to further amplification. Fibroblasts grown in the presence of the inflammatory cytokine interleukin 1β secreted significantly more PGE_2 than control cells in response to histamine stimulation, and this correlated with an increased ability to transmit a signal for chloride secretion to T_{84} cells in response to bradykinin, histamine, or hydrogen peroxide (Hinterleitner et al. 1991). It remains to be demonstrated whether such epithelial-fibroblast interactions are enhanced in chronic inflammation, where significant fibrosis is frequently encountered.

5 Summary

Since its initial description in 1984, the T_{84} cell line has come to play a preeminent role in the investigation of intestinal ion transport at the cellular level. This is largely because the use of this model offers a significant number of experimental advantages. First, the cell line represents a homogeneous population of secretory epithelial cells. Thus, secretory processes are not masked, as they are in the native intestine, by simultaneously occurring electrolyte absorption. This has allowed the precise definition of the stoichiometry, electrophysiology, and biochemistry of the membrane transport pathways that comprise the chloride secretory mechanism. Second, the model consists solely of epithelial cells, without the influence of endogenous hormones and neurotransmitters and other cell types normally resident in the intestinal mucosa. This feature of the model makes it useful for tracing pathways of cellular and molecular interactions that control secretion. Third, the cells offer the possibility of a theoretically, essentially unlimited supply of experimental material. The T_{84} cell line has already contributed to the discovery of the protein defective in most cases of cystic fibrosis, CFTR, and should also permit further detailed molecular analysis of other proteins involved in chloride secretion. Finally, the model is amenable to chronic manipulation in culture. This should allow us to understand how the epithelium alters its properties to compensate for a change in the osmotic, inflammatory, microbial, or hormonal environment of the gut.

It should not be forgotten that the T_{84} cell line is of tumor origin, and therefore may not be truly representative of the properties of the native intestinal crypt epithelium. The process of malignant transformation may cause cells to lose differentiated properties, or, less frequently, to acquire new characteristics which are not shared by their normal counterparts. These reservations notwithstanding, provided that we are cautious about extrapolating conclusions from this cell type, and take into account complementary information derived from intact tissue and in vivo experiments, investigations such as those described here, as well as those using novel electrophysiological and molecular techniques, should remain of value. It is to be expected that studies of the T_{84} cell will continue to contribute to our knowledge of intestinal transport, and ultimately to our ability to treat diarrheal diseases and other conditions where electrolyte transport is disregulated, such as cystic fibrosis.

Acknowledgments. Studies from the author's laboratory have been supported by grants from the National Institutes of Health (AI 24992, DK 28305). I thank Ms. Glenda Wheeler for her expert editorial assistance in the preparation of this manuscript. I also thank the following former and present colleagues whose helpful discussions and/or technical expertise have contributed to some of the studies described here: Kiertisin Dharmsathaphorn, M.D., Gregory Beuerlein, Patrick Huott, M.D., Udom Kachintorn, M.D., James A. McRoberts, Ph.D., Alexis E. Traynor-Kaplan, Ph.D., and Mana Vajanaphanich, M.D.

References

Anderson MP, Rich DP, Gregory RJ, Smith AE, Welsh MJ (1991) Generation of cAMP-activated Cl^- currents by expression of CFTR. Science 251: 679–682

Barrett KE (1989a) Immune-related intestinal secretion: control of colonic chloride secretion by inflammatory mediators. In: MacDermott RP (ed) Inflammatory bowel disease: current status and future approach. Elsevier, Amsterdam, pp 377–382

Barrett KE (1989b) A role for arachidonic acid in the induction of epithelial chloride secretion. FASEB J 3: A281 (Abstr)

Barrett KE (1990) A diglyceride lipase inhibitor, RG80267, delineates two types of cyclic mucleotide-dependent epithelial chloride secretagogues. FASEB 4: A550 (Abstr)

Barrett KE (1991) Immune-related intestinal chloride secretion. III. Acute and chronic effects of mast cell mediators on chloride secretion by a human colonic epithelial cell line. J Immunol 147: 959–964

Barrett KE, Dharmsathaphorn K (1991) Secretion and absorption: small intestine and colon. In: Yamada T (ed) Textbook of gastroenterology. Lippincott, Philadelphia, pp 265–294

Barrett KE, Cohn JA, Huott PA, Wasserman SI, Dharmsathaphorn K (1990) Immune-related intestinal chloride secretion. II. Effect of adenosine on T_{84} cell line. Am J Physiol 258: C902–C912

Barrett TA, Musch MW, Chang EB (1990) Chemotactic peptide effects on intestinal electrolyte transport. Am J Physiol 259: G947–G954

Bern MJ, Sturbaum W, Karayalcin SS, Berschneider HM, Powell DW (1989) Immune system control of rat and rabbit colonic electrolyte transport. Role of prostaglandins and enteric nervous system. J Clin Invest 83: 1810–1820

Berridge MJ (1975) The interaction of cyclic nucleotides and calcium in the control of cellular activity. Adv Cyclic Nucleotide Res 6: 1–98

Berschneider HM, Powell DW (1989) Fibroblasts mediate intestinal secretory response to bradykinin, H_2O_2 and serotonin. Gastroenterology 96: A41 (Abstr)

Cartwright CA, McRoberts JA, Mandel KG, Dharmsathaphorn K (1985) Synergistic action of cyclic AMP and calcium-mediated chloride secretion in a colonic epithelial cell line. J Clin Invest 76: 1837–1842

Cliff WH, Frizzell RA (1990) Separate Cl^- conductances activated by cAMP and Ca^{2+} in Cl^--secreting epithelial cells. Proc Natl Acad Sci USA 87: 4956–4960

Cohn JA (1987) Vasoactive intestinal peptide stimulates protein phosphorylation in a colonic epithelial cell line. Am J Physiol 253: G420–G424

Cohn JA (1990) Protein kinase C mediates cholinergically regulated protein phosphorylation in a Cl^--secreting epithelium. Am J Physiol 258: C227–C233

Cohn JA, Nairn AC, Marino CR, Melhus O, Kole J (1992) Characterization of the cystic fibrosis transmembrane conductance regulator in a colonocyte cell line. Proc Natl Acad Sci USA 89: 2340–2344

Dharmsathaphorn K, Madara J (1990) Established intestinal cell lines as model systems for electrolyte transport studies. In: Fleischer S, Fleischer B (eds) Methods in enzymology, vol 192. Biomembranes, part W. Cellular and subcellular transport: epithelial cells. Academic Press, San Diego, pp 354–389

Dharmsathaphorn K, Pandol S (1986) Mechanism of chloride secretion induced by carbachol in a colonic epithelial cell line. J Clin Invest 77: 348-354

Dharmsathaphorn K, Mandel KG, McRoberts JA, Tisdale LD, Masui H (1984) A human colonic tumor cell line that maintains vectorial electrolyte transport. Am J Physiol 246: G204–G208

Dharmsathaphorn K, Mandel KG, Masui H, McRoberts JA (1985) Vasoactive intestinal polypeptide-induced chloride secretion by a colonic epithelial cell line: direct participation of a basolaterally-localized Na^+, K^+, Cl^- cotransport system. J Clin Invest 77: 348–351

Dharmsathaphorn K, Cohn J, Beuerlein G (1989a) Multiple calcium-mediated effector mechanisms regulate chloride secretory responses in T_{84} cells. Am J Physiol 256: C1224–C1230

Dharmsathaphorn K, Huott PA, Vongkovit P, Beuerlein G, Pandol SJ, Ammon HV (1989b) Cl^- secretion induced by bile salts. A study of the mechanism of action based on a cultured colonic epithelial cell line. J Clin Invest 84: 945–953

Drumm ML, Pope HA, Cliff WH, Rommens JM, Marvin SA, Tsui LC, Collins FS, Frizzell RA, Wilson JM (1990) Correction of the cystic fibrosis defect in vitro by retrovirus-mediated gene transfer. Cell 62: 1227–1233

Fitz JG, Cohn JA (1991) VIP-stimulated T_{84} cell chloride conductance is blocked by intracellular perfusion with a protein kinase inhibitor. Gastroenterology 100: A685 (Abstr)

Gregory RJ, Cheng SH, Rich DP, Marshall J, Paul S, Hehir K, Ostedgaard L, Klinger KW, Welsh MJ, Smith AE (1990) Expression and characterization of the cystic fibrosis transmembrane conductance regulator. Nature 347: 382–386

Halm DR, Dawson DC (1983) Cation activation of the basolateral sodium-potassium pump in turtle colon. J Gen Physiol 82: 315–329

Hinterleitner TA, Berschneider HM, Powell DW (1991) Fibroblast-mediated Cl^- secretion by T_{84} cells is amplified by interleukin-1β. Gastroenterology 100: A690 (Abstr)

Huott PA, Liu W, McRoberts JA, Giannella RA, Dharmsathaphorn K (1988) The mechanism of *E. coli* heat stable enterotoxin in a human colonic cell. J Clin Invest 82: 514–523
Kachintorn U, Traynor-Kaplan AE, Dharmsathaphorn K (1990a) Characterization of inositol phosphate production in T_{84} cells stimulated with histamine and carbachol. FASEB J 4: A1778 (Abstr)
Kachintorn U, Vajanaphanich M, Traynor-Kaplan A, Thastrup O, Dharmsathaphorn K (1990b) The effect of thapsigargin on $[Ca^{++}]_i$ and Cl^- secretory response in T_{84} monolayers. Clin Res 38: 102A (Abstr)
Kachintorn U, Vongkovit P, Vajanaphanich M, Dinh S, Barrett KE, Dharmsathaphorn K (1992) Dual Effects of a phorbol ester on calcium-dependent chloride secretion by T_{84} epithelial cells. Am J Physiol 262: C15–C22
Kirk KL, Halm DR, Dawson DC (1980) Active sodium transport by turtle colon via an electrogenic Na-K exchange pump. Nature 287: 237–239
Lin M, Nairn AC, Guggino SE (1991) Cyclic GMP-dependent protein kinase (PKG) opens Cl^- channels in T_{84} cells. Gastroenterology 100: A694 (Abstr)
Mandel KG, Dharmsathaphorn K, McRoberts JA (1986a) Characterization of a cyclic AMP-activated Cl^- transport pathway in the apical membrane of a human colonic epithelial cell line. J Biol Chem 261: 704–712
Mandel KG, McRoberts JA, Beuerlein G, Foster ES, Dharmsathaphorn K (1986b) Ba^{++} inhibition of VIP and A23187 stimulated Cl^- secretion by T_{84} cell monolayers. Am J Physiol 250: C486–C494
McRoberts JA, Barrett KE (1989) Hormone-regulated ion transport in T_{84} colonic cells. In: Matlin KS, Valentich JD (eds) Functional epithelial cells in culture. Liss, New York, pp 235–265
McRoberts JA, Beuerlein G, Dharmsathaphorn K (1985) Cyclic AMP and Ca^{++} activated K^+ transport in a human colonic epithelial cell line. J Biol Chem 260: 14163–14172
Nash S, Stafford J, Madara JL (1987) Effects of polymorphonuclear leukocyte transmigration on the barrier function of cultured intestinal epithelial monolayers. J Clin Invest 80: 1104–1113
Nash S, Parkos C, Nusrat A, Delp C, Madara JL (1991) In vitro model of intestinal crypt abscess. A novel neutrophil-derived secretagogue activity. J Clin Invest 87: 1474–1477
Nishizuka Y (1988) The heterogeneity and differential expression of multiple species of the protein kinase C family. Biofactors 1: 17–20
Parkos CA, Arnaout MA, Madara JL (1991) Polymorphonuclear leukocyte (PMN) transmigration through intestinal epithelium is dependent on a CD11b/CD18-mediated event. Gastroenterology 100: A607 (Abstr)
Ringe D, Petsko GA (1990) A transport problem? (editorial). Nature 346: 312–313
Riordan JR, Rommens JM, Kerem B-S, Alon N, Rozmahel R, Grzelczak Z, Zielenski J, Lok S, Plavsic N, Chou J-L, Drumm ML, Iannuzzi MC, Collins FS, Tsui L-C (1989) Identification of the cystic fibrosis gene: cloning and characterization of complementary DNA. Science 245: 1066–1073
Rogers KV, Goldman PS, Frizzell RA, McKnight GS (1990) Regulation of Cl^- transport in T_{84} cell clones expressing a mutant regulatory subunit of cAMP-dependent protein kinase. Proc Natl Acad Sci USA 87: 8975–8979
Shorofsky SR, Field M, Fozzard HA (1982) The cellular mechanism of active chloride secretion in vertebrate epithelia: studies in intestine and trachea. Philos Trans R Soc Lond (Biol) 299: 597–607
Vajanaphanich M, Kachintorn U, Dharmsathaphorn K, Traynor-Kaplan A (1990) The role of phosphatidic acid (PA) in Cl^- secretion in T_{84} colonic epithelia. Gastroenterology 98: A559 (Abstr)
Vajanaphanich M, Kachintorn U, Barrett KE, Traynor-Kaplan AE (1991a) Agents which modulate epithelial secretion alter accumulation of lipids in the PI pathway. Gastroenterology 100: A707 (Abstr)

Vajanaphanich M, Kachintorn U, Traynor-Kaplan AE, Barrett KE (1991b) Evidence for reciprocal control of epithelial Cl^- secretion by protein kinase C and calcium/calmodulin dependent protein kinase. Gastroenterology 100: A707 (Abstr)

Wagner JA, Cozens AL, Schulman H, Gruenert DC, Stryer L, Gardner P (1991) Activation of chloride channels in normal and cystic fibrosis airway epithelial cells by multifunctional calcium/calmodulin-dependent protein kinase. Nature 349: 793–796

Wasserman SI, Barrett KE, Huott PA, Beuerlein G, Kagnoff MF, Dharmsathaphorn K (1988) Immune-related intestinal Cl^- secretion. I. Effect of histamine on the T_{84} cell line. Am J Physiol 254: C53–C62

Welsh MJ, Smith PL, Frizzell RA (1982) Chloride secretion by canine tracheal epithelium. II. The cellular electrical potential profile. J Membr Biol 70: 227–238

Weymer A, Huott P, Liu W, McRoberts JA, Dharmsathaphorn K (1985) Chloride secretory mechanism induced by prostaglandin E_1 in a colonic epithelial cell line. J Clin Invest 76: 1828–1836

Wong SM, Tesfaye A, DeBall MC, Chase HS Jr (1990) Carbachol increases basolateral K^+ conductance in T_{84} cells. Simultaneous measurements of cell [Ca] and gK explore calcium's role. J Gen Physiol 96: 1271–1285

Worrell RT, Butt AG, Cliff WH, Frizzell RA (1989) A volume-sensitive chloride conductance in human colonic cell line T_{84}. Am J Physiol 256: C1111–C1119

Chapter 11

Ion Transport in HT_{29} Colonic Carcinoma Cells

Karl Kunzelmann, Monika Tilmann and *Rainer Greger*[1]

Contents

1 Introduction

The permanent cell line HT_{29} was established by Fogh and Trempe in 1975 from an adenocarcinoma of human colon (Fogh and Trempe 1975). In the following years, the cell line was examined and characterized with respect to many different biochemical and biophysical properties. HT_{29} cells, besides other colonic carcinoma cell lines such as T_{84} and CACO-2, serve as a very useful model in studying growth and differentiation of epithelial cells (Rousset 1986; Le Bivic et al. 1988; Hekmati et al. 1990). Moreover, the presence of a number of different hormone receptors was demonstrated in the membranes of HT_{29} cells. Second messenger systems activated by receptor binding were examined and are still under investigation (cf. Table 1). In all instances, the occupation of the receptor was linked to a change in the ion conductance properties of the HT_{29} cell. In terms of membrane transport proteins, in addition to the (Na^+/K^+)-ATPase, the presence of the

[1]Physiologisches Institut der Albrecht-Ludwigs-Universität, Hermann-Herder-Straße 7, W-7800 Freiburg, Germany

Advances in Comparative and Environmental Physiology, Vol. 16

Table 1. Regulation of HT_{29} function by hormones and agonists

Receptor	Cellular events	References
Vasointestinal peptide	cAMP increase	
	Act. of $Na^+/2Cl^-/K^+$ carrier	Laburthe et al. (1978)
	Act. of Na^+/K^+-ATPase	Turner et al. (1990)
	Act. of Cl^- channels	
Neurotensin	Ca^{2+} increase	Lohrmann et al. (1991)
	IP_3 increase	Bozou et al. (1989)
	Act. of K^+ channels	Amar et al. (1986)
	Act. of Cl^- channels	Wu et al. (1991)
		Leipziger et al. (1991)
α_2-Receptor	Inhibition of cAMP pathway	Cantiello and Lanier (1989)
		Bouscarel et al. (1985)
β-Receptor	Act. of Cl^- channels	Lohrmann et al. (1991)
		Greger and Kunzelmann (1991)
Muscarinic	Ca^{2+} increase	Lohrmann et al. (1991)
		Nitschke et al. (1993b)
	Act. of Cl^- channels	Greger and Kunzelmann (1991)
		Rechkemmer et al. (1991)
	Act. of K^+ channels	Ziss and Hegel (1988)
		Wu et al. (1991)
Purinergic (type 2)	Ca^{2+} increase	Lohrmann et al. (1991)
		Kunzelmann et al. (1992b)
	Act. of K^+ channels	Wu et al. (1991)
		Leipziger et al. (1991)

Act. = activation

following systems has been shown in HT_{29} cells: the Na^+/H^+ exchanger (Cantiello and Lanier 1989); the $Na^+/2Cl^-/K^+$ cotransporter (Kim et al. 1988); a K^+ conductance (Wu et al. 1991); and a Cl^- conductance (Ziss et al. 1987; Lohrmann et al. 1991; Kunzelmann et al. 1992a). Because of their polar growth, which is developed under certain culture conditions, HT_{29} cells are also useful for studying electrolyte transport, and they are commonly looked at as a paradigm of the colonic crypt cell. Very recently, HT_{29} cells have also been examined with respect to their expression of the cystic fibrosis transmembrane conductance regulator (CFTR), the gene product of the cystic fibrosis gene locus (Rommens et al. 1989). CFTR was shown to be related to the Cl^- conductance in epithelial cells (Rommens et al. 1989; Sood et al. 1990; Montrose-Rafizadeh et al. 1991). For this reason and because they possess hormonally regulated chloride channels, HT_{29} cells have become a very important model system in cystic fibrosis research (Kunzelmann et al. 1989c; Anderson and Welsh 1990).

After a short summary on the morphological aspects and some other characteristics of these cells, the present short review will focus mostly on the Cl^- conductance and its regulation.

2 HT_{29} Cells in Culture

2.1 Differentiation of the Cells

HT_{29} is a rather undifferentiated epithelial cell line when kept in media containing a high glucose concentration (Zweibaum et al. 1985). In contrast, treatment of the cells with sodium-butyrate (Augeron and Laboisse 1984) or replacing D-glucose in the media by D-galactose (Remy et al. 1984; Zweibaum et al. 1985) induces a morphological differentiation of the cells: cell polarity arises and tight junctions develop. The apical cell membrane forms microvilli and possesses membrane proteins characteristic for the luminal part of the cell membrane. Moreover, the grade of differentiation correlates with the biochemical properties of these cells such as the membrane lipid composition and the presence of glycosaminoglycane (Simon-Assman et al. 1987; Reynier et al. 1991). A number of different subclones were established which differentiate in glucose-free media either as absorptive or mucous cells (Huet et al. 1987; Montrose and Murer 1989; Kreusel et al. 1990; Montrose-Rafizadeh et al. 1990).

2.2 CFTR and Cl^- Conductance

It has also been shown that the intracellular content of mRNA coding for the CFTR was significantly increased (9- to 18-fold) when the cells were kept under glucose-free conditions (Sood et al. 1990; Montrose-Rafizadeh et al. 1991). The appearance of a calcium-induced but not of a forskolin-induced Cl^- efflux was correlated with the cytosolic content of CFTR-mRNA (Montrose-Rafizadeh et al. 1991). In the HT_{29} clone used in this study the forskolin-induced Cl^- flux was low in cells containing either high or low concentrations of CFTR-mRNA. Therefore, the question of whether the CFTR-mRNA content correlates with the magnitude of the cAMP-induced Cl^- conductance remains open. Induction of a Cl^- conductance by the cAMP-dependent pathway was shown in undifferentiated (Ziss and Hegel 1988) (high D-glucose) as well as in differentiated (Kreusel et al. 1990) (D-glucose-free) HT_{29} cells and in cells kept in media containing a low D-glucose/high D-galactose concentration (Kunzelmann et al. 1992a; Lohrmann et al. 1991). Therefore, the property of cAMP-dependent Cl^- secretion, however, does not exclusively depend on the cell culture conditions. More studies are needed to correlate the biophysiological properties and the morphological differentiation of HT_{29} cells. It is probably an oversimplification to assume that HT_{29} cells in general reflect the properties of the crypt cells of the intact colon.

3 Transport Properties of HT_{29} Cells

3.1 Hormone Receptors and Regulatory Pathways

The presence of a number of different hormone receptors, which activate chloride (NaCl) transport, have been demonstrated in HT_{29} cells (cf. Table 1). VIP, which was shown to activate the $Na^+/2Cl^-/K^+$ carrier, also increases the Cl^- conductance (authors' unpubl. data); (Fig. 1), most probably due to an increase in the cytosolic cAMP concentration (Laburthe et al. 1978). β-Adrenergic stimulation by isoproterenol also induces a Cl^- conductance (Kunzelmann et al. 1992a; Lohrmann et al. 1991). In recently performed experiments we found that the effect of isoproterenol can be mimicked by forskolin, membrane-permeable derivatives of cAMP or inhibition of the phosphodiesterase by isobutylmethylxanthine (IBMX) (Kunzelmann et al. 1992a). These results indicate that the cAMP pathway is involved in the activation of the Cl^- conductance. It was shown that the increase in cytosolic cAMP induced by VIP is reduced by the activation of α_2-adrenergic receptors (Bouscarel et al. 1985; Paris et al. 1985). The mechanism by which this occurs is not fully understood as yet.

Activation of another group of receptors present in HT_{29} cell membranes increases the cytosolic Ca^{2+} concentration. This was shown for carbachol, ATP and neurotensin (Leipziger et al. 1991; Rechkemmer et al. 1991). The effect of carbachol on cytosolic Ca^{2+} activity could be inhibited by pirenzipine, indicating the presence of a muscarinic receptor. Carbachol depolarized the cell membrane potential in impalement studies (Lohrmann et al. 1991) as well as in patch-clamp studies (Greger and Kunzelmann 1991). This depolarization was due to an activation of a Cl^- conductance (Greger and Kunzelmann 1991; Lohrmann et al. 1991). ATP and neurotensin, which are known cotransmitters, also strongly depolarized the cell membrane potential. In the case of neurotensin the depolarization was very transient (Fig. 2). Again, for both agonists the depolarization was due to activation of a Cl^- conductance (Greger and Kunzelmann 1991; Lohrmann et al. 1991). The nucleotide receptor responsible for the ATP effect could be characterized as a P_2 receptor according to the potency sequence for different nucleotides (Leipziger et al. 1991; Lohrmann et al. 1991). In the case of neurotensin, a parallel activation of the phosphatidylinositol turnover was demonstrated (Amar et al. 1986; Bozou et al. 1989), strongly suggesting an IP_3-induced Ca^{2+} release from intracellular Ca^{2+} stores.

These results indicate that HT_{29} cells express muscarinic, purinergic (P_2) and neurotensin receptors. Agonist binding to these receptors causes an increase in the cytosolic Ca^{2+} and a depolarization of the cell membrane potential. With all three agonists, the depolarization is followed by a transient hyperpolarization, reflecting most probably the activation of Ca^{2+}-activated K^+ channels (Wu et al. 1991).

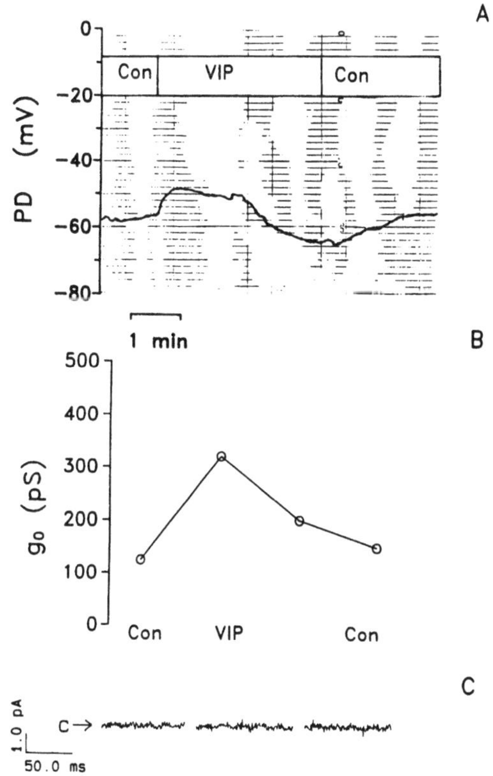

Fig. 1A,C. Effect of the vasointestinal polypeptide (VIP; 10^{-9} mol/l) on a HT_{29} cell. The method used was the CAN method (Greger and Kunzelmann 1991). The membrane potential (**A**) was depolarized by VIP. This depolarization was followed by a hyperpolarizing phase. Simultaneously, the input conductance of the cell-attached membrane patch was reversibly increased (**B**). No single channel activity was observed in the voltage-clamp mode of the patch-clamp amplifier at a clamp voltage of $Vc = +40$ mV. *C* → indicates the zero current level. Current traces are filtered at 300 Hz. These data indicate that (i) VIP acts by increasing the Cl^- conductance and that (ii) the ion channels responsible for this increase in conductance are too small to be identified as single channel events

Besides the cAMP- and Ca^{2+}-dependent activation, a Cl^- conductance can also be induced by cell swelling. Whenever the osmolarity of the extracellular solution is reduced, HT_{29} cells activate chloride and potassium channels which facilitate the outflow of KCl and therefore the reduction of the intracellular osmolarity. This leads to shrinkage of the cell and is labelled regulatory volume decrease (RVD). Up to now, the cellular mechanisms of channel activation in the RVD are not clear. The intracellular Ca^{2+} is increased in hypotonically swollen cells (Nitschke et al. 1993a), suggesting that Ca^{2+} is the second messenger involved. Unpublished data from the authors' laboratory suggest that the Cl^- currents induced

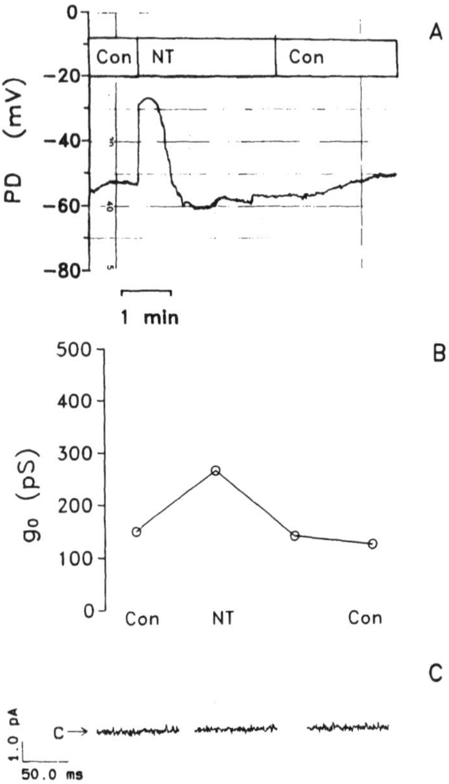

Fig. 2A,C. Effect of neurotensin (NT; 10^{-9} mol/l) on a HT_{29} cell. The method used was the CAN method (Greger and Kunzelmann 1991). The membrane potential (**A**) was depolarized by NT, which was followed by a hyperpolarizing phase. Simultaneously, the input conductance of the cell-attached membrane patch was reversibly increased (**B**). No single channel activity was observed in the voltage-clamp mode of the patch-clamp amplifier at a clamp voltage of Vc = +40 mV. *C* → indicates the zero current level. Current data were filtered at 300 Hz. These data indicate that (i) NT acts by increasing the Cl^- conductance and that (ii) the ion channels responsible for this increase in conductance are too small to be identified as single channel events

either by ATP (increase in cytosolic Ca^{2+}) or forskolin (increase in cytosolic cAMP) are not additive to the current induced by the hypotonic bath solution. This suggests that RVD is due to activation of Cl^- channels which can also be activated by cAMP or Ca^{2+}. In further preliminary experiments we were not able to inhibit the hypotonically induced depolarization and the respective conductance increase in the cell membrane by nordihydroguaiaretic acid, an inhibitor of the lipoxygenase (Fig. 3), suggesting that the metabolism of the lipoxygenase pathway is not involved in the activation of Cl^- channels during RVD. In another study, transient shrinkage of HT_{29} cells was induced by a Ca^{2+} ionophore and a more stable shrinkage

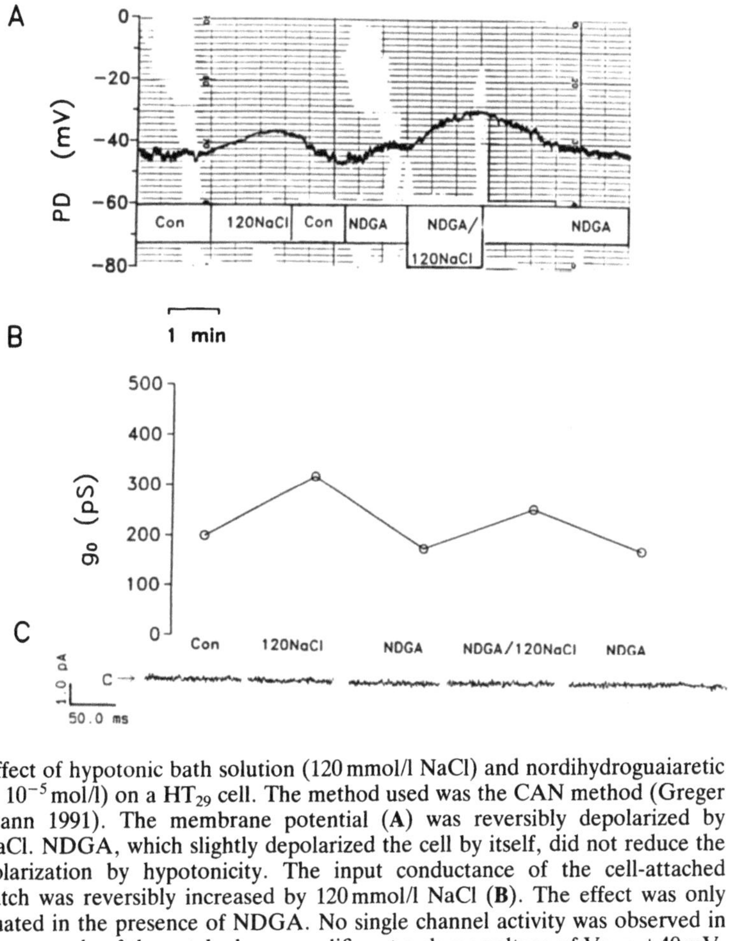

Fig. 3A,C. Effect of hypotonic bath solution (120 mmol/l NaCl) and nordihydroguaiaretic acid (NDGA, 10^{-5} mol/l) on a HT_{29} cell. The method used was the CAN method (Greger and Kunzelmann 1991). The membrane potential (**A**) was reversibly depolarized by 120 mmol/l NaCl. NDGA, which slightly depolarized the cell by itself, did not reduce the induced depolarization by hypotonicity. The input conductance of the cell-attached membrane patch was reversibly increased by 120 mmol/l NaCl (**B**). The effect was only slightly attenuated in the presence of NDGA. No single channel activity was observed in the voltage-clamp mode of the patch-clamp amplifier at a clamp voltage of Vc = +40 mV. *C* → indicates the zero current level. Current data were filtered at 300 Hz. These data indicate that (i) hypotonic swelling induces a Cl^- conductance, that (ii) small Cl^- channels, not resolved as single channel events, are responsible for this conductance and that (iii) NDGA does not abolish this response

by forskolin (Montrose 1990). The latter suggests an involvement of cAMP during volume regulation.

In yet another study it was shown that Cl^- conductance in membrane vesicles of HT_{29} cells can be activated directly by GTP-gamma-S (Tilly et al. 1991). In patch-clamp experiments the activation of a 20-pS inwardly rectifying Cl^- channel by GTP-gamma-S was found. The regulation of Cl^- channels by G-proteins could play a role in channel activation during the RVD. From our own recent data this appears, however, unlikely as we never observed the activation of a 20-pS channel, neither in cell-attached membrane patches of HT_{29} cells exposed to hypotonic bath solution nor in cell-excised membrane patches exposed to GTP or GTP-gamma-S (Kunzelmann et al. 1992b). More experiments are required to define the mechanisms of Cl^- conductance increase in RVD. The question whether Cl^- channels in HT_{29} cells can also be activated by a cGMP-dependent kinase or by protein kinase C is the subject of current studies.

At this stage it appears justified to summarize that a Cl^- conductance can be activated by three different mechanisms, namely (1) an increase in the intracellular cAMP; (2) an increase in the cytosolic Ca^{2+}; and (3) a hypotonically induced cell swelling. In a latter section, we shall discuss the single channel events basically responsible for these different Cl^- conductances.

3.2 The Various Membrane Transport Properties

HT_{29} cells possess the $Na^+/2Cl^-/K^+$ cotransporter in their basolateral membranes which is inhibited by bumetanide (Franklin et al. 1989). It has been shown that treatment of the cells with phorbol esters decreases the number of cotransporters in HT_{29} cells, suggesting the involvement of protein kinase C in the regulation of the $Na^+/2Cl^-/K^+$ cotransport. Similarly, stimulation of the cells by VIP increases the transport rate of the cotransporter, which serves as the main pathway for Cl^- uptake into the cell and its accumulation at concentrations above its electrochemical equilibrium. Cl^- leaves the cell, driven by its electrochemical gradient, via apical Cl^- channels, whereas K^+ recycles basolaterally probably via calcium-dependent K^+ channels (Wu et al. 1991). The Na^+ ions will be pumped out of the cell by the (Na^+/K^+)-ATPase (Greger and Schlatter 1984; Turner et al. 1990; cf. Fig. 4). The Na^+/H^+ antiporter has also been found in these cells (Cantiello and Lanier 1989). Besides its function in maintaining the cellular pH homoeostasis, it was assumed to be involved in cell growth, as it has been shown for other cell types (Rothenberg et al. 1983; Rotin et al. 1989). However, the amiloride-induced growth retardation in HT_{29} cells does not seem to correlate with cytosolic pH changes (Szolgay-Daniel et al. 1991). Moreover, unlike in other cells, the activity of the Na^+/H^+ exchanger does not appear to be regulated by the α_2-adrenergic pathway (Cantiello and Lanier 1989).

4 Cl^- Channels

4.1 Outwardly Rectifying Cl^- Channels

In single channel patch-clamp studies an intermediate conductance, outwardly rectifying Cl^- channel (ICOR) of around 50 pS was found in membranes of HT_{29} cells (Hayslett et al. 1987). The channel activity, i.e., its open probability and the current amplitude of the channel, was independent

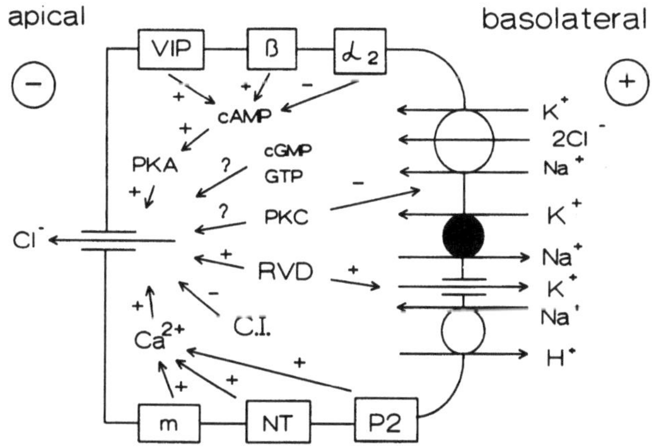

Fig. 4. Model for the transport of electrolyte in a polarized HT_{29} cell. *Circles* indicate ion carriers and pumps ((Na^+/K^+)-ATPase). = ion channel. *Boxes* indicate receptors for hormones. *VIP* Vasointestinal peptide; *NT* neurotensin; *m* muscarinic; P_2 type-2 purinergic receptor. Apical Cl^- channels can be activated by an increase in cytosolic Ca^{2+} and cAMP and by regulatory volume decrease (*RVD*). Cl^- channels are inhibited by a cytosolic inhibitory molecule (*C.I.*). Activation of Cl^- channels by GTP, protein kinase C (*PKC*) and the cGMP-dependent kinase has been suggested and is currently under further experimental examination

of changes in the cytosolic calcium concentration and pH, respectively. Furthermore, the channel could be blocked by a number of compounds, such as 5-nitro-2-(3-phenylpropylamino)-benzoate (NPPB), indanyloxyacetic acid (IAA94/95) and different stilbene derivatives (DIDS, DNDS) (Tilmann et al. 1991). It is exclusively permeable to small anions with a sequence: $J^- \geqslant Br^- \geqslant Cl^- > NO_3^- > HCO_3^- > SO_4^{2-}$ = gluconate = isethionate = 0. Since the channel was found in cell-excised as well as in cell-attached patches, it was suggested to be responsible for the Cl^- secretion in these cells. This type of channel is also present in other Cl^- secreting epithelial cells, e.g., T_{84} and respiratory epithelial cells (Kunzelmann et al. 1991b), and it was postulated that the regulation of this channel is defective in cystic fibrosis (CF) (Schoumacher et al. 1987; Li et al. 1988). However, the incidence of the channel in HT_{29}, T_{84} or respiratory cells was much higher in cell-excised patches when compared to cell-attached recordings. Moreover, and unlike others (Welsh 1986; Frizzell et al. 1987), we found no correlation between the frequency of cell-attached ICOR channels and stimulation of the cells with β-adrenergic compounds (authors' unpubl. data). The observation of excision activation of ICOR channels led us to the assumption that some cytosolic compound inhibits this channel in the intact cell. In a larger series of experiments we could show that, indeed, a small amphiphilic molecule, which is as yet unidentified and which we named cytosolic inhibitor (CI), is responsible for this channel inhibition in HT_{29} cells as well as in other epithelial cell types (Krick et al. 1991; Kunzelmann et al. 1991b). CI, besides its effects on the ICOR channel, also reduced the baseline current in excised membrane patches of HT_{29} cells. This Cl-sensitive baseline current was identified as the synchronous operation of small, non-resolvable Cl^- channels (Hansen et al. 1991). The relevance of the cytosolic inhibition of the ICOR and small Cl^- channels is not clear at the moment but it could reflect an antagonistic pathway for Cl^- channel regulation. Further studies have to show whether Cl can be modified by stimulation of the cell with agonists, which are known to increase the Cl^- conductance. Also, it is not clear at this stage whether the Cl content of CF tissues is altered.

4.2 Small Conductance Cl^- Channels

In our opinion one of the most intriguing and puzzling observations has been the missing correlation between cellular stimulation by secretagogues and the appearance of the ICOR channel. To resolve this puzzle we recently developed a modification of the patch-clamp technique. We labelled this modified technique the cell-attached nystatin (CAN) technique. With CAN, the cell membrane potential and the properties of the cell-attached membrane can be measured simultaneously. This is achieved by a partial permeabilization of the cell-attached membrane by low concentrations of nystatin in the filling solution of the pipette (Greger and Kunzelmann 1991). Using the CAN technique, we observed a depolarization and the activation of a Cl^- conductance in HT_{29} cells by hormones increasing the cytosolic Ca^{2+} and cAMP concentration or by hypotonically induced cell swelling (cf. above). In none of these series of experiments did we observe the activation of an ICOR channel or another Cl^- channel large enough to be resolved as a single channel current (Figs. 1–3). In contrast, the input conductance of the cell-attached cell membrane was increased significantly by all of these manoeuvres, but to a variable degree (Greger and Kunzelmann 1991; Kunzelmann et al. 1992a). These observations clearly indicate that the depolarization of the cell and the increase in a macroscopic Cl^- conductance were paralled by the activation of small Cl^- channels with a conductance of less than 8 pS. Therefore, one can conclude that not the ICOR channel is responsible for the Cl^- conductance in HT_{29} cells but rather other types of small Cl^- channels, which cannot be resolved as single channel currents in patch-clamp experiments. This conclusion is supported by the fact that very potent inhibitors of the ICOR channel are ineffective in inhibiting the Cl^- conductance induced by the different agonists (authors' unpubl. observ.). The functional role of the ICOR channel is not yet clear and remains a matter of further experimentation.

5 K^+ Channels and Other Ion Channels

5.1 K^+ Channels

Much less is known about the properties of K^+ channels in HT_{29} and other Cl^--transporting cells and epithelia. It is clear a priori that K^+ conductance is required for (1) the recycling of K^+, taken up by the (Na^+/K^+)-ATPase; (2) the recycling of K^+, taken up by the $Na^+/2Cl^-/K^+$ cotransporter; (3) the maintenance of cell volume, e.g., in the RVD. Comprehensive analyses are available for the rectal gland of *Squalus acanthias*, the thick ascending limb of the loop of Henle, the proximal tubule, the amphibian diluting segment and respiratory epithelial cells (Gluck et al. 1982; Gögelein and Greger 1987; Greger et al. 1987; Kunzelmann et al. 1989a,b). Two major classes of K^+ channels have been observed. One of them is the Ca^{2+}-dependent, big K^+ channel and the other is an intermediate conductance, Ca^{2+}-independent K^+ channel. In a few patch-clamp experiments we addressed this question regarding HT_{29} cells. In preliminary experiments we rarely found Ca^{2+}-dependent K^+ channels. Surprisingly, the mean conductance of the K^+ channels found falls in the range of the intermediate conductance K^+ channel. These channels were blocked by barium (authors' unpubl. data). It is generally believed that this type of K^+ conductance in epithelial cell gains importance mostly when cytosolic Ca^{2+} is increased by hormonal agonists.

5.2 Other Channels

Resting HT_{29} cells usually have a membrane voltage of -50 to $-60\,mV$ which makes it unlikely that the chloride and potassium conductances suffice to explain it. In fact, the fractional conductances for K^+ and Cl^- do not add up to unity. This has prompted experiments in our group to test for other possibly contributing ions. Since amiloride does not hyperpolarize these cells, the presence of the classical amiloride-sensitive Na^+ channel appears unlikely. On the other hand, almost complete removal of Na^+ and replacement by large cations hyperpolarize these cells. Hence, we believe that an amiloride-insensitive conductance for small cations contributes to the membrane voltage.

6 Conclusion

Many of the data summarized in this brief review suggest that HT_{29} cells can serve as a model for Cl^- secretory processes in epithelia. These cells resemble most probably the crypt cells of the colonic mucosa. They possess transport systems for the basolateral uptake and the apical secretion of Cl^-

ions. This seems to be equally true for poorly differentiated cells grown in D-glucose-containing media and for some subclones grown under specific culture conditions. The correlation between morphological differentiation of the cells, their expression of CFTR and of certain biochemical properties, on the one hand, and the functional properties, on the other, is not well understood at present and will be the subject of further studies.

The transcellular ion transport in HT_{29} cells can be activated in different ways: by hormones increasing the cytosolic Ca^{2+} or cAMP concentration and by hypotonically induced cell swelling. It is very important to realize that small conductance Cl^- channels rather than ICOR channels are responsible for the Cl^- secretion in these and other epithelial cells.

Acknowledgments. The data from the authors' laboratory presented in this review have been continuously supported by: DFG G480/10, GIF I-86-100.1/88 and BMFT 01GA 8816.

References

Amar S, Kitabgi P, Vincent J-P (1986) Activation of phosphatidylinositol turnover by neurotensin receptors in the human colonic adenocarcinoma cell line HT29. FEBS Lett 201: 31–36

Anderson MP, Welsh MJ (1990) Fatty acids inhibit apical membrane chloride channels in airway epithelia. Proc Natl Acad Sci USA 87: 7334–7338

Augeron C, Laboisse CL (1984) Emergence of permanently differentiated cell clones in a human colonic cancer cell line in culture after treatment with sodium butyrate. Cancer Res 44: 3961–3969

Bouscarel B, Cortinovis C, Carpene C, Murat JC, Paris H (1985) α2-Adrenoreceptors in the HT29 human colon adenocarcinoma cell line: characterization with [3H] clonidine; effects on cyclic AMP accumulation. Eur J Pharmacol 107: 223–231

Bozou J-C, Rochet N, Magnaldo I, Vincent J-P, Kitabgi P (1989) Neurotensin stimulates inositol trisphosphate-mediated calcium mobilization but not protein kinase C activation in HT29 cells. Biochem J 264: 871–878

Cantiello HF, Lanier SM (1989) α2-Adrenergic receptors and the Na/H exchanger in the intestinal epithelial cell line HT-29. J Biol Chem 264: 16000–16007

Fogh J, Trempe G (1975) New human tumor cell lines. In: Fogh J (ed) Human tumor cells in vitro. Plenum Press, New York, pp 115–141

Franklin CC, Turner JT, Kim HD (1989) Regulation of $Na^+/K^+/Cl^-$ cotransport and $[^3H]$bumetanide binding site density by phorbol esters in HT29 cells. J Biol Chem 264: 6667–6673

Frizzell RA, Schoumacher RA, Shoemaker RL, Halm DR (1987) Chloride channel regulation in secretory epithelial cells. Pediatr Pulmonol Suppl 1: 24–25

Gluck D, Kelly S, Al-Awqati Q (1982) The proton translocating ATPase responsible for urinary acidification. J Biol Chem 257(16): 9230–9233

Gögelein H, Greger R (1987) Properties of single K^+ channels in the basolateral membrane of rabbit proximal straight tubules. Pflügers Arch Eur J Physiol 410: 288–295

Greger R, Kunzelmann K (1991) Simultaneous recording of the cell membrane potential and properties of the cell attached membrane of HT29 colon carcinoma and CF-PAC cells. Pflügers Arch Eur J Physiol 419: 209–211

Greger R, Schlatter E (1984) Mechanism of NaCl secretion in the rectal gland of spiny dogfish (*Squalus acanthias*). I. Experiments in isolated in vitro perfused rectal gland tubules. Pflügers Arch Eur J Physiol 402: 63–75

Greger R, Gögelein H, Schlatter E (1987) Potassium channels in the basolateral membrane of the rectal gland of the dogfish (*Squalus acanthias*). Pflügers Arch Eur J Physiol 409: 100–106

Hansen CP, Roch B, Kunzelmann K, Greger R (1991) Inhibition of epithelial chloride channels by epithelial cytosols. Pflügers Arch Eur J Physiol 419: R101

Hayslett JP, Gögelein H, Kunzelmann K, Greger R (1987) Characteristics of apical chloride channels in human colon cells (HT_{29}). Pflügers Arch Eur J Physiol 410: 487–494

Hekmati M, Polak Charcon S, Ben-Shaul Y (1990) A morphological study of a human adenocarcinoma cell line (HT29) differentiating in culture. Similarities to intestinal embryonic development. Cell Differ Dev 31: 207–218

Huet C, Sahuquillo-Merino C, Coudrier E, Louvard D (1987) Absorptive and mucus-secreting subclones isolated from a multipotent intestinal cell line (HT-29) provide new models for cell polarity and terminal differentiation. J Cell Biol 105: 345–357

Kim HD, Tsai Y-S, Franklin CC, Turner JT (1988) Characterization of Na^+-K^+-Cl^- cotransport in cultured HT29 human colonic adenocarcinoma cells. Biochim Biophys Acta 946: 397–404

Kreusel KM, Fromm M, Lempart U, Sorgenfrei D, Hegel U (1990) Epithelial monolayers of differentiated HT-29 cells exhibit crypt-like chloride secretion. Pflügers Arch Eur J Physiol 415: R33 (Abstr)

Krick W, Disser J, Hazama A, Burckhardt G, Frömter E (1991) Evidence for a cytosolic inhibitor of epithelial chloride channels. Pflügers Arch Eur J Physiol 418: 491–499

Kunzelmann K, Pavenstädt H, Beck C, Ünal Ö, Emmrich P, Arndt HJ, Greger R (1989a) Characterization of potassium channels in respiratory cells I. General properties. Pflügers Arch Eur J Physiol 414: 291–296

Kunzelmann K, Pavenstädt H, Greger R (1989b) Characterization of potassium channels in respiratory cells II. Inhibitors and regulation. Pflügers Arch Eur J Physiol 414: 297–303

Kunzelmann K, Pavenstädt H, Greger R (1989c) Properties and regulation of chloride channels in cystic fibrosis and normal airway epithelial cells. Pflügers Arch Eur J Physiol 415: 172–182

Kunzelmann K, Gerlach L, Fröbe U, Greger R (1991a) Bicarbonate permeability of epithelial chloride channels. Pflügers Arch Eur J Physiol 417: 616–621

Kunzelmann K, Tilmann M, Hansen CP, Greger R (1991b) Inhibition of epithelial chloride channels by cytosol. Pflügers Arch Eur J Physiol 418: 479–490

Kunzelmann K, Grolik M, Kubitz R, Greger R (1992a) cAMP dependent activation of small conductance Cl^- channels in HT_{29} colon carcinoma cells. Pflügers Arch Eur J Physiol (in press)

Kunzelmann K, Kubitz R, Grolik M, Warth R, Greger R (1992b) Small conductance Cl^- channels in HT29 cells: activation by Ca^{2+}, hypotonic cell swelling and 8-Br-cGMP. Pflügers Arch Eur J Physiol (in press)

Laburthe M, Rousset M, Boissard C, Zweibaum A, Rosselin G (1978) Vasoactive intestinal peptide: a potent stimulator of adenosine 3′:5′-cyclic monophophate accumulation in gut carcinoma cell lines in culture. Proc Natl Acad Sci USA 75: 2772–2775

Le Bivic A, Hirn M, Reggio H (1988) HT-29 cells are an in vitro model for the generation of cell polarity in epithelia during embryonic differentiation. Proc Natl Acad Sci USA 85: 136–140

Li M, McCann JD, Liedtke CM, Nairn AC, Greengard P, Welsh MJ (1988) Cyclic AMP-dependent protein kinase opens chloride channels in normal but not cystic fibrosis airway epithelium. Nature 331: 358–360

Lohrmann E, Cabantchik ZI, Greger R (1991) The membrane potential (PD) and the conductance properties of HT29 Cells. Pflügers Arch Eur J Physiol 418: R326 (Abstr)

Montrose MH (1990) Linkage between second messenger levels and cell volume in cultured intestinal cells (HT29-C1). FASEB J 4: A448 (Abstr)
Montrose MH, Murer H (1989) Polarized response to propionate in cultured human colonic epithelial cells (HT29/C1). Organization of membrane polarity in epithelial cells, Arolla (Abstr)
Montrose-Rafizadeh C, Guggion WB, Montrose MH (1990) Expression of CFTR is modulated by cell differentiation in a cloned human intestinal cell line. Pediatr Pulmonol Suppl 5: 192 (Abstr)
Montrose-Rafizadeh C, Guggino WB, Montrose MH (1991) Cellular differentiation regulates expression of Cl^- transport and cystic fibrosis transmembrane conductance regulator mRNA in human intestinal cells. J Biol Chem 266: 4495–4499
Nitschke R, Leipziger J, Greger R (1993a) Intracellular Ca^{2+} transients in HT_{29} cells induced by hypotonic cell swelling. Pflügers Arch (in press)
Nitschke R, Leipziger J, Greger R (1993b) Agonist induced intracellular Ca^{2+} transients in HT_{29} cells. Pflügers Arch (in press)
Paris H, Bouscarel B, Cortinovis C, Murat JC (1985) Growth-related variation of α2-adrenergic receptivity in the HT29 adenocarcinoma cell-line from human colon. FEBS Lett 184: 82–86
Rechkemmer GR, Bökenkamp D, Jeromin A (1991) Intracellular calcium concentration in the human colonic tumor cell line HT-CL19A. Effects of carbachol, forskolin and vasoactive intestinal polypeptide. Pflügers Arch Eur J Physiol 418: R59 (Abstr)
Remy L, Marvaldi J, Rua S, Secchi J, Lechene de la Porte P (1984) The role of intracellular lumina in the repolarization process of a colonic adenocarcinoma cell line. Virchows Arch 46: 297–305
Reynier M, Sari H, d'Anglebermes M, Ah Kye E, Pasero L (1991) Differences in lipid characteristics of undifferentiated and entrocytic-differentiated HT29 human colon cells. Cancer Res 51: 1270–1277
Rommens JM, Iannuzzi B-SK, Drumm ML, Melmer G, Dean M, Rozmahel R, Cole JL, Kennedy D, Hidaka N, Zsiga M, Buchwald M, Riordan JR, Tsui L-C, Collins FS (1989) Identification of the cystic fibrosis gene: chromosome walking and jumping. Science 245: 1059–1065
Rothenberg P, Glaser L, Schlesinger P, Cassel D (1983) Activation of Na^+/H^+ exchange by epidermal growth factor elevates intracellular pH in A 431 cells. J Biol Chem 258: 12644–12653
Rotin D, Steele-Norwood D, Grinstein S, Tannock I (1989) Requirement of the Na^+/H^+ exchanger for tumor growth. Cancer Res 49: 205–211
Rousset M (1986) The human colon carcinoma cell lines HT-29 and Caco-2: two in vitro models for the study of intestinal differentiation. Biochimie 68: 1035–1040
Schoumacher RA, Shoemaker RL, Halm DR, Tallant EA, Wallace RW, Frizzell RA (1987) Phosphorylation fails to activate chloride channels from cystic fibrosis airway cells. Nature 330: 752–754
Simon-Assman P, Bouziges F, Daviaud D, Haffen K, Kedinger M (1987) Synthesis of glucosaminoglycans by undifferentiated and differentiated HT29 human colonic cancer cells. Cancer Res 47: 4478–4484
Sood R, Auerbach W, Shannon W, Buchwald M (1990) Regulation of expression of CFTR with in vitro differentiation of human intestinal epithelial cells. Pediatr Pulmonol Suppl 5: 194 (Abstr)
Szolgay-Daniel E, Carlsson J, Zierold K, Holtermann G, Dufau E, Acker H (1991) Effect of amiloride treatment on U-118 MG and U-251 MG human glioma and HT-29 human colon carcinoma cells. Cancer Res 51: 1039–1044
Tilly BC, Kansen M, Gageldonk PGM, van den Berghe N, Galjaard H, Bijman J, De Jonge HR (1991) G-proteins mediate intestinal chloride channel activation. J Biol Chem 266: 2036–2040
Tilmann M, Kunzelmann K, Fröbe U, Cabantchik I, Lang HJ, Englert HC, Greger R (1991) Different types of blockers of the intermediate conductance outwardly rectifying chloride channel (ICOR) in epithelia. Pflügers Arch Eur J Physiol 418: 556–563

Turner JT, Franklin CC, Bollinger DW, Kim HD (1990) Vasoactive intestinal peptide stimulates active K^+ transport and $Na^+/K^+/Cl^-$ cotransport in HT-29 cells. Am J Physiol 258: C266–C273

Welsh MJ (1986) An apical-membrane chloride channel in human tracheal epithelium. Science 232: 1648–1649

Wu H, Franklin CC, Kim HD, Turner JT (1991) Regulation of calcium-activated potassium efflux by neurotensin and other agents in HT-29 cells. Am J Physiol 260: C35–C42

Ziss W, Hegel U (1988) Human colon cancer cells (HT-29) exhibit ion transport properties of intestinal crypt cells. Pflügers Arch Eur J Physiol 411: R81 (Abstr)

Ziss W, Fromm M, Sorgenfrei D, Hegel U (1987) Effect of chloride step changes on the membrane potential in the human colonic carcinoma cancer cell line HT-29. Pflügers Arch Eur J Physiol 408: R32 (Abstr)

Zweibaum A, Pinto M, Chevalier G, Dussaulx E, Triadou N, Lacroix B, Haffen K, Brun J-L, Rousset M (1985) Enterocytic differentiation of a subpopulation of the human tumor cell line HT-29 selected for growth in sugar-free medium and its inhibition by glucose. J Cell Physiol 122: 21–29

Subject Index

MIX
Papier aus verantwortungsvollen Quellen
Paper from responsible sources
FSC® C105338

If you have any concerns about our products,
you can contact us on
ProductSafety@springernature.com

In case Publisher is established outside the EU,
the EU authorized representative is:
Springer Nature Customer Service Center GmbH
Europaplatz 3, 69115 Heidelberg, Germany

Printed by Libri Plureos GmbH
in Hamburg, Germany